Practical
Rock
Mechanics

Applied Geotechnics Series

Series Editor: William Powrie,
University of Southampton, United Kingdom

PUBLISHED TITLES

David Muir Wood, *Geotechnical Modelling*
Hardback ISBN 978-0-415-34304-6 • Paperback ISBN 978-0-419-23730-3

Alun Thomas, *Sprayed Concrete Lined Tunnels*
Hardback • ISBN 978-0-415-36864-3

David Chapman *et al.,Introduction to Tunnel Construction*
Hardback ISBN 978-0-415-46841-1 • Paperback ISBN 978-0-415-46842-8

Catherine O'Sullivan, *Particulate Discrete Element Modelling*
Hardback • ISBN 978-0-415-49036-8

Steve Hencher, *Practical Engineering Geology*
Hardback ISBN 978-0-415-46908-1 • Paperback ISBN 978-0-415-46909-8

Martin Preene *et al.,* *Groundwater Lowering in Construction*
Hardback • ISBN 978-0-415-66837-8

Steve Hencher, *Practical Rock Mechanics*
Paperback • ISBN 978-1-4822-1726-1

Mike Jefferies *et al.,* *Soil Liquefaction, 2nd ed*
Hardback • ISBN 978-1-4822-1368-3

Paul F. McCombie *et al.,* *Drystone Retaining Walls: Design, Construction and Assessment.*
Hardback • ISBN 978-1-4822-5088-6

FORTHCOMING

Zixin Zhang *et al.,* *Fundamentals of Shield Tunnelling*
Hardback • ISBN 978-0-415-53597-7

Christoph Gaudin *et al.,* *Centrifuge Modelling in Geotechnics*
Hardback • ISBN 978-0-415-52224-3

Kevin Stone *et al.,* *Weak Rock Engineering Geology and Geotechnics*
Hardback • ISBN 978-0-415-56071-9

Practical Rock Mechanics

Steve Hencher

*Hencher Associates, Limited
University of Leeds,
UK and Halcrow, Hong Kong*

CRC Press
Taylor & Francis Group
Boca Raton   London   New York

CRC Press is an imprint of the
Taylor & Francis Group, an **Informa** business

A SPON PRESS BOOK

CRC Press
Taylor & Francis Group
6000 Broken Sound Parkway NW, Suite 300
Boca Raton, FL 33487-2742

Printed on acid-free paper
Version Date: 20160225

International Standard Book Number-13: 978-1-4822-1726-1 (Paperback)

Visit the Taylor & Francis Web site at
http://www.taylorandfrancis.com

and the CRC Press Web site at
http://www.crcpress.com

Contents

Preface

My good friend and co-teacher of the MSc engineering geology at Leeds University for many years, Alastair Lumsden, is the source of many wonderful anecdotes and philosophies, as his numerous past students will attest. One of his best stories related to the four-yearly assessment by the Natural Environment Research Council (NERC) of MSc courses. NERC, a government-funded body, provided grudging crumbs of support to anything remotely useful in the university earth sciences environment. The argument went along the lines that if training or research were actually useful, then the industry should pay for it in some way. They have since stopped supporting such courses at all.

Anyway, one of the other four key university players in the provision of MSc courses in engineering geology in the United Kingdom at the time (Leeds, Imperial College, Durham and Newcastle) was criticised for their course being overly-mathematical. Two very eminent geotechnical engineers ran the course. They took umbrage at the criticism and wrote to complain to the NERC, and provided three pages of calculations to prove it. If you do not find this tale even faintly amusing, put this book down carefully. Otherwise, read on.

This book was partly inspired following requests to present a course on rock mechanics as part of an MSc study programme at Hong Kong University and the same at Imperial College, London. The course in Hong Kong was to be suitable for geologists and others with little background in engineering or mechanics. Such students often struggle with the concepts of stress and stress transformation, especially where rock mechanics is taught as a predominantly mathematical subject.

When considering the possible course texts to adopt, I judged that most available textbooks on rock mechanics have been written from a mechanical point of view and that rocks are usually treated essentially as uniform materials albeit with a few micro-cracks. Furthermore, structural geology, that is also based on fundamental mechanical principles and is extremely important to practical engineering and mining, particularly concerning the formation and characterisation of rock discontinuities, has been rather skimmed over in most rock mechanics textbooks.

One of my fellow researchers at Imperial College in the 1970s, Grant Hocking, once shocked me into awed silence by declaring that he had worked his way through every equation in Jaeger and Cook's *Fundamentals of Rock Mechanics* (1969). All such equations are, of course, important for thorough understanding but I remain unconvinced that such intense mathematical treatment is really necessary for most rock mechanics as it applies generally in engineering practice. As illustrated later, the vast majority of failures in geotechnical engineering are the result of misinterpreted or overlooked geological or hydrogeological factors – not a matter of getting the sums wrong: cue the insightful cartoon of Frank and Ernest that is sellotaped inside my first edition of Hoek and Bray (1974).

Frank and Ernest

This book is aimed at teaching from first principles, assisting the understanding of the practical application of rock mechanics theory and knowledge to major rock engineering projects. It is also intended to inspire and encourage practitioners to use the full gamut of geological skills to ensure that ground models and designs are correct and realistic, produced cost-effectively and to avoid too many surprises. The text sometimes strays somewhat from what has become the traditional practice area of 'rock mechanics' but, no apologies, if you do not deviate from the path occasionally, you might miss the best views!

Acknowledgements

Thanks to Dr. Marco Redaelli of CH2M, UK who drafted the case study summarising the investigation and thinking regarding the North Anchorage for the Izmit Bay Crossing suspension bridge (Chapter 7). This case study describes the work done in close co-operation with the designers for the bridge (notably Brian Foged of COWI and their geotechnical consultants, especially Bob Mackean of GeoDesign). Thanks are due to contractors IHI and owners KGM, NOMAYG for permission to publish details and photographs from the project.

Mike Palmer, also of CH2M, UK provided much of the background information on the Kishinganga tailrace tunnel (Chapter 9).

Much of the practical guidance on tunnel support measures in Section 9.6.6 is based on recommendations by and discussions with tunnel guru Nick Swannell, also of CH2M, UK.

Thanks to my helpful reviewers professor Andrew Malone for Chapters 1 and 2, Drs. Bob Fowell for Chapters 2 and 6, Jared West for Chapter 4 on hydrogeology, Andy Pickles for Chapter 7 on foundations and Laurie Richards for Chapter 9 on underground excavations. In each case, they made useful comments and drew attention to errors and omissions but they have not seen the final versions; all the remaining weaknesses and schoolboy howlers are mine alone.

Many figures were inspired by published diagrams and I hope that I have acknowledged the original sources adequately. I have found particularly useful the clearly presented introduction to *Rock Mechanics Principles in Engineering Practice* by John Hudson (1989) and the well-illustrated *Structural Geology* by Davis and Reynolds (1996).

Thanks to my son, Sam, who helped with some of the line drawings and to Marji, my wife, who checked some of the grammar and references and put up with this yet again.

Specific permissions

The Frank and Ernest cartoon in the preface is used with the permission of Sara Thaves and the Cartoonist Group. All rights reserved.

Figures 1.10 and 4.11 Rock joint Fracman models are courtesy of Dr. Bill Dershowitz of Golder Associates, Canada.

Figure 1.16 Flac model with permission of Nick Swannell, CH2M, UK.

Figures 2.16–2.19 have been kindly provided by Drs. Branko Damjanac and Jim Hazzard of Itasca Corporation, USA.

Figure 2.20 is courtesy of Dr. Bob Pine and Davide Elmo and reproduced with kind permission of Professor Giovanni Barla, the chief editor of *Rock Mechanics and Rock Engineering.*

The underlying image to Figure 3.1 is reproduced with courtesy of Laurie Kovalenko: Tectonics Observatory, California Institute of Technology, USA.

Figure 3.2 is reproduced with courtesy of Oliver Heidbach at the World Stress Map Project, Potsdam, Germany (ref: Heidbach et al., 2010).

Figures 3.17, 3.18, 3.20–3.22 are courtesy of Dr. Graham Mcleod of the University of Leeds, UK.

Figure 3.26 photograph and description are courtesy of Dr. Bruce Yardley, who is also of the University of Leeds.

Figure 3.47a, photo courtesy of Professor Terry Engelder, Pennsylvania State University, USA.

Figures 3.49 and 3.50 are from Rawnsley et al. (1992), reprinted with permission of Elsevier.

Figure 3.67 reproduced with kind permission of Siegaverde, Spain.

Figures 3.71–3.77 photographs and data for this case example reproduced with kind permission of John Morhall.

Figures 5.1, 5.57 and 5.58 reproduced with kind permission of Benoit Wentzinger and David Starr of Golder Associates, Brisbane, Australia. With acknowledgements to Sinclair Knight Merz (now Jacobs Engineering) and Department of Transport and Main Roads, Queensland. Construction by Leighton Contractors.

Figures 5.11 and 5.12 and discussion, courtesy of Dr. Seifko Slob, Holland (and thanks to Dr. Robert Hack).

Figure 5.13 and the accompanying text, courtesy of Dr. Thomas Seers, Manchester University.

Figures 5.35 and 5.36 are reproduced with permission of the head of the Geotechnical Engineering Office and the director of the Civil Engineering and Development, the Government of the Hong Kong Special Administrative Region.

Figure 5.52 The Geological Strength Index chart is reproduced with permission of Dr. Evert Hoek.

Figure 5.59a and b reproduced with permission from Measurand, Canada.

Figures 6.2 and 6.3 are courtesy of Jun Long Shan, the University of Leeds, UK.

Figure 6.4 is courtesy of Dr. Bill Murphy, the University of Leeds, UK.

Figures 7.15, 7.21 and 7.22 are reproduced with courtesy of IHI, contractors for the construction of Izmit Crossing.

Figures 9.15a and b and 9.23 are courtesy of Charlie Watts, New Zealand.

Author

Steve Hencher is research professor in engineering geology at the University of Leeds, UK, and an honorary professor in the Department of Earth Sciences at the University of Hong Kong and works as an independent consultant. Up until recently, he was the director of Halcrow (China) and head of Geotechnics in the Hong Kong Office for about 10 years. Halcrow is now part of CH2M. Previously, he worked with Bechtel on the design of the high-speed railway in South Korea and for the Geotechnical Control Office of the Hong Kong Government. His early years were spent working for WS Atkins on sites throughout the United Kingdom, including 2 years on-site for the construction of Drax Power Station and in North Africa. He also spent 12 intervening years teaching the MSc in engineering geology and conducting research at the University of Leeds.

Chapter 1

Introduction to rock mechanics

Confidence is what you have before you understand the problem.

Woody Allen

1.1 INTRODUCTION

In geology, 'rock' may be defined as a natural aggregate of minerals (Whitten and Brooks, 1972), which is a very broad definition that encompasses all naturally-formed soil and rock as otherwise distinguished in engineering practice. In geotechnical engineering, the term 'rock mechanics' is usually confined to the stronger and lithified end of the spectrum – the state when the mineral grains have become bonded and cemented together over geological time. In sedimentary rocks, this is partly due to the secondary effect of minerals such as calcite, silica and iron oxides that are transported and then deposited in the pore space. These secondary minerals cement the sedimented mineral grains to one another. Strong electrochemical bonds also exist directly between individual mineral grains in all rocks. In sands, grains in contact deform under high pressure to form strong bonds ('pressure solution' and 'interpenetration'); in igneous and metamorphic rocks strong bonds are formed on cooling and the transition from liquid to solid. This bonding, together with cementation, results in intact rocks having tensile and cohesive strength, which is their defining characteristic (see Section 1.6 for definitions).

1.2 DIFFERENTIATING BETWEEN SOIL AND ROCK

First, let us consider what is meant by soil, in geotechnical engineering.

In a review of good and bad practice in geotechnical engineering, Atkinson (2002) suggests a way to judge the validity of a textbook on ground engineering is to pose the question: '*does the book distinguish between cohesive and frictional soils (wrong) or between fine-grained and coarse-grained soils (correct)?*' Knappett and Craig, in the eighth edition of Craig's *Soil Mechanics* (2012), make almost no reference to cohesion other than to define it as '*the term used to describe the strength of a clay sample when it is unconfined, being due to negative pressure in the water filling the void space. This strength would be lost if the clay were immersed in a body of water*'.

The implication is that soil, by definition, has no true cohesion and, certainly, the discipline of 'soil mechanics' is primarily concerned with the properties and behaviour of assemblages of mineral grains that lack strong physical bonds or cementation.

It follows that any materials that have significant true cohesion and/or tensile strength should be defined as rock rather than soil, and that is essentially the way that rock is defined here. This book deals with both the 'hard' and 'soft' end of rock mechanics, which includes weakly-cemented sediment on its way towards becoming strong, lithified rock as well as weathered igneous rock that has very little strength left compared to the original fresh rock.

The transition from 'soil' to 'rock' is usually taken at a uniaxial compressive strength of about 1 MPa.* This strength represents a material that is so weak that it can be readily scratched by thumb nail and broken down by hand, but has some tensile strength. To put this value in context, the compressive strength at which a piece of rock (55 mm diameter) can be broken by hand (a subjective but useful test) is about 12.5 MPa (British Standards Institution, 1999; Geotechnical Control Office, 1988). By comparison, concrete strength typically ranges from about 20 to 90 MPa. The compressive strength of intact fresh rock can exceed 400 MPa.

1.3 MECHANICS OF FAILURE

There are only two modes of intact rock failure: tension (pulling apart), and shearing. Shearing is differentiated into in-plane and out-of-plane modes as illustrated in Figure 1.1. In rock mechanics, as it applies to structural geology, a distinction also needs to be made between 'direct' shear and 'simple' shear. Direct shear is where two blocks are sheared in opposite directions either side of a plane (Figure 1.1, mode II), for example, where two sides of a transcurrent fault are forced to move in opposing directions. In simple shear, a volume of rock is deformed through an angle. These different modes of shearing are associated with the development of particular geological structures as discussed further in Chapter 3 and illustrated in Figure 1.2.

1.4 CLASSIFICATION OF INTACT ROCK

Despite these fundamental modes of failure identified for a rock, confusion can arise because intact rock is rarely classified by its tensile strength and never according to its shear strength although both these parameters are very important in practice. Rock is generally classified instead with reference to its 'compressive strength'. This is because rocks are commonly tested under compression either as freestanding cylindrical samples (unconfined) or in a special apparatus where a confining stress can be applied to the sample using pressurised fluid. Despite the conditions of the test being compressive, the actual mode of failure of the

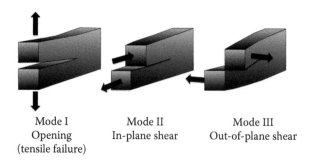

Mode I
Opening
(tensile failure)

Mode II
In-plane shear

Mode III
Out-of-plane shear

Figure 1.1 Three modes of fracturing. (After Atkinson, B.K. 1987. *Fracture Mechanics of Rock*, Academic Press, 1–26.)

---

* The terms and units are defined in Chapter 2 and conversion factors are given in a table towards the end of the book.

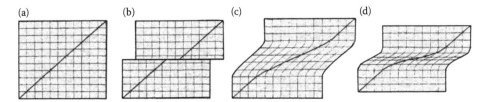

Figure 1.2 Different types of shear zone. (Modified from Ramsay, J.G. 1980. *Journal of Structural Geology*, **2**, 83–99.) (a) Original undeformed rock. (b) Brittle fracture (mode II) with development of specific shear discontinuity (fault). Shear zone walls show no signs of deformation but might be brecciated (fragmented). (c) Heterogeneous, simple shear zone (typical of ductile – high-temperature, high confining pressure conditions). No discontinuity. Rock walls remote from shear zone show no signs of strain; deformation within the shear zone is by simple shear (angular deformation with no volume change). Typically, the proportions of different mineral grain species such as quartz, calcite and feldspar remain essentially constant across the shear zone and into the parent rock although there may well be differences in grain size and fabric. (d) Heterogeneous simple shear with volume change (also ductile). This is typical of low-to-medium grade metamorphic rocks. Relatively mobile minerals such as quartz and calcite may go into solution and migrate from the shear zone resulting in volumetric loss; there is a proportional increase within the shear zone of less mobile components such as clay, chlorite and mica.

rock specimen is through a combination of tension, at least at the micro-mechanics scale, and shearing as explained in Chapter 2.

1.5 COMPRESSIVE STRENGTH OF WEAK* ROCK

Compressive strength of 1 MPa is towards the weakest end of rock strength but is not insignificant for engineering design as will be illustrated by an example. Figure 1.3 is a schematic diagram (2D) representing an axially-loaded compression test on a cylindrical core sample of a very weak rock. Failure occurs when the vertical stress, σ_c, is 1 MPa. In this example,

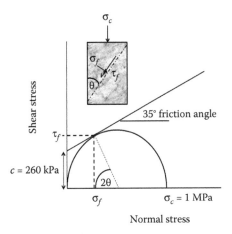

Figure 1.3 Representation of a uniaxial compressive strength test on a sample of rock so weak that you could crumble it by hand.

---

* Note that terms such as 'weak' and 'strong' are used by the geotechnical community to define rock with specific ranges of 'compressive' strength as discussed in Chapter 5. Here, the words are used with their normal, non-technical and relative meanings.

there is no confining stress on the sides of the cylinder. Although the rock sample is under uniaxial compression from its ends, it will most likely fail by shearing along an inclined plane as illustrated. How stresses vary through a rock body and how this stress state can be expressed as a 'Mohr's circle', will be explained in detail later; accept for the moment that the semi-circle shown in Figure 1.3 represents the stress state at different orientations on an infinite series of planes through the sample. On the ends of the cylinder at failure, the stress is 1 MPa; on the sides, it is zero. At various angles, $\theta°$ (pronounced as *theta*), measured from the vertical axis, there are stresses acting normally and along any inclined plane and these will be represented by some point on the Mohr circle. The normal stress acting orthogonally to the failure plane* is σ_f (*sigma* at failure) and the shear stress parallel to the failure plane is termed as τ_f (*tau*). These values can be measured directly from the graph in Figure 1.2, provided the angle of inclination of the plane on which failure occurred is known.

1.6 ORIGINS OF SHEAR STRENGTH IN INTACT ROCK

The shear strength of an intact rock is derived from three sources:

Cohesion: If two pieces of rock are glued together, a 'tensile' strength will resist attempts to pull the pieces apart. If direct shear is applied to the glued sample (parallel to the joint), the glue will provide 'cohesive' resistance. If the glued area is doubled, the cohesive force will also be doubled. The unit for cohesion is force per unit area – kN/m^2 or kPa as discussed in Chapter 2. Cohesion is independent of the level of applied normal stress.

Friction: Friction contributes to the shear resistance between any two surfaces and is derived from interactions and attractions between surfaces. Friction, like cohesion originates from electro-chemical interactions (e.g. Bowden and Tabor, 1956). Unlike cohesion in a rock, however, the bonds causing frictional resistance are temporary and provide no routinely-measurable tensile resistance to pulling apart. Friction, unlike cohesion, increases proportionally to the confining force and does not vary with apparent contact area of the sample.† These are known as Amontons' first and second laws respectively (Amontons, 1699).

Friction not only acts on surfaces in contact but acts internally, for example within a body of sand where the frictional resistance is derived from many contacts between the sand grains carrying the applied loads. Friction also acts internally in a sample of rock. Whereas sand has no true cohesion – its shear strength is purely frictional (and 'dilational' – see below) – in rock, both cohesion and friction resist shear at the same time. Once the rock is fractured and shear displacement occurs, the original cohesion from long-term bonding is lost and the remaining shear strength is purely frictional‡ as for sand.

The 'angle of friction', ϕ (*phi*) is the tilt angle at which a block of some material will slide on a plane of the same material. A 'coefficient of friction' (μ, *mu*) can be defined

---

* Actually, for this 2D case, in ideally isotropic rock, there would be two failure planes developed, one to the left as shown, and the other to the right. These would make a 'conjugate set' and the acute angle between these planes (2θ) is known as the 'dihedral' angle.

† The 'apparent contact' area in macroscopic and measurable directly, e.g. length times width of a house brick. The actual contact points on a mineral grain or rock-joint surface are microscopic – the true contact area over these points of interaction, that together carry the load, increases directly with normal load, essentially elastically but sometimes plastically given the very high contact stresses.

‡ Frictional resistance for a rock joint needs to be broken down into different components as discussed in Chapter 6.

as the ratio between shear (τ) and normal stress (σ) at failure in a cohesionless material such as sand.

The angle of friction for a cohesionless material is related to the coefficient of friction and stresses on the plane as follows:

$$\tan\phi = \mu = \tau/\sigma \tag{1.1}$$

Dilatency: When dense sand, intact rock or a rough rock discontinuity is sheared at a relatively low normal stress, it will open up (dilate). Work carried out against the normal load as the sample dilates causes additional resistance to shear. Since the dilatency effect varies with normal load, as does friction, for many analyses in civil engineering, dilatency is expressed as an apparent increase in friction angle. At very high loads, for example, deep in the Earth's crust, the dilatency effect is totally suppressed.

1.7 SHEAR STRENGTH PARAMETERS FOR THE SAMPLE IN FIGURE 1.3

Knowing the value for θ (Figure 1.3) from observation of the geometry of failure in a test, values for σ_f and τ_f can be calculated readily (in this case, normal stress, $\sigma_f = 500\,(1 - \cos 2\theta)$ kPa; shear resistance, $\tau_f = 500 \times \sin 2\theta$ kPa). The relative contributions to the shear resistance from friction and cohesion are however unknown, but an educated guess can be made.

A fundamental and remarkable fact is that the internal angle of the friction of rock masses at a very large scale is usually somewhere between 30° and 40°. For Anderson's (1951) theory of faulting discussed in Chapter 3, he adopted 30° as the internal friction angle for the Earth's crust. This value is verified by the geometry of many large-scale geological structures (Price, 1966). For rock joints at high stress, the friction angle is typically 38° (Byerlee, 1978).

So, for the rock sample depicted in Figure 1.3, let us assume a friction angle of 35°. Then, using trigonometry, the cohesional strength at zero normal stress can be calculated to be 260 kN/m² (260 kPa).

To put this into context, cohesion of 260 kPa is slightly higher than the shear strength of a stiff cohesive* soil. 'Very stiff' or 'hard' clay, the uppermost category in soil classification, is defined with 'undrained shear strength' >150 kPa (British Standards Institution, 1999). Remember again, that rock with a uniaxial compressive strength of 1 MPa is extremely weak. Few engineers would consider it as rock at all.

1.8 STABILITY OF A CUT SLOPE IN WEAK ROCK

Consider an instance where a slope is to be constructed in this 'rock'. If it was required to be cut steeply at 75° beside a road because of limited land availability, would it be stable? If the geology had simply comprised cohesionless sand with a friction angle of 35°, then that would be the limiting angle, the 'angle of repose', and it would be necessary to cut it even more shallowly to allow for some degree of safety.[†]

---

* There is a strong debate in the soil mechanics literature as to whether soil is or can be regarded as cohesive especially for design purposes: see Atkinson (2002) and Schofield (2006) and the discussions by Burland (2008) and others in the same volume. For the purposes of this broad, comparative discussion, we will adopt the common approach that the undrained shear strength of 'clay' is essentially the equivalent of cohesion.

† Allowance might also need to be made for groundwater pressure and other loading conditions that would reduce the acceptable cut angle for the slope well below 35°.

Using sophisticated analytical software (in this case FLAC/SLOPE, by Itasca Corporation), it is possible to calculate the height at which a slope cut at 75° into the hypothetical weak rock will fail. Failure would theoretically occur where the shear forces encouraging failure are equal to the shear resistance in the ground. The analysis in Figure 1.4 indicates that the slope might be cut to almost 150 m height before it would fail, which is perhaps intuitively surprising for such a weak material.

The results from a series of theoretical trial analyses for the same weak rock, varying the slope height systematically, are presented in Figure 1.5. The results demonstrate that even a 60 m-high slope would have a factor of safety (FoS) >1.6. This means that the available shear strength in the rock mass exceeds the shearing stress (derived from gravity) along the most adversely oriented potential shear plane by a factor of 1.6. To place this number in context, for slope stability design, an FoS of 1.2–1.4 is generally accepted as safe enough depending upon the degree of confidence in the ground model and consequence associated with potential failure (Geotechnical Control Office, 1984a).

If the slope was fully saturated with water, it would be far less stable theoretically but analysis shows that a slope of 10 m height could still be cut at 75° quite safely in the same uniform weak rock with the shear strength parameters derived earlier. The important lesson is that even a small cohesive strength component is extremely important for stability in all rock engineering. The problem with this theoretical discussion however is that most rock masses, including many over-consolidated and partially-cemented 'soils', contain natural fractures (discontinuities) and these affect the way that we must consider their stability and other behaviour as discussed below. The presence and influence of natural discontinuities is one of the most important differences between 'rock mechanics' and 'soil mechanics' approaches to geotechnical engineering.

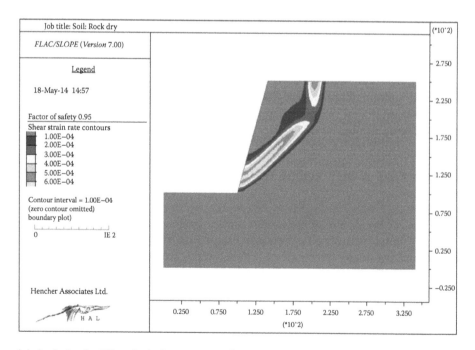

Figure 1.4 Analysis of a 150-m-high slope, cut at 75° in isotropic rock with uniaxial compressive strength of only 1 MPa. Zero pore water pressure. The calculated FoS is slightly <1 and the zone of shearing is clearly identified.

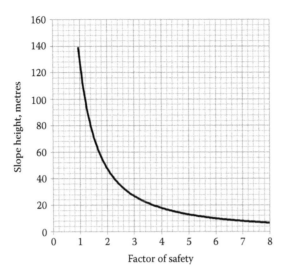

Figure 1.5 Height versus FoS for a slope cut at 75° in weak intact rock with shear strength parameters c = 260 kPa, phi = 35°. No water pressure. Analyses were conducted using FLAC/SLOPE.

1.9 DISCONTINUITIES IN ROCK MASSES

1.9.1 Introduction and relationship to geological history

Rock masses, even very weak rocks, typically contain a network of fractures that has developed throughout their geological history. Such fracture networks commonly control rock mechanical behaviour at the large scale including strength, deformability and permeability. Brittle fractures develop as the result of overstressing of the rock material. A fracture may cease propagation at a particular location as the driving stress is diminished so that the fracture is arrested and this limits the persistence (lateral extent) of any discontinuity. The broad stress field stemming from such process as burial or cooling is then redistributed so that subsequent fractures are initiated elsewhere within the rock mass. Experiments show that the development of a fracture is gradual (Chapter 2); at each stage of fracture some stress is shed and will be distributed into the adjacent rock mass, long before full fracture development. As a consequence, many rock masses contain a network of incipient geological weaknesses. Some of these will eventually develop to become fully-mechanical fractures with zero tensile strength, especially where the rock mass is weathered and unloaded close to the Earth's surface, but many will be present as relatively benign weaknesses.

Under higher confining stress and at high temperature at depth in the Earth's crust, the overstressing may result in plastic folding and general deformation rather than fracturing. Between the brittle and plastic (ductile) extremes, the rock may be both deformed and fractured in response to overstressing. The situation can be highly complex reflecting the longevity of geologial history. At many locations, it can be demonstrated that some joints formed prior to folding, some formed at the same time as folding and some formed after folding. To understand and explain the fracture and folding patterns requires detailed geological detective work.

1.9.2 Fracture development

As discussed in Chapters 2 and 3, the stress tensor at any point in the Earth's crust can be characterised in terms of three principal stress directions (maximum, intermediate and minor). These principal stress directions and relative magnitudes largely control the broad geometry of discontinuity formation although locally, fracture development can be extremely complex; the stress concentrations initiating fracture are dependent on the rock material fabric and any contained flaws (such as fossils, concretions or just bedding intersections) and the influence of pre-existing fractures. An example is shown in Figure 1.6.

In the example shown, there are clearly sets of sub-parallel, intersecting and sometimes curved fractures that might be explained with reference to the geological history at the location in terms of various periods of burial, compression and extension as discussed in more detail in Chapter 3. Some of the longer fractures are associated with regular patterns of short, oblique 'pinnate' joints and might be interpreted as associated with shearing (Hancock, 1985). In the foreground, a series of discrete, mostly four- or five-sided blocks can be made out – almost like mud cracks; the sides are nearly vertical as seen in nearby exposures and Rawnsley (1990) interprets these as 'isotropic hydraulic joints'. This well-exposed and much-studied location has many other geologically significant and indicative structures, including slickensided bedding-parallel calcite veins, cross-cutting dykes filled with sediment, broken and fractured zones and polygonal fold structures. Many of these features have been variously attributed to syn-sedimentary deformation, earthquakes (for the sedimentary dykes), natural 'fracking' from the migration of hydrocarbon gas and conjugate thrusting associated with various tectonic episodes. This complexity might seem inconsistent with the apparently simple geology, with regularly spaced strata gently dipping as seen in the background in Figure 1.7. These points are not made gratuitously but just to emphasise the importance of careful, 'pure' geological study to unravel features that may or may not prove to be of importance as a prerequisite to understanding rock mechanical behaviour in an engineering context. I refer the reader to the Woody Allen quote at the start of this chapter.

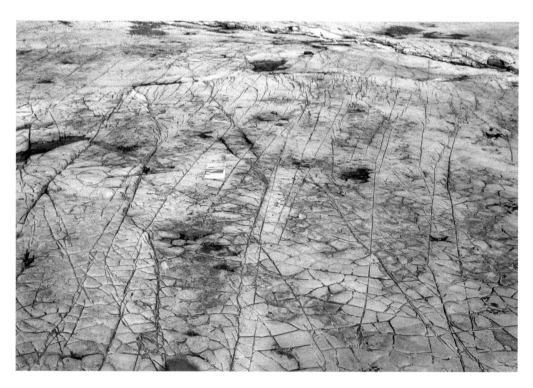

Figure 1.6 Fracture network in strong to extremely strong Jurassic dolomite exposed in a wave-cut platform at Kimmeridge Bay, Dorset, UK. Map (240 mm length), slightly left of centre, for scale.

Figure 1.7　General view to the northeast across Kimmeridge Bay showing the general dip of strata with light coloured, mostly continuous dolomite bands and black organic shales. In the foreground, some of the minor folds (underlain by slickensided calcite sheets), within one of the dolomite horizons have been attributed by various authors to both sedimentary and tectonic origins. West (2014) provides a web page with a very well-illustrated discussion of the geology and history of this location and others along the Dorset Coast, with detailed references.

1.9.3 Joints

The most common natural visible and extensive fractures in rock masses are joints, which are defined as natural fractures with no visible shear displacement. There is an often-expressed misperception that because joints are common, they will be encountered in all rock masses, that they are 'pervasive'. It is true that they are a common visible feature in many rocks, especially in the zone of surface weathering to depths of perhaps 100 m and it is also true that joints due to overstressing could develop at any depth in the upper 10 km or so of the Earth's crust where brittle fracturing dominates. However, rock masses are encountered, especially at a depth, where there is no visible jointing or where joints are very few and far between as discussed in Chapter 3.

The stress states that result in the formation of joints may develop during the burial of sediments, can be due to unloading (e.g. from erosion or from the melting of ice sheets) and from relief of 'residual stresses' locked into the rock, initiated by hydraulic pressures (water, gas or oil) or associated with tectonic activity. The resulting fracture networks can extend over tens of square kilometres or can be very localised.

Most joints form as tensile features (mode I) but also form associated with shearing and as hybrid features (Hancock, 1985) and are discussed in detail in Chapter 3. Joints are found not only in ancient rocks that have undergone long geological histories but also in recent, weak, young and un-lithified geological masses as illustrated in Figures 1.8 and 1.9. Rawnsley (1990) measured the characteristics of 154 joints similar to those in Figure 1.8 and concluded that these are tension fractures that probably owe their origins to the weight of ice several thousand years ago although other possibly contributing factors include unloading, desiccation and extension of near-parallel joints from the underlying bedrock. Figure 1.9 shows systematic joint sets in relatively young colluvium (described as a 'very stiff' soil by Fyfe et al., 2000).

Figure 1.8 Vertical set of joints orthogonal to the cliff face in glacial boulder clay. Robin Hood's Bay, Yorkshire, UK. Rucksack for scale.

1.9.4 Faults

Faults, like joints, are the result of overstressing of the rock but are distinct from joints in that they exhibit measurable shear displacement. Faults are predominantly associated with mountain-building episodes and tectonic movements as discussed in some detail in Chapter 3. They can occur as single features or more commonly as fault systems. Faults are often highly important to rock-engineering projects because they are often associated with large zones of weakened rock. They also act as barriers or passageways for fluids, cause general disruption to the geology, and are associated with sudden releases of energy (earthquakes). An example of a disruptive fault (inactive) is illustrated in Figure 1.10.

1.10 THE IMPORTANCE OF DISCONTINUITIES TO STABILITY

The importance of the adverse influence of discontinuities to stability (and to mass strength generally for all geotechnical engineering) is obvious just from field observation (e.g. Figure 1.11) but is also clear through the examination of stability data for slopes.

Figure 1.12 is a plot of height versus slope angle data from large open-pit mines. The plot includes examples of measured geometries of both stable and unstable slopes.

Figure 1.9 Joints through colluvium on hillside. The colluvium is probably of Pleistocene age (about 1 million years old) according to Fyfe et al. (2000). A set of parallel and orthogonal joints (coated with white kaolin, black manganese dioxide and red iron oxides) can be seen cutting through the colluvium, which comprises mixed weathered granite and volcanic rocks. The location is below Fei Ngo Shan, New Territories, Hong Kong. Geological hammer for scale. The joints probably developed in response to the weight of overburden pressure (now removed).

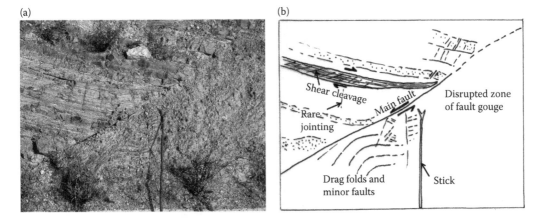

Figure 1.10 (a) Roadside cutting above Morenos, Portugal. (b) Interpretation. Fractured, faulted and folded Carboniferous schist (with alternating sandstone and shaley horizons). One major fault is quite obvious but there has also been slippage along schistosity with a well-defined fracture cleavage confined to one shaley horizon. The degree of jointing is remarkably low. To the right of the walking stick, the rock below the main fault has lost a recognisable structure compared with the folded zone to its left. Clearly, the history here is complicated – not a simple single faulting episode.

Figure 1.11 Large rock failure involving sliding on an irregular fracture network of joints and faults through limestone and ophiolite melange, Dibba Road, Dubai.

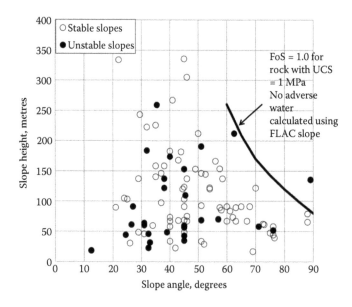

Figure 1.12 Data on large open-pit mine slopes in 'hard' rock. (Redrawn from Hoek, E. and Bray, J.W. 1974. *Rock Slope Engineering*, Institution of Mining and Metallurgy, 309pp.)

The first observation regarding the data in Figure 1.12 is the remarkable lack of any separation between the geometry of stable and unstable slopes. Some slopes are apparently stable at angles of almost 90°, up to about 75 m in height, and at 45° up to about 350 m in height. The second observation is that many slopes have failed at much shallower angles. Many of these are at lower heights than analysis has shown that they should be stable for even the weakest intact rock as discussed earlier (see Figure 1.5) and evident from the FoS = 1.0 line superimposed on Figure 1.12 which is for rock that could readily be broken into pieces by hand. This apparent conundrum is largely explained by the weakening influence of discontinuities and it follows that the important exercises of characterisation of fractured rock masses and determining properties of discontinuities are major themes throughout this book. The other explanation for many slopes failing at lower angles is the influence of 'environmental' effects such as adverse water pressure and seismic shaking although in most cases it is geological factors – the weakening effect of joints and faults – that provide the fundamental cause.

A similar exercise to the one reported in Hoek and Bray was carried out in the early 1980s for slopes cut into weathered rock profiles in Hong Kong – the 'CHASE' study (Figure 1.13).

For the CHASE study, almost 200 slopes were examined in considerable detail with the ground profile characterised up the centre of slopes in inspection trenches. For stable slopes (showing no failure scars), the steepest and highest slopes were selected for study. Failed slopes were selected for study whatever their original geometry. The exercise was aimed at deriving some empirical rules that could be used for designing safe slopes, thereby avoiding extensive and difficult sampling, laboratory testing and analysis in the weathered rock often encountered in Hong Kong.

As for the earlier example of open-pit slopes in 'hard' rock, Figure 1.13 shows poor differentiation between stable and unstable slopes.

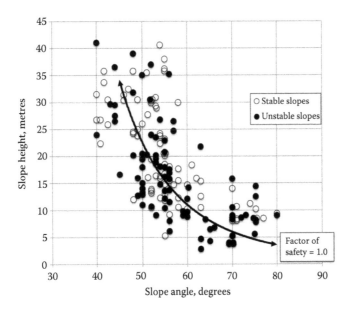

Figure 1.13 Geometry of 'stable' and 'failed' slopes investigated for the CHASE study. FoS = 1.0 curve is for material with cohesion of 7 kPa and phi = 37° but no adverse water pressure. (Redrawn from Geotechnical Control Office. 1984b. *CHASE: Cutslopes in Hongkong Assessment of Stability by Empiricism.* Unpublished report, Geotechnical Control Office, Hong Kong, 5 Volumes.)

The trial FoS = 1 curve superimposed on the data in Figure 1.13 is for the analysis of soil-like material with equivalent uniaxial compressive strength of about 30 kPa, which is considerably weaker than the weak rock discussed earlier. Such strength is also considerably lower than that of most intact weathered rock making up cut slopes in Hong Kong. Despite this, in Figure 1.13, there are many failures that plot below the FoS = 1.0 line. This partly reflects the fact that water pressure is not incorporated in this particular FoS calculation (a factor which triggers most landslides in Hong Kong) but again relates to the important weakening effect of rock discontinuities even in weathered rock profiles.

1.11 EARLY LESSONS AND THE RELEVANCE OF ROCK MECHANICS

The main lessons from these early discussions are

1. The important role of cohesive bonding to the strength of intact rock. Even a rock that can be broken by hand will stand steeply in a slope or stably in a tunnel if unfractured and will also carry high loads in foundations.
2. Rock behaviour, strength, deformability and fluid flow, are often predominantly governed by the fracture network of discontinuities. Much of this book will concentrate on such discontinuities, their origins and how to measure and characterise them.

1.12 APPLICATION OF ROCK MECHANICS

Rock mechanics is the theoretical and applied science of the mechanical behaviour of rock and of the way that rock responds to forces. The same mechanical and physical principles that apply in geotechnical engineering equally apply to structural geology (mountain building, faulting and joint formation) even if the scales and physical conditions including time, pressure and temperature are often rather different (Figure 1.14).

Figure 1.15 illustrates how a rock mass exposure in a quarry might be examined and prove relevant to a wide range of engineering disciplines at a project scale including quarrying, civil engineering, mining and oil and gas reservoir engineering as well as to structural geology itself.

1.13 HISTORY OF THE SUBJECT AREA

People have been constructing and mining in rock since pre-history. The nature of the early man-made structures using rock demonstrates a working understanding of strength, deformation and stress distribution for thousands of years (Gordon, 1978). The use and strength of rock arches in major structures goes back to at least Roman times (Figure 1.16) and has been used ever since to form elegant structures (Figure 1.17). The terminology used in Roman masonry bridge construction is still employed in tunnel liner design where the 'arching effect' within the rock mass is highly significant as discussed in Chapter 9.

1.14 ROCK MECHANICS AS A SCIENTIFIC DISCIPLINE

Rock mechanics has been identified as a separate scientific discipline in engineering only recently. Hoek (2014a,b) gives a concise history of the development of rock engineering. He identifies that a large part of the impetus to develop rock mechanics as a study area

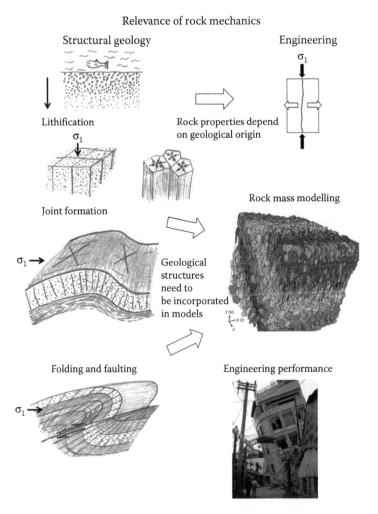

Relevance of rock mechanics

Figure 1.14 Rock mechanics principles are fundamentally important to both structural geology and to engineering. An understanding of the origin of rock material and rock mass characteristics is important for the creation of ground models whether for civil engineering, mining or petroleum reservoir exploitation.

came from the failure of the Malpasset Dam foundations in France in 1959 and the overtopping of the Vajont Dam in Italy from a rock landslide that failed into the reservoir in 1963. Similarly, motivation was given to engineering geology and soil mechanics studies by the failure of the St. Francis Dam in Southern California in 1928 (Krynine and Judd, 1957).

The International Society for Rock Mechanics (ISRM) was established in 1962, developing from an earlier society formed in the 1950s; its first president was Professor Leopold Müller from Austria. Rock mechanics issues received relatively little attention in geotechnics textbooks of that era (compared to soil mechanics) although evidence reveals a long gestation period. For the First International Rock Mechanics Congress in Lisbon in 1966, eight broad topics were identified. These were

1. Exploration of rock masses
2. Description of rocks with a view to their physical and mechanical behaviour

Geological issues
- Origin and age of rocks
- Relations between underlying and overlying strata
- Folding, faulting, jointing
- Dykes and mineralisation
- Weathering history

Oil, gas, water issues
- Source rocks for hydrocarbons
- Rock fracture network connectivity
- Structures and permeability contrasts

Quarrying issues
- Intact rock properties
- Block size
- Blasting
- Crushability

Civil engineering issues
- Intact and rock mass properties
- Stability
- Underground excavation
- Stability in foundations

Figure 1.15 Rock mechanics issues at a single exposure. Note the strong link between different disciplines and how geological factors are fundamentally important to all engineering applications. Dry Rigg Quarry in Silurian sedimentary rock, near Horton in Ribblesdale, Yorkshire, UK. A dump truck for scale.

Figure 1.16 View upwards towards one of the 167 arches supporting the Roman aqueduct at Segovia, Spain. This structure has survived the best part of 2000 years. There is no cement mortar. The rocks forming the arch are called voissoirs; the one at the top of the arch is the keystone. The Romans evidently understood that the arch takes the vertical load from gravity, turns this into lateral loads between the individual blocks and runs them around the arch ring to be carried by the abutments. During the Roman era, each of the three tallest arches displayed a sign in bronze letters, indicating the name of its builder along with the date of construction.

Figure 1.17 The medieval (probably thirteenth century) Puente de Roman, at Cangas do Onis, Northern Spain. The structure is formed of rock blocks and founded on rock foundations.

3. Properties of rocks and rock masses
4. Residual stresses in rock masses
5. Comminution
6. Natural and excavated slopes
7. Underground excavations and deep borings
8. Behaviour of rock masses as structural foundations

There were 241 papers from 32 different countries, and many of the major papers remain useful today and form the foundation for modern rock mechanics practice. All of the research areas are still important.

There are two major, long-established journals devoted to rock mechanics (the *International Journal of Rock Mechanics and Mining Sciences* and *Rock Mechanics and Rock Engineering*) and a relatively newcomer, the *Journal of Rock Mechanics and Geotechnical Engineering*. Rock mechanics issues and engineering in rock are also reported in a very wide range of other journals, including those concerning structural geology, petroleum and oil engineering, environmental, earthquake and general geotechnical engineering, and this reflects the wide-ranging applicability of the basic principles of mechanics to the study of rocks. Indeed, many of the principles regarding forces, stress and material behaviour apply throughout all engineering disciplines and material sciences. Much of the stress analysis in rock mechanics, at least for isotropic situations,* is derived from elasticity theory in general engineering science and physics. One difference is that in rock mechanics practice, structural geology and rock engineering, compressive stress is taken as positive, and tensile stress is taken as negative (the opposite convention is adopted in much engineering stress analysis). This is appropriate because stress in the Earth's crust is usually compressive and

---

* Hudson and Harrison (1997) differentiate between a CHILE (continuous, homogeneous, isotropic and linearly elastic) concept for materials that '*are traditionally required for modelling*' and DIANE (discontinuous, inhomogeneous, anisotropic and non-elastic), which is the '*rock with which the engineer has to deal*'. They comment, '*There is little chance of any modelling based on CHILE assumptions being realistic*'.

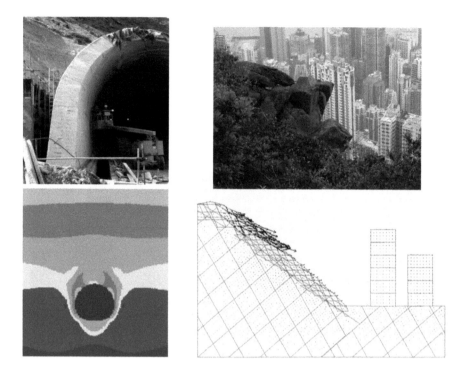

Figure 1.18 To the left, isotropic stress analysis for tunnel liner design (FLAC) and to the right, rock slide potential using the discontinuum model (UDEC). Both software packages are by Itasca Corporation. (Left figure: From Swannell, N.G. and Hencher, S.R. 1999. *Proceedings of the 10th Australian Tunnelling Conference*, Melbourne, March.)

the conflict is not a practical problem. The other major difference is that in rock mechanics, we are often dealing with irregularly fractured and heterogeneous rock masses to which relatively simple elastic theory is not appropriate.

One of the major achievements in rock mechanics over the last 50 years is the development of sophisticated computer software that can simulate the behaviour of such complex rock masses under load and model fluid flow through them (Figure 1.18). One of the major challenges that remain is that methods of characterising rock masses (the structural geology end of rock mechanics) has not really kept up with numerical modelling capabilities for analysing engineering situations. A stochastic (statistical) approach is often used in setting up geological models. This approach has been described as *'informed guesswork'*. There is also an argument that the rock mechanics (engineering end) fraternity has not paid as much attention to developments in structural geology as might have been done. Hopefully, this book will help redress that balance.

1.15 LOAD CHANGES

Much of rock mechanics practice involves prediction of deformation or strain in response to changing load. Deformation must be kept within tolerable limits for structural performance and safety.

Changes in load may be natural or anthropogenic. Examples of natural load changes include burial by sediment, changes in fluid pressure due to groundwater rise or fall and erosion. The melting of glaciers since the last ice age 20,000 years ago has led to an overall isostatic rebound (uplift) in Scotland, Scandinavia and N. Canada of perhaps 800 m that is continuing at about 10 mm per year. This is similar to the rates of tectonic uplift in the Himalayas. Since the last glacial maximum, sea levels have risen by between 110 and 175 m (Pirazzoli, 1996) with obvious consequences for coastal retreat as well as river gradients and patterns. Parts of the east coast of Yorkshire are being eroded at a rate of about 1 m per year.

Load change may be essentially static (constant) or dynamic as in the case of an earthquake or wave loading. Fluid pressures are often important, for example, increasing due to rainfall and causing a rise in groundwater table or decreasing as oil and gas are extracted or where groundwater is exploited for water supply. Increasing water pressure can reduce the available shear strength and cause landslides; reducing fluid pressure can have the opposite effect with increased effective stress causing consolidation settlement, even in rock.

An example of intense anthropogenic-loading conditions can be found in the construction of a suspension bridge with a very high vertical compressive loading in the foundations below the towers and huge tension forces in the cables that need to be resisted by anchor blocks at each end of the cables. In the case of Tsing Ma Bridge in Hong Kong, the two anchorage concrete blocks weigh 200,000 and 250,000 tonnes to resist a maximum tensile pull from each of the two suspension cables of 50,000 tonnes. A case example of the procedures necessary to ensure a valid design for rock anchorage of a major suspension bridge, subject to a very high earthquake loading in Turkey is presented in Chapter 7. For the Incheon 2nd Crossing, the longest span cable-stay bridge in South Korea, loading tests were conducted on individual bored piles founded in rock with loads applied in excess of 30,000 tonnes. This is described more fully in Chapter 7. Tunnels, cut slopes and mining generally cause unloading of the rock mass which leads to deformation driven by changes in gravity and other *in situ* stresses as the lateral or vertical support is removed.

Chapter 2

Fundamental mechanics

All science is either physics or stamp collecting...

Ernest Rutherford

2.1 DEFINITIONS

The concepts of stress and strain are fundamental to all aspects of geotechnical engineering and it is important to define some terms before discussing how these are used.

2.1.1 Force and load

Force and *load* are synonymous and are expressed in the metric system in newtons (N) or multiples of newtons. For most rock mechanics applications, the units kilonewtons (10^3 N) and meganewtons (10^6 N) are appropriate at project scale; giganewtons (10^9 N) come into play where describing stiffness or deformability of rock.

One newton is the force produced by 1 kg (mass) accelerated by 1 m/s². On Earth, the acceleration due to gravity is about 9.81 m/s², so 1.0 kg mass weighs about 10 N.

The weight 10 N is the equivalent of 1 kg force or 1 kgf. If one purchased 1 kg of apples comprising 10 apples, each apple would weigh about 1 newton.

One tonne is 1000 kg (mass); the term is also used colloquially as the weight of 1000 kg (should actually be tonne-force). One-tonne-force is 9.81 kN on Earth, and generally this is approximated with sufficient accuracy in calculations as 10 kN. One cubic metre of water weighs approximately 1.0 tonne (10 kN). One cubic metre of unweathered granite weighs about 2.7 tonnes (27 kN); the same volume of completely weathered granite containing many voids might weigh 1.2 tonnes – not much heavier than water. A cubic metre of sand might weigh about 1.6 tonnes (16 kN).

Force is a *vector* in that it has magnitude and direction. It is the product of mass, a *scalar* quantity with only magnitude (no direction), and acceleration, a vector, with magnitude and direction (you need to define towards which direction the acceleration acts). In the case of weight it is directed towards the centre of the Earth with the mass accelerated by gravity.

Other examples of scalar quantities include temperature, time and density (Mg/m³) – they are described completely by single values and have no defined direction.

2.1.2 Stress

Stress is force divided by area. The average vertical stress due to the weight of a 10-m high column of granite would be about 270 kN/m². This is also expressed as 270 kPa (kilopascals). *Pressure* is also defined as force divided by area but is usually used to describe the state of stress in a gas or liquid, and sometimes essentially interchangeably with stress as in the term 'overburden pressure' to mean the vertical stress at depth in the ground.

Stress is defined as a *tensor* in that it has magnitude and direction but must also be defined with reference to the plane on which it is acting. The natural stress state in the ground will usually differ according to the direction that we measure or experience it. For example, at some point in a pile of sand, the vertical stress will be due to the vertical weight of the sand above that point; lower stresses will also act radially towards that point, due to lateral expansion caused by the gravitational loading* (the horizontal stress would probably be about 0.3 times the vertical stress). That stress could be measured physically with an instrument such as a pressure gauge. Measured stresses would differ with direction within the pile of sand, all due to the same gravitational loading as explained later; so, measuring and defining the state of stress in a soil or rock mass is not a simple matter.

The most common symbol generally used for stress is σ (sigma) with different suffixes used to denote relative magnitude and direction. Sigma is also used to denote maximum strength (stress at failure) as in σ_c to represent compressive strength and σ_t for tensile strength. Compressive stress is taken as positive, tensile stress as negative. Other suffixes used include σ_v for vertical, and σ_h for horizontal. The overall stress state in a body can be expressed according to three orthogonal, '*principal stresses*', σ_1, σ_2 and σ_3, in decreasing order of magnitude, as explained later. In the ground, maximum or minimum stresses are typically either vertical or horizontal but also might be inclined, depending on the topographic and geological setting. Stresses may also be described relative to arbitrary axes, x, y and z, as in σ_x, σ_y and σ_z.

The stress acting at right angles to a plane is defined as the '*normal*' stress, σ_n. Stress acting parallel to a plane is defined as the '*shear*' stress, τ (tau) (Figure 2.1).

For a mug of tea on a table, the force acting on the table is due to the weight of the mug (mass times gravitational acceleration); the normal stress underneath the mug is its weight divided by its basal area. This is the way that gross engineering stress is calculated. In fact, the base of the mug is probably not flat, so the actual stress acting between the mug and table is carried on a smaller area of contact and is much higher. If we were to look at a microscopic level, we would see that the actual contact points are tiny and stresses very high. This does not matter generally, when designing bridges (or mugs) but when we come to deal with friction, we will see that it does, if the phenomenon is to be understood scientifically rather than just empirically.

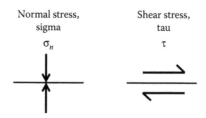

Figure 2.1 Normal and shear stresses relative to a given plane.

---

* The cause is the Poisson effect. If a rubber band is stretched it becomes noticeably thinner; if a rubber cylinder (or pile of sand) is compressed from its ends it will bulge in the middle. The inverse ratio of strain in the major stress direction to the strain at right angles in known as Poisson's ratio and is typically about 0.3 for rock and soil.

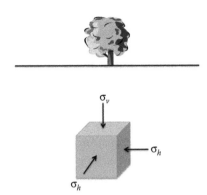

Figure 2.2 Cube with three principal stresses acting on principal planes.

If the table is horizontal, there is no force causing the mug to slide – there is no shear-stress component relating to the gravitational weight of the mug – no shear 'traction'. A normal stress with no shear component is called a *principal stress* and the plane on which that stress is applied is called a *principal plane*. All unsupported excavation or natural surfaces, for example a tunnel wall or a slope face, are principal planes by definition in that there is no disturbing shear traction in the steady state.

More generally, Figure 2.2 shows an elementary cube representing a single point in the ground, with stresses acting on all six faces.

With the cube oriented as shown, the vertical stress, σ_v is due to gravity; the stresses acting on the vertical sides of the cube, σ_h, relate to the deformability of the ground and are squeezing or 'confining' stresses. The cube is in balance and there are no shear stresses acting on the sides of the cube. In this case, all of the normal stresses are also principal stresses and the planes are all principal planes. The stresses may all have different values, in which case the maximum principal stress is σ_1, the lowest stress is σ_3, and the intermediate stress is σ_2. If the stresses were all the same, in every direction, the stress state would be termed 'hydrostatic' as it is at a point at depth in a lake or swimming pool.

2.1.3 Stress transformation

In a given stress field, which can be defined by the three principal stress magnitudes and their orientations, the stress conditions will vary according to the plane under consideration, as illustrated in Figure 2.3 (see Box 2.1). In the left-hand drawing, for a horizontal plane, the force, *F*, acts at right angles and there is no shear stress. For an inclined plane, drawn

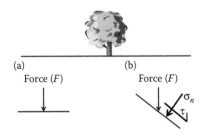

Figure 2.3 (a) Force in the ground and (b) stresses (normal and shear), generated on a particular plane in the ground, by that force.

BOX 2.1

Tricky concepts: The stress state applies throughout a rock mass but varies according to the direction we sample it (and from location to location). If we cut across the stress field in one direction at one point (say horizontal), we would measure different normal and shear stresses acting on that horizontal plane than if we cut through it, at the same point, in a different direction. Because the stress field acts on a body in every direction, any measurement needs to be defined both by magnitude and direction of sampling, and these are the qualities that make stress a 'tensor'. It differs from a 'vector' such as force or velocity, in that a vector has magnitude and a unique direction. It also differs from properties such as temperature or speed that have magnitude but no directional property and are called 'scalar'.

through exactly the same point at the same time, the stress state needs to be defined both by the normal stress acting across the plane, and the shear stress acting parallel to the plane. In three dimensions it becomes more complex; so, if we rotate the cube in Figure 2.2 so that none of the planes are principal planes, then the stress state needs to be defined by six independent shear-stress measurements and three normal stresses acting at right angles to each of the planes, as illustrated in Figures 2.4 and 2.5.

As shown in Figure 2.4, in the generalised case, there are nine stress components acting on an element of rock. The horizontal plane in Figure 2.4 (not necessarily horizontal in reality) is denoted as z, the planes orthogonal to plane z as x and y. There are two subscripts to each of the stresses. The plane on which the stress acts defines the first subscript; the second subscript denotes the direction of stress. The stresses normal to the cube faces are σ (sigma); σ_{xx} acts normally to plane x but also in an x direction.

Shear stresses τ (tau) act on each of the planes in this generalised case, and can be defined in two orthogonal directions tangential to each plane. Thus τ_{zy} and τ_{zx} act tangentially to plane z (the first subscript), and in directions y and x, respectively.

Given that for a stable situation (avoiding rotation of the elemental cube), opposing pairs of shear stress need to cancel each other out, the state of stress in the ground, as a general rule, requires only (only!) six separate measurements – three normal stresses and three shear stresses, as depicted in matrix form in Figure 2.5. This analysis is important if one is trying to measure the 3-dimensional (3D) state of stress in the ground and methods to do so

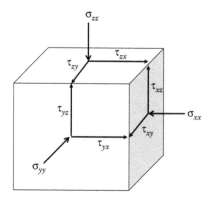

Figure 2.4 Stresses acting on a generalised cube at a point in the rock mass.

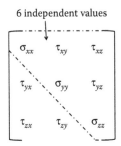

6 independent values

Figure 2.5 Stress matrix for the stresses acting on the elemental cube in Figure 2.4.

are discussed in Chapter 5. Commonly used methods, such as hydraulic-fracturing, only provide three principal stress measurements, and then are interpreted on the assumption that one principal stress direction is vertical. It takes very special equipment to measure the total stress state at a point, and even then the data may only be locally relevant because of geological complexity. Hudson (1989) and Hudson and Harrison (1997) clearly explain stress analysis and its measurement and are recommended reading. For those who wish to delve more deeply, then Jaeger et al. (2007) and Farmer (1983) offer fuller discussion.

If 3D analysis is important because of the geometry of the situation, for example at the intersection of tunnels, then there are software programmes that can be used to calculate the stress conditions, including Examine3D (Rocscience) and FLAC3D (Itasca). Where the 3D geological conditions are important, software like 3DEC (Itasca) can be used to analyse the situation. Fortunately, for many practical engineering and structural geological analyses, 2D analysis is adequate. One of the advantages of 2D analysis is that the models are relatively easy to set up, so that many simple analyses can be conducted quickly and relatively cheaply, to explore the relative importance of various assumptions in the ground models (Starfield and Cundall, 1988). For most engineering projects, one assumes that gravity is one of the principal stress directions, usually the major one, but this is not always the case, as discussed in more detail in Chapter 3.

2.2 MOHR CIRCLE REPRESENTATION OF STRESS STATE

For most projects, the situation can be adequately considered in two dimensions, chiefly in the plane containing both the maximum σ_1 and minimum σ_3 principal stresses – that is, at right angles to σ_2. It is in this plane that failure generally occurs in nature, as in the Anderson (1951) theory of major faulting discussed in Chapter 3.

In Figure 2.6, a force, F, is applied to the top of the elemental cube with no lateral force. This is termed 'uniaxial loading'. The vertical stress $\sigma_v = F/A$ where A is the cross-sectional area of the cube. There is no shear stress acting on the upper cube surface so σ_v is a principal stress – it is also the maximum principal stress, σ_1 because the lateral stresses are zero. Consider a plane inclined at $\theta°$ relative to the applied force. By resolution of forces, the component of F acting normal to the inclined plane, F_n, is given by

$$F_n = F \cdot \sin\theta \tag{2.1}$$

However, the area of the inclined plane A' is larger than area A:

$$A' = A/\sin\theta \tag{2.2}$$

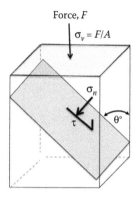

Figure 2.6 Calculating stresses on an inclined plane.

The normal stress, σ_n, acting on the inclined plane is

$$F_n/A' = (F \cdot \sin\theta)/(A/\sin\theta) = (F/A)\sin^2\theta \tag{2.3}$$

so that:

$$\sigma_n = \sigma_v \cdot \sin^2\theta \tag{2.4}$$

Considering the shear stress, τ, acting on the plane, the component of vertical force acting tangential to the inclined plane is

$$F_t = F \cdot \cos\theta \tag{2.5}$$

As before:

$$F_t/A' = (F/A) \cdot \cos\theta \cdot \sin\theta \tag{2.6}$$

that is

$$\tau_v = \sigma_v \cdot \cos\theta \cdot \sin\theta \tag{2.7}$$

If one carries out the same exercise for a force acting horizontally (again, uniaxial conditions) then the normal stress component calculated is

$$\sigma_n = \sigma_h \cdot \cos^2\theta \tag{2.8}$$

where σ_h is the horizontal principal stress.

In a biaxial situation, where $\sigma_1 = \sigma_v$ and $\sigma_3 = \sigma_h$ (or the other way around), the normal stress on the inclined plane will be the sum of the resolved stresses so that

$$\sigma_n = \sigma_1 \cdot \sin^2\theta + \sigma_3 \cdot \cos^2\theta \tag{2.9}$$

Carrying out the same analysis as in Equation 2.2, above, for the shear stress component derived from the horizontal stress ($\sigma_h = \sigma_3$),

$$\tau_h = \sigma_3 \cdot \cos\theta \cdot \sin\theta \tag{2.10}$$

But this shear component acts in the opposing direction to that derived from resolving σ_v (σ_1) so that the total shear stress tangential to the inclined plane is

$$\tau = (\sigma_1 - \sigma_3) \cdot \cos\theta \cdot \sin\theta \tag{2.11}$$

From trigonometry, Equations 2.4 and 2.6 can be written following the double-angle rule for 2θ so that:

$$\sigma_n = \frac{\sigma_1 + \sigma_3}{2} - \frac{\sigma_1 - \sigma_3}{2} \cdot \cos 2\theta \tag{2.12}$$

and

$$\tau = \frac{\sigma_1 - \sigma_3}{2} \sin 2\theta \tag{2.13}$$

It is a convenient fact that the complete 2D stress state at a single point can then be represented graphically by a circle with axes of shear stress and normal stress, with diameter $= \sigma_1 - \sigma_3$. This is known as a Mohr stress circle (actually, only semicircles are necessary for analysis) and is illustrated in Figure 2.7.

The Mohr circle represents all the stress states acting within the sample in 2-D, simultaneously. If the sample were to fracture at some increased value in σ_1 we could determine the stress state on that plane simply by measuring the inclination of the failure plane and consulting our Mohr plot. If we do this for several different samples at different confining stresses (σ_3), we can plot out a strength envelope for the intact rock as illustrated in Figure 2.8 and discussed in more detail in Chapters 3 and 5. In this figure, the strength envelope is drawn as a straight line following the Mohr–Coulomb failure criterion so that:

$$\tau = \sigma \tan\varphi + c \tag{2.14}$$

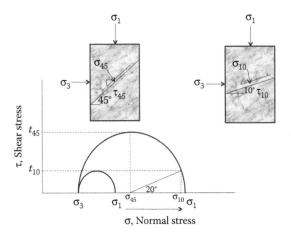

Figure 2.7 Mohr representation of changing stress state in a sample. The two semi-circles represent the stress change as σ_1 increases from one value to another. The minimum stress, σ_3, is kept constant. The two rectangles illustrate the stress state on two arbitrary planes through the stressed sample, inclined at 10° and 45° to the horizontal, and how these can be read from the Mohr circle representation. (Note that, as explained in Section 1.5, the dihedral angles (2θ) between the maximum principal stress direction and conjugate inclined surfaces, are 160° and 90°, respectively, for these planes.)

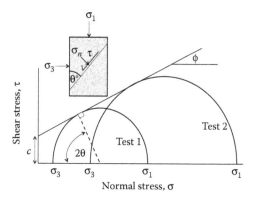

Figure 2.8 Mohr circle presentations of stress states in two different samples, at failure. The tangential point on each circle is the critical stress state at which failure occurs (other stress conditions within each sample are stable, being below the failure envelope).

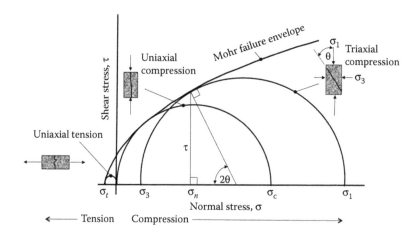

Figure 2.9 Generalised strength envelope for intact rock.

Figure 2.9 shows a more typical generalised, curved failure envelope for intact rock. At the left, there is a circle representing the uniaxial tensile strength of the rock. Also shown are Mohr circles for the uniaxial compressive situation and for a triaxial condition, where the rock is confined by a lateral stress and the mode of failure is by shear on an inclined plane.

2.3 STRESS CONCENTRATION IN UNDERGROUND OPENINGS

Cutting a hole in a pre-existing stress field will cause the stress conditions to change as illustrated in Figure 2.10.

The stress changes can be calculated following elastic theory, as per Figure 2.11.

In the example illustrated at a depth of 25 m, the induced stress conditions would not be very significant for an excavation in strong rock. In weaker rock such as over-consolidated London Clay or chalk, however, such stress concentrations might be significant and, indeed,

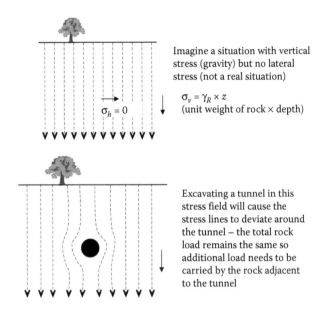

Imagine a situation with vertical stress (gravity) but no lateral stress (not a real situation)

$\sigma_v = \gamma_R \times z$
(unit weight of rock × depth)

$\sigma_h = 0$

Excavating a tunnel in this stress field will cause the stress lines to deviate around the tunnel – the total rock load remains the same so additional load needs to be carried by the rock adjacent to the tunnel

Figure 2.10 Pre-existing lines of stress will deviate (flow) around any underground opening, which itself, of course, cannot carry any stress.

for the early stages of the Heathrow Express collapse (in London Clay), cracking along the length of the tunnel occurred in the invert (Muir Wood, 2000). See also Chapter 9.

Many underground excavations are irregular in shape, with different elements of a project, such as cross-excavations or parallel tunnels grouped closely together, and, probably, with irregular and anisotropic ground conditions. Stress conditions and concentrations would usually be calculated using computer-based numerical methods, as discussed earlier.

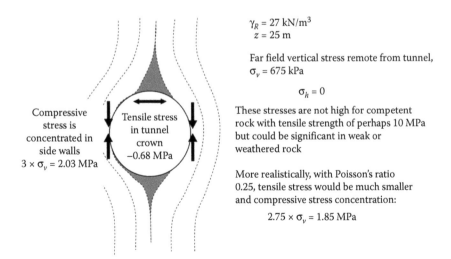

$\gamma_R = 27 \text{ kN/m}^3$
$z = 25 \text{ m}$

Far field vertical stress remote from tunnel,
$\sigma_v = 675 \text{ kPa}$

$\sigma_h = 0$

These stresses are not high for competent rock with tensile strength of perhaps 10 MPa but could be significant in weak or weathered rock

Compressive stress is concentrated in side walls
$3 \times \sigma_v = 2.03 \text{ MPa}$

Tensile stress in tunnel crown
-0.68 MPa

More realistically, with Poisson's ratio 0.25, tensile stress would be much smaller and compressive stress concentration:

$2.75 \times \sigma_v = 1.85 \text{ MPa}$

Figure 2.11 For this imaginary situation with zero horizontal stress, the vertical stress will be concentrated in the side walls of the tunnel by three times. In the crown and invert of the tunnel, tensile stress will develop (−0.68 MPa for this particular calculation). Calculations follow Megaw and Bartlett (1981).

2.4 STRESSES BELOW FOUNDATIONS

Design and construction of foundations on rock are addressed in Chapter 7. In terms of stress, the normal assumption based on elastic theory is that there is a bulb of stress beneath a footing that diminishes with depth, and laterally away from the footprint of the foundation. Generally, at a depth of about 1.5 times the foundation width, the vertical stress is reduced by about 90% so that almost all the deformation of the ground (causing settlement for the structure) will be in the zone immediately beneath the footing. Where the rock is strongly anisotropic, the stress 'bulb' will be deformed to follow the structure; this will not usually be of concern unless the rock is very weak and deformable, or where there is a specific potential failure mechanism as discussed in Chapter 7.

Many rocks are sufficiently strong and stiff so that settlement calculations are only a minor concern. That is not the case where the rock is very weak through fracturing or weathering, and varying conditions across the founding area may need to be dealt with either by local excavation or strengthening of the foundation itself.

2.5 EFFECTIVE STRESS

The concept of effective stress is very important to all geotechnical engineering: '*effective stress rules!*' In the case of dry sand, all the stress (from an overlying building or just from self-weight) will be carried by the contacts between sand grains, and these contacts will generate friction. However, if the sand is saturated, then any fluid pressure will reduce the stress acting between grains, with a consequent reduction in friction and in overall mass strength. In the case of partially saturated sand, there will be an increased effective stress due to suction effects, as the sand dilates and water menisci between grains are deformed. Terzaghi (1925) first identified the principal of effective stress, in the context of soil mechanics. Hubbert and Rubey (1959), recognising that extremely high fluid pressures were encountered sometimes in oil wells with pressure '*as high above the ground as the well was deep*', and noting Terzaghi's work on effective stress, identified that high fluid pressures could also be invoked to explain the apparent gliding movements of blocks of rock, km in thickness, on almost horizontal planes. This work by Hubbert and Rubey '*completely transformed the way in which geologists think about joints, shear fractures and faults*' (Davis and Reynolds, 1996), and is discussed in more detail in Chapter 3.

It can be seen that, in rock mechanics, water and other fluid pressures are very important to mechanical behaviour. This is especially true for the shear strength of rock discontinuities. Generally, pore pressure in the rock matrix is less important than it is to soil but there are exceptions, especially in water, oil and gas extraction, where matrix porosity and pressure within the pores play a vital role, both in terms of extraction, and reservoir behaviour on depletion, as the *in situ* pressure is reduced.

2.6 ROCK DEFORMATION AND BEHAVIOUR

2.6.1 Elastic behaviour and Young's modulus

Many rocks behave essentially elastically when loaded lightly. This means that:

1. Deformation increases linearly with applied load
2. When the load is removed, the rock returns to its original shape

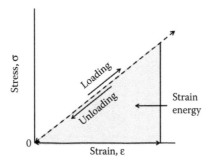

Figure 2.12 Schematic representation of ideal elastic behaviour.

This concept is illustrated in Figure 2.12.

The area under the stress–strain curve is the strain energy – essentially stored energy that could be released suddenly, like a coiled spring. This simple concept has wide application and is the crux of the commonly accepted mechanism for how fault-slip releases damaging seismic waves: rock is gradually strained due to ground movements (Chapter 3). The stored strain energy is suddenly released when the strength of the fault is exceeded – this is essentially the elastic rebound theory of Reid (1910).

The ratio stress/strain (σ/ε) is known as the modulus of elasticity or *Young's modulus* (E). Strain (ε) is a number (ratio of change of length over original length), so Young's modulus has the same units as stress, σ (kN/m^2).

If the whole rock volume is compressed by a uniform stress (hydrostatically), then the stress/volumetric strain is called the *bulk modulus* (G).

2.6.2 Rock behaviour

Rock responds to change in stress in different ways according to whether or not it is 'confined' by surrounding rock. Its response also depends on temperature, pressure and rate of loading. Even the strongest and most brittle rocks at the Earth's surface flow and deform like modelling clay at great depth in the Earth's crust (Figures 2.13 and 2.14).

Figure 2.13 Plastically deformed gneiss with late-stage quartz-rich pegmatite vein from cooling fluids under great pressure, Island of Coll, West Scotland. Binoculars, length 140 mm, for scale.

Figure 2.14 Plastic folding in hornblendic gneiss. Note also that some black bands are segmented with well-developed boudinage structure. These are relatively brittle horizons that have been stretched and deformed within the less competent, more plastic horizons. Island of Coll, West Scotland. A4 map folder for scale.

2.6.2.1 Brittle fracture and Griffith crack theory

When a structure or sample is broken by loading in tension the failure is not caused by the direct action of the force pulling on the electro-chemical bonds between atoms. The theoretical strength of the bonds is far higher than strength measured in tensile tests. Instead, failure results from strain energy being stored in the body as it is loaded and *'the sixty-four thousand dollar question, whether the structure actually breaks at any particular juncture, depends upon whether or not it is possible for this strain energy to be converted into fracture energy so as to form a new crack'* (Gordon, 1978).

Brittle fracture is defined as sudden failure without prior plastic yielding (Figure 2.15). Glass is perhaps the most brittle material we are familiar with; it has high tensile strength (much higher than rock or concrete) but a very low 'work of fracture', which is the energy per unit area required to extend a fracture through the intact material. For glass this energy requirement is of the order of 1 J/m^2. As Gordon (1978) puts it:

> *As a matter of fact, 1 J/m^2 is a rather pathetically small amount of energy. It is a sobering thought that, on the simplest theory, the strain energy which could be stored in one kg of tendon would 'pay' for the production of 2,500 square metres of broken glass surface – which accounts for the effects of bulls in China-shops.*

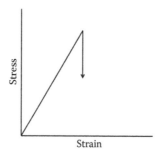

Figure 2.15 Brittle behaviour.

Rocks have a higher work of fracture than glass (by a factor of perhaps 4) and so they are less prone to shattering, but this is a factor of thousands less than the energy required to extend a fracture through mild steel. Through comparison with glass and steel, most rocks are generally described as essentially brittle at the low temperatures of the Earth's surface but intact rock specimens are not fully brittle in reality (i.e. not quite as shown in Figure 2.15).

Rocks typically contain many flaws such as pores and mineral grain boundaries. These weaknesses act as the loci for tensile stress concentration at crack points in both tensile and compressive remote stress fields (e.g. Mandl, 2005). The failure process, either in tension or compression, involves adversely oriented micro-cracks linking up and propagating, often tangentially under a hybrid tensile and shear stress field to form a macroscopic fracture (e.g. Hoek and Bieniawski, 1965; Engelder, 1999).

According to Griffith's (1921) crack theory, a critical length of crack exists at which the strain energy stored in the rock volume (3D) adjacent to the crack tip balances the energy required for a planar crack (2D) to propagate. Once the crack extends slightly beyond this point then the crack can develop uncontrollably as the strain energy is released. This does not necessarily happen in reality partly because the driving stress may be relieved, especially where the fracture growth is being driven by water pressure, but nevertheless, the extension of minor flaws to become major discontinuities is the generally accepted mechanism of formation and can be observed in laboratory tests and simulated in numerical models as we shall see.

As discussed in Chapter 3, the redistribution of stress as fractures develops in a rock mass, and transfer of that stress to other parts of the rock body, helps to explain the systematic nature of many joint systems (e.g. Ladeira and Price, 1981). The local propagation of a single fracture relieves the stress locally (releases strain energy adjacent to the fracture) but the large-scale stress field retains its general orientation and driving forces. New fractures will tend to develop with the same general geometry as earlier ones but at some critical distance. Furthermore, the fact that the shedding of stress probably occurs before a fracture has fully developed may help to explain the incipient nature and limited persistence of most geological discontinuities as discussed in more detail in Chapter 3.

2.6.2.2 Failure of rock

Figure 2.16 shows a numerical model comprising almost 10,000 spherical particles, simulating the mineral grains in a rock specimen. Each particle has its own stiffness defined, and is bonded to its neighbours with inter-particle properties modelled fairly realistically in terms of friction, cohesion, tensile strength and stiffness, with variance in bond tensile strength and cohesion as might be expected in a real rock.

Figure 2.17 shows the results from a simulated unconfined strength test on the sample in 2.16. Details of test procedures are presented in Chapter 6.

Figure 2.18 shows the test at a stage before peak strength has been reached. Some of the particle bonds are failing in tension, others in shear, and the loci of damage are beginning to define linear directions consistent with shear failure. Hazzard and Damjanac (2013) calculated the likely acoustic emissions (noise) that would be associated with formation of microcracks and these are growing exponentially with strain in the model. Such acoustic emissions also accompany real rock failure in the laboratory and in the field and a practical application of this phenomenon is monitoring by sensitive instruments to identify gradual failure mechanisms and to provide warnings, for example, of sudden rock bursts in highly stressed rock in deep mines.

Post-peak, full fractures develop as illustrated in Figure 2.19. In real tests these are typically either vertical (essentially tensile failure – opening mode) or follow shear directions,

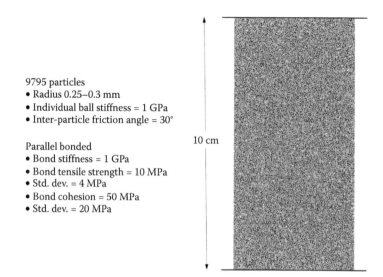

9795 particles
• Radius 0.25–0.3 mm
• Individual ball stiffness = 1 GPa
• Inter-particle friction angle = 30°

Parallel bonded
• Bond stiffness = 1 GPa
• Bond tensile strength = 10 MPa
• Std. dev. = 4 MPa
• Bond cohesion = 50 MPa
• Std. dev. = 20 MPa

10 cm

Figure 2.16 Simulation of gradual breakage of rock cylinder under uniaxial compression, using discrete element code PFC (Itasca).

even though the mechanism is still largely due to localised tensile failure triggered at flaws (e.g., Engelder, 1999).

While the test simulation illustrated in Figure 2.17 shows essentially constant stress–strain (modulus of elasticity) up to peak strength and failure, carefully conducted uniaxial compressive strength tests on intact rock show more complex behaviour in detail as illustrated in Figure 2.20. On initial loading there is some axial shortening of the specimen with relatively little lateral strain (reduced Poisson effect compared to what is to come). The stress–strain ratio is relatively low and this is attributed to an initial phase during which flaws, microcracks and pores are compressed. After this initial phase, the sample becomes stiffer (higher stress–strain ratio) and the rock is behaving essentially elastically. At some

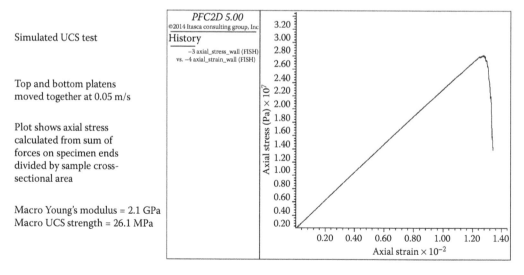

Simulated UCS test

Top and bottom platens moved together at 0.05 m/s

Plot shows axial stress calculated from sum of forces on specimen ends divided by sample cross-sectional area

Macro Young's modulus = 2.1 GPa
Macro UCS strength = 26.1 MPa

Figure 2.17 Details of simulated unconfined compressive strength test with stress–strain diagram.

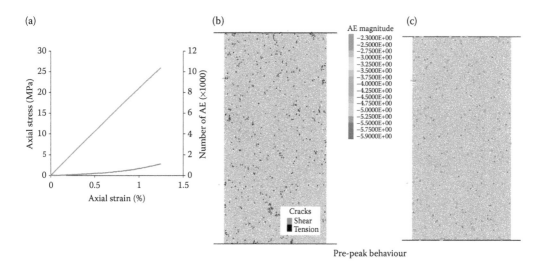

Figure 2.18 Pre-peak behaviour with localised failure of bonds in tension and shear. (a) Stress–strain curve in blue (upper line). Number of acoustic emissions (lower line) increasing with strain exponentially. (b) Development of microcracks some of which are lining up at angles of about 45 to 80 degrees relative to vertical loading direction (mostly in upper part of model). (c) Acoustic emissions match loci of microfracture development.

stress level, typically between 40% and 60% of the peak strength, new cracks will start to open – some tensile (Mode I) where oriented roughly parallel to the axial loading and others shearing (Mode II) on grain boundaries and other flaws oriented obliquely to the axial load. This crack formation is accompanied by acoustic emissions. At about 70–90% of the peak strength, the total volume of the rock specimen will begin to increase (dilation) and this volume growth is accompanied by the propagation of larger fractures through the coalescence of microfractures and a strong increase in acoustic noise (Hoek and Martin, 2014).

Gradual and progressive failure is also the norm for more complex rock masses containing incipient discontinuities of different strengths and orientations, as illustrated in Figure 2.21.

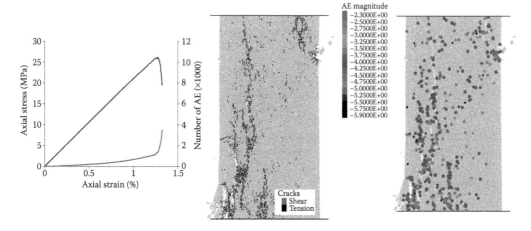

Figure 2.19 Failed sample with vertical and inclined (shear) directions dominating (see Figure 2.18 for explanation of individual graphs and images).

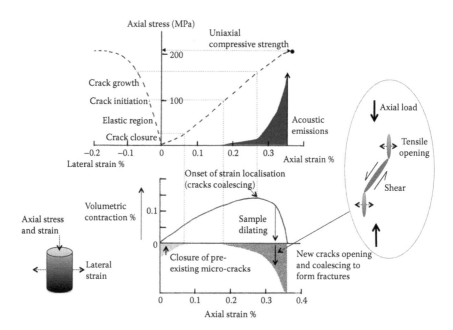

Figure 2.20 Detailed behaviour of a rock sample under uniaxial loading to failure. (Redrawn and based on Hoek, E. and Martin, C.D. 2014. Fracture initiation and propagation in intact rock – A review. *Journal of Rock Mechanics and Geotechnical Engineering*, **6**, 287–300.)

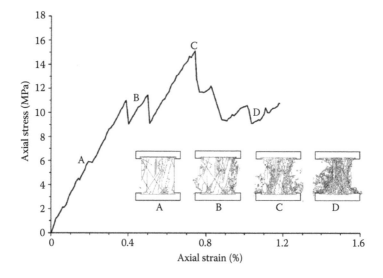

Figure 2.21 Simulation of gradual pillar collapse using finite element computer software ELFEN. (From Pine, R.J. et al. 2006. *Rock Mechanics and Rock Engineering*, 39(5), 395–419.)

2.6.2.3 Plasticity

A material is described as plastic or ductile where it continues to yield and absorb energy without loss of strength as in Figure 2.22.

Plastic behaviour is typical of metals that deform considerably before they fracture. This property is exploited in the design of steel structures to withstand earthquakes, where they

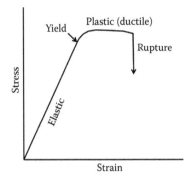

Figure 2.22 Plastic behaviour.

are designed to work elastically under most loading. But it would be difficult or extremely expensive to design the structure to withstand a huge earthquake; in that event, the structure will be designed to deform permanently and, in doing so, will absorb energy, without collapse. Structural engineers calculate the 'ductility demand' from the shaking associated with any huge earthquake, and ensure that the structure has sufficient 'redundancy' to withstand the loading without failing. If necessary, they will add in 'redundant' additional steel beams and bracing, and control the vibrational response of the structure to move to a less damaging natural frequency during the ground-shaking.

As stated earlier, all rocks behave plastically at high temperature and pressure and, at the extreme, will melt and recrystallise. Some rocks behave plastically even at low temperature and pressures, the chief examples being salts such as halite and gypsum. These rocks deform and flow naturally within the Earth's crust, forming diapiric structures where, because of their low density, they rise as large bodies through denser sedimentary sequences, causing deformation in the surrounding rocks and, sometimes forming suitable traps for oil and gas (because of their low permeability). Because salts flow, rather than fracture, they attract quite a lot of consideration for nuclear waste disposal, the downside being that they can be mobile. Other rocks that show plastic characteristics at relatively low pressure and temperature are chalk, limestone and marble.

2.6.2.4 Poisson's ratio

As discussed earlier, when a sample is placed in tension or in compression, any change in axial length is matched by a change in lateral dimension. If you load a cylindrical sample vertically along its axis, its circumference increases. The ratio of change (lateral divided by axial strain) is the Poisson's ratio. For most intact rock, the value is somewhere between 0.25 and 0.35. The value is used in various ways for numerical analysis but, generally, any increase in vertical stress (say beneath a building) will be accompanied by an increase in horizontal stress because, whilst the rock would like to expand because of the Poisson effect, in nature it is restrained or confined by the surrounding rock mass.

2.7 DIRECT SHEAR

Direct shear applies where two halves of a sample are sheared in opposing directions either side of a distinct plane. Direct shear is employed as a testing method, as discussed in Chapter 6, and is applicable to conditions such as the sliding of a rock block above an adversely oriented joint, through the rock mass.

In a triaxial or uniaxial compression test on material that can be regarded as homogeneous and isotropic (such as many remoulded soils), one of the usual modes of failure is along conjugate shear surfaces inclined at an angle $45 + \phi/2°$ relative to horizontal, where $\phi°$ is the material friction angle. In soil mechanics, direct shear tests are carried out to mimic the stress state along the failure surface through the triaxial test specimen – by applying a normal stress and then increasing shear stress until the material fails. There are various criticisms of the test (for soil), in that the stress state is not fully known, as explained by Atkinson (1990). However, in discontinuous rock, which is strongly anisotropic, shear surface geometry will usually be constrained by the pre-existing weak planes. In this case, it makes very good sense to carry out direct shear tests, along the discontinuities. Again, this is explained in detail in Chapter 6.

2.8 SIMPLE SHEAR AND ASSOCIATED ROCK STRUCTURES

As explained in Chapter 1 simple shear occurs essentially as an angular deformation of a body. The classic analogue model to illustrate this is to deform a stack of playing cards as shown in Figure 2.23. The rough circle drawn on the side of the cards is drawn out into an ellipse, extending along its main axis, and contracting laterally. This occurs a great deal in nature, especially where a relatively weak stratum is sandwiched between two 'more competent' beds, and field examples of strain ellipse are much-studied in structural geology, to determine the direction and amount of strain in a tectonic belt (e.g., Ramsay, 1967).

A wide range of structures is developed in association with simple shear strain, including extensional tension gashes (Figures 2.24 and 2.25) and axial plane cleavage (Figures 2.24 and 2.26). McClay (1987) provides excellent guidance on the many geological structures associated with real rock deformation and their interpretation. Fractures may develop in other directions, including Riedel shears that are a conjugate set of fractures relating to the stress field within the strain ellipsoid (Tchalenko, 1968). This pattern of fracturing is not only seen in thin beds, but on a scale of kilometres in fault zones (Scholz, 1990).

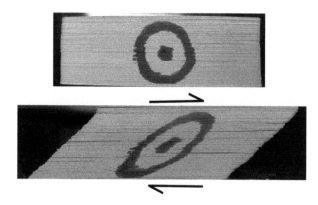

Figure 2.23 Simple shear, demonstrated with a stack of cards. Note that the circle is deformed to an ellipse, while the thickness of shear band remains constant.

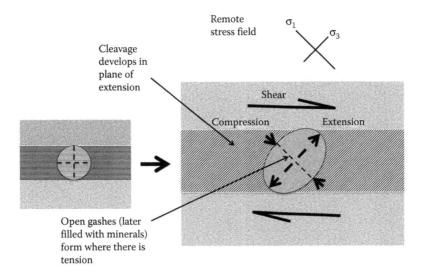

Figure 2.24 Simple shear in a rock stratum, induced by a remote stress field. Circle is deformed to an ellipse, with lengthening along one axis and shortening in the other.

Figure 2.25 Tension gashes in shear zone. Extension leads to the opening up of fairly regularly spaced gashes. As deformation has continued, the fabric has rotated and further infill taken place (see discussion in Price and Cosgrove, 1990). Gashes have been infilled with secondary mineralisation – in this case, calcite. Dry Rigg Quarry, Horton-in-Ribblesdale, Yorkshire. Geological hammer for scale.

Figure 2.26 Shear zone in relatively weak stratum between more competent beds (Dryrigg Quarry). The competent beds show many fractographic features typical of tensile fractures, including arrest marks and fringes with hackle fractures oblique to the main joint planes. In the shear zone, original bedding can be seen parallel to the main boundaries, but a cleavage has developed in the main direction of extension.

2.9 SURFACE FEATURES ON ROCK FRACTURES

As fractures propagate, the stress state changes from that which initiated the fracturing, away from some weakness point. Further, as the fracture meets some pre-existing boundary, the stress field may be affected, leading to change in geometry. This is particularly true of tensile fractures. Fractography is the science that deals with the description, analysis and interpretation of fracture surface morphologies, and links them to the causative stress, mechanisms and subsequent evolution of fractures (Ameen, 1995). According to Ameen, all fractures develop fractographic features, though some are microscopic. However, many can be seen by the naked eye. Kulander and Dean (1995) report experiments to produce typical fracture geometries using glass slides and rods, and similar features are produced just by breaking rock with a hammer and chisel, as illustrated in Figure 2.27. Similar features can be produced on dry, fine soil that breaks in a brittle manner, for example, on the faces of mud cracks.

Many tensile fractures exhibit 'hackle fringes' comprising 'twist hackle faces' and shorter 'hackle steps', as illustrated in Figure 2.28, supposedly in response to a twist imposed by the remote stress field (Roberts, 1995). In Figure 2.29, the near-vertical joint has developed as a tensile fracture originating from some point on the ripple-marked bedding surface, with numerous arrest ribs. There is a hackle fringe to the right, probably indicating the influence of another, pre-existing, cross joint.

(b)

(a)

Figure 2.27 Fractography features on tensile fractures in rocks, broken by impact with hammer and chisel. (a) Feather/plume markings on limestone block radiating from point of impact. Euro coin for scale. (b) Concentric plume markings radiating from point of impact in micritic limestone cobble (foot for scale). Note also the strong rib marks – also known as arrest or hesitation lines each marking a 'hesitation' or 'velocity discontinuity' during propagation, although, given the mode of formation of the cobbles in (b), the fluctuation in velocity must have been of extremely short duration. Both photographs, Tavira, Portugal.

Figure 2.28 Example of hackle fringe developed in orthogonally-jointed Carboniferous sandstone, above Accrington, Lancashire, UK. Block is about 1 m high.

Figure 2.29 Tensile fracture with arrest rib marks and hackle fringe, almost orthogonal to ripple-marked bedding surface in sandstone, Tsao-lin landslide, Taiwan. Black lens cap (52 mm) on bedding surface for scale.

2.10 CONCLUSIONS TO THIS SECTION

Many aspects of rock mechanics theory and observation are important to all aspects of structural geology as well as engineering rock mechanics. Interpretation of geological features in terms of their geological origins and stress histories plays an important role in developing ground models that make sense, rather than relying solely on measurement and statistical treatment, which is common practice. This theme will be developed further throughout this book.

Chapter 3

Geological processes and the nature of rock masses

The world was changing, as if with one piece missing the whole thing had come loose and was running wild.

Ross Macdonald
The Underground Man (1971)

3.1 INTRODUCTION

There can be no 'soil mechanics' without soil and no 'rock mechanics' without rock. Soil and rock are geological materials and to practice rock mechanics in geotechnical engineering, petroleum engineering, mining or nuclear waste studies, it is important to remember that we are dealing with natural geological materials and masses. These rocks owe their characteristics to a long history, which includes the mechanics and chemistry of formation together with subsequent stress history, cementation and weathering. Rocks are neither isotropic nor elastic, even though, in engineering practice, they are generally treated as such for analysis. Rocks contain flaws ranging from microscopic cracks within minerals through to joints at the scale of metres and faults that extend for kilometres.

The principles of physics and mechanics apply equally to engineering practice and to geology. An understanding of the processes that have given rise to the site-specific nature of rock masses is extremely important if investigations and analyses are to be cost-effective and failure to be avoided.

3.2 EARTH STRESSES

3.2.1 Plate tectonics

Geophysical studies have established that the solid crust of the Earth is a thin layer, often likened to the shell of an egg. Beneath the oceans, the solid basaltic crust is dense (29 kN/m^3), relatively young (less than 150 million years) and thin – a few km. compared to the Earth's radius, which exceeds 6350 km. Under the continents the crust is lighter, with the density of granite (27 kN/m^3), thicker (up to 40 km) and some parts are almost as old as the Earth itself (4.6 billion years). The rigid crust and upper mantle is split into seven or eight major plates, and several minor plates that drift above the plastic asthenosphere, driven by hot convective currents and gravitational imbalances. The boundaries of the plates are defined clearly in oceanic areas by the epicentres of major earthquakes and volcanic activity, as illustrated in Figure 3.1. In continental areas, the plates are often less readily defined, and damaging earthquakes sometimes occur unexpectedly at some distance from established

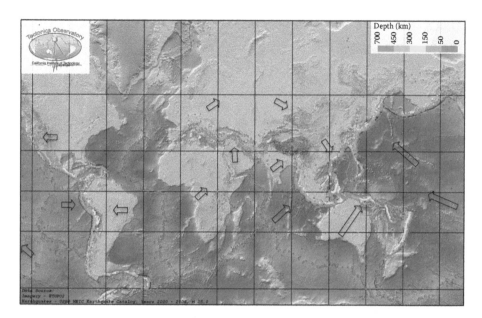

Figure 3.1 Red dots are shallow earthquakes, becoming deeper through orange, yellow and green, marking the Benioff zone where one colliding plate descends within the Earth's mantle. Major plate boundaries are clearly delimited by earthquake epicentres although some are rather diffuse particularly through the Middle East and north of the Indian subcontinent. The arrows mark some of the general directions of plate movement and relative magnitude (various sources).

plate boundaries, as in Boston, United States, in 1755 and the Newcastle, Australia, earthquake of 1989.

Where plates move apart, as along the mid-Atlantic ridge that runs through Iceland, hot magma wells up in volcanoes and solidifies to create new crust. Where plates collide, one forces its way above the other; the other plate that is forced downwards heats up and is re-melted. The overriding plate lifts and, essentially, forms a crumple zone as in the mountain chains running north–south down the west coast of the Americas. The re-melted crust rises to form new major igneous intrusive bodies on cooling. Elsewhere, plates grind past each other along transform faults, as discussed later. Rates of movement are actually quite high, compared to many other geological processes and measurable using GPS – typically, a few cm. per year.

It is remarkable that the now-accepted theory of plate tectonics only became established scientifically about 50 years ago. Arthur Holmes, an inspirational geologist, wrote in 1965:

> *If the Pacific floor is migrating radially outwards and passing beneath the continental crust, the basaltic layer of the oceanic crust must be continually renewing itself. It is hard to see how this could happen except by the ascent from the mantle of new material that takes place of the old as the latter is carried away or pushed aside.*

That is what we now know is happening. The theory immediately gave a unifying explanation for geological observations painstakingly made over centuries. Plate movements help to explain how mountains have formed, igneous rocks are emplaced, and huge sedimentary basins develop and down-warp over millions of years. It was quite a revolution of thought. Now that we have a working understanding of these major geological processes, we have a better vision of the conditions in which rock-engineering projects need to be conducted.

We now understand why the sea floor is not flat, as had been assumed by engineers of the Atlantic Telegraph Company who laid a cable between Europe and North America. The connection broke because, rather than being laid across flat land, it was suspended between peaks higher than the Alps, along the mid-Atlantic ridge (Jones, 1999). If anyone is in any doubt as to the severity and challenges for construction in the conditions associated with tectonic regions, an examination of Hoek (2000) and Carter (2011) will put them back on course. Interestingly, such onerous conditions for construction, caused by the vagaries of geology, whilst usually associated with active plate margins, can also occur in apparently inactive areas. For example, in Hong Kong, hundreds of miles from any known plate boundary, hot water was suddenly encountered during construction of the Butterfly Valley tunnel in north Kowloon, under such pressure that it buckled 40 mm thick steel linings (Robertshaw and Tam, 1999).

3.2.2 Earth stresses: Prediction, measurement and significance to engineering projects

Near to the surface of the Earth, much of the stress is gravitational due to self-weight. Vertical stress can be estimated reasonably accurately by multiplying depth by unit weight of the rock (typically 27 kN/m^3) (Brown and Hoek, 1978). Horizontal stress is more difficult to predict. Close to the Earth's surface, horizontal stresses often exceed vertical stress, and, in tectonically active regions or areas where residual stress is locked in after erosion of mountain chains, horizontal stress can exceed vertical by up to 15 times, with severe consequence for construction, such as squeezing and deformation. Elsewhere, horizontal stresses can be much lower, again with consequences for construction, as at Pergau Dam, Malaysia where unexpectedly open joints and voids led to high water inflows and necessitated the re-design of shafts and tunnel linings (Murray and Gray, 1997).

Fluid pressure typically increases essentially linearly with depth (depth times unit weight of water, which is about 10 kN/m^3), so is less than 50% of vertical rock stress. Sometimes, however, water pressure is found to be much higher than anticipated in geologically-young sedimentary basins, where fluids are confined and compressed by sediment compaction, and in tectonically active areas. Fluid stress can approach and even exceed the rock overburden pressure (Hubbert and Rubey, 1959). This has geological consequences such as the formation of mud volcanoes, the development of over-thrust faults that can extend many kilometres, and for the development of hydraulic joint networks in sedimentary rock, as discussed later.

3.2.3 Measurement of stress

Orientation and magnitude of stresses are measured at depth in boreholes, using hydraulic fracturing techniques, and by over-coring sometimes for major underground engineering works or mines. Hydraulic fracturing involves pumping water into a borehole, between inflated packers, until a new rock fracture is created and water pressure drops. The resulting fracture follows the major principal stress direction, and magnitude can be estimated through consideration of the water pressure to create the fracture and to maintain fluid flow (Hubbert and Willis, 1957). The orientation of the fracture can be determined using oriented cameras or from rubber packers inflated against the walls of the borehole. Stress conditions are also deduced from natural borehole fracturing (breakouts), and from the interpretation of earthquakes. The data is compiled and made freely available in a World Stress Map, as illustrated in Figure 3.2. Data is, of course, influenced by local geological structure, which limits extrapolation. On a regional scale, however, trends are consistent with overall plate movements (Heidbach et al., 2010).

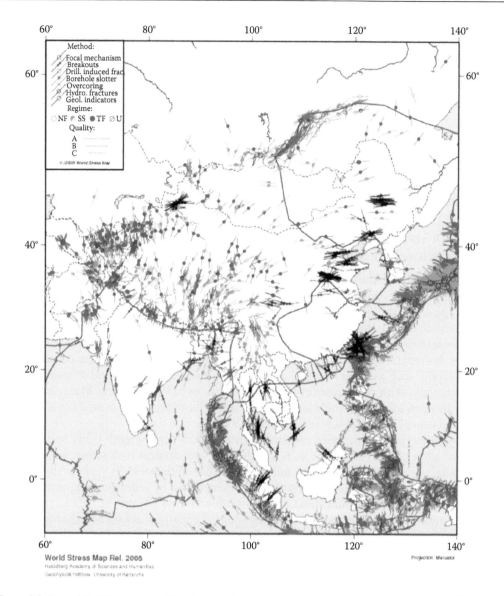

Figure 3.2 Part of the World Stress Map, for East Asia, showing major principal stress directions. Clustering of thrust mechanisms (TF) compared with normal faulting (NF) and strike-slip (SS) is interpretable with respect to plate movements and interactions. Global crustal stress pattern based on the World Stress Map database release 2008.

3.3 FAULTS

3.3.1 Significance of faults to ground engineering

Faults are fractures in the Earth's crust distinguished by the fact that there has been displacement on either side. These can be on a scale of metres, up to many kilometres extending through the Earth's crust. Disintegrated rock within the fault zone is generally known as fault gouge. In active fault zones, near the Earth's surface, the gouge will generally be relatively loose and granular, due to the predominantly brittle failure mechanisms that operate at relatively low temperatures and pressure; in older faults, the fault rock may be cemented or weathered and, in faults that have been active at depth, the gouge may be crushed to mylonite (meaning 'milled rock') that has deformed plastically at very high temperature and pressure. Faults have major consequences for geology and for engineering and mining projects, for example:

1. Faults may throw beds up against each other that have very different rock-mass properties. In mining, faults can result in truncation and separation of the ore body; conversely faults themselves may be mineralised and the target of mining activities. In coal-bearing strata faulting often leads to repetition of productive seams which greatly enhances the value of a mine.
2. Faults can be highly permeable leading to excessive inflow of groundwater into tunnels, migration of oil and gas, and potentially rapid escape of radionuclides from a nuclear waste repository or disposal site.
3. Faults that are cemented, are infilled with clay gouge or were formed at great depth where faulting was ductile/plastic may, conversely, act as barriers to fluid flow; these are known as 'fault seals' in oil and gas engineering. Such faults lead to hydrogeological compartmentalisation of the broad geological structure.
4. Fault rock close to the surface can be very weak and, therefore, may collapse in tunnels and other underground excavations. Where the fault rock is thick, there may be long delays to tunnelling, while the fault rock is grouted to reduce permeability, and strengthened, or while other engineering measures (such as pipe pile canopies) are adopted to allow tunnelling to proceed. Fault zones encountered unexpectedly can bury or lead to the inundation of tunnel-boring machines, resulting in extensive cost and delay, while the situation is rectified.
5. Because of their extent and, often, planarity, faults form extensive weak planes on which sliding failure can occur in slopes or underground excavations.
6. Movement on faults results in earthquake-related shaking and tsunamis.

The nature and distribution of faults can be understood at least in part through structural geological study.

3.3.2 General

Structural geologists make much use of physical models to mimic structures that we see in rock masses. The models do not necessarily meet the strict rules of similitude favoured for rock mechanics experiments (Fumagalli, 1968), but they can be very revealing.

Figure 3.3 illustrates results from a sand box experiment carried out by Hubbert (1951). Layers of differently coloured sand or dry powdered clay are placed in a box, on either side of a vertical plate connected to a screw. When the screw is advanced to the right, it simultaneously applies horizontal compression to the right whilst releasing the passive horizontal support to the left. Shallow, overthrust faulting and folding occurs to the right of the vertical plate, and normal faulting to the left (these terms will be explained shortly).

Anderson (1951) recognised that the Earth's surface can be assumed to be a principal plane with no shear traction; that being so, one principal stress direction would be vertical, and the other two horizontal. The relative magnitudes of these three principal stresses give rise to the main forms of faulting: normal, thrust and strike-slip.

3.3.3 Normal faults

Dip of geological structures is usually measured from the horizontal, as discussed in Chapter 5, but for faults, structural geologists sometimes use the term 'hade', which is the dip angle

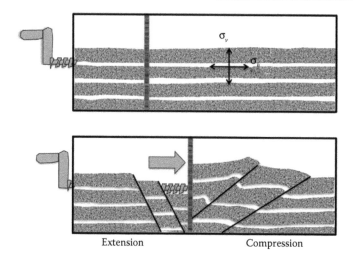

Figure 3.3 Sand box experiment used by Hubbert (1951) to illustrate the development of thrusts where horizontal stress exceeds vertical (to the right) and normal faulting where vertical stress exceeds horizontal (to the left). (Figure based on Davis, G.H. and Reynolds, S.J. 1996. *Structural Geology of Rocks and Regions*, 2nd edition. John Wiley & Sons, 776pp.)

measured from the vertical. A normal fault, by definition, 'hades towards the down-throw side'.* The down-thrown side of the fault is called the 'hanging wall'; the rock beneath the fault is the 'footwall'. An area down-thrown between conjugate sets of normal faults is known as a 'graben' and the whole structure a 'rift'. Famous examples of this type of structure include much of the Rhine Valley and the East African Rift Valley.

Following Anderson's analysis, normal faulting occurs where the vertical stress is the major principal stress ($\sigma_v = \sigma_1$) and the minimum principal stress (σ_3) is horizontal, and acts at right angles to the direction of fault propagation, as shown in Figure 3.4.

Given the stress conditions and evidence from Hubbert's simple experiment in Figure 3.3, one would expect to find normal faults where plates are moving apart or where they did so at some stage in geological history – in areas of 'extensional tectonics' – as is typically the case. Anderson argued that given an internal friction angle for rock of 30°, normal faults

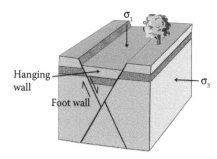

Figure 3.4 Normal faults. The mechanics are the same as seen in a typical triaxial rock laboratory experiment with the formation of conjugate shear surfaces. (Drawing based on Ramsay, J.G. 1967. *Folding and Fracturing of Rocks*, McGraw-Hill, 568pp.)

---

* The term 'normal' fault derives from coal mining where this type of steeply dipping fault was the most commonly encountered. To find the displaced coal seam, miners proceeded downwards and out towards the hade direction of the fault. A 'reverse' fault is the opposite ('hades towards the upthrow side').

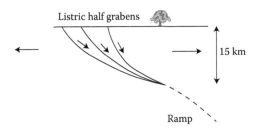

Listric half grabens

15 km

Ramp

Figure 3.5 Imbricate set of listric normal faults ('half grabens' because only one side of a graben structure is illustrated).

should dip at 60° to the horizontal. Price (1966) discusses the mechanics in detail and notes that, indeed, most normal faults dip at 60°–65°. At depth, normal faults often become shallower and curved and are then termed 'listric' (Figure 3.5).

Extensional tectonic regimes often mark the development of huge, gradually subsiding basins bounded by a series of roughly parallel, curved normal faults. Movement on these 'growth' faults, over millions of years, control rate and location of sedimentation; an understanding of such syn-sedimentary faulting is very important to oil and gas exploration (e.g., Leeder, 1999; Gawthorpe and Leeder, 2000).

3.3.4 Thrust faulting

Thrust faulting is essentially the antithesis of normal faulting, as illustrated in Figure 3.3. Thrusting occurs in compressional tectonic regimes where plates collide. In this case, one of the horizontal stresses is the maximum principal stress, as illustrated in Figure 3.6.

Anderson calculated that thrusts close to the Earth's surface should dip at about 30° and this is commonly the case. Thrusts are often associated with major folding, with 'nappe' folds thrust over the footwall, in some cases for kilometres. Often, there are series of thrusts that 'ramp' up, cutting across beds and then linking up, shearing through weak strata or exploiting pre-existing discontinuities such as bedding surfaces. As for normal faulting, roughly parallel thrusts form complex 'imbricate' structures.

Some major thrusts occur at much shallower angles than in Anderson's models, as in the Moine thrust belt in western Scotland, where rock has been displaced up to 40 km, sometimes at dip angles of only 10° (McClay and Coward, 1981). Evidence from the extreme disintegration and plastic flow structure of the rock (mylonite) indicates that some of the thrusting occurred at great depth (at least 10–12 km). Hubbert and Rubey (1959) proposed and demonstrated, as structural geologists do, with the help of full and, later, empty beer cans, that shallow over-thrusting could be the result of reduced effective stress (and, therefore, reduced friction) due to high fluid pressure.

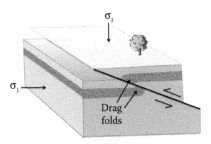

σ_3

σ_1

Drag folds

Figure 3.6 Stress regime for thrust faulting.

3.3.5 Reverse faults and inversion tectonics

The Anderson dynamic models for faulting do not have a set of stress conditions that explain the development of reverse faults, that is, with the same style as thrust faults, hading towards the up-thrown side, but dipping at steeper angles relative to the horizontal (>45°).

Studies show that many reverse faults are actually reactivated normal faults. The normal faults were formed during extension, but when plate movements reversed direction, and previously opening oceanic basins were again closed, the normal faults were reactivated as weaknesses, but now with a reverse direction. This process is termed 'inversion tectonics' and serves as a reminder to non-geologists and geologists alike, just how long is the geological history of the Earth. The current plate system is just the latest version. The mountains of north Wales (on the eastern shore of the Atlantic Ocean) largely comprise sedimentary rocks with a cumulative thickness of more than 10 km that were deposited over a period of more than 100 million years in an earlier ocean (Rayner, 1967). This ocean was finally closed by plate convergence about 400 million years ago, and the rocks folded and faulted during a mountain-building episode called the Caledonian Orogeny. The Atlantic Ocean of today began to open again with South America starting to drift away from Africa about 190 million years ago. Rifting in the North Atlantic started about 90 million years ago. As part of this opening, huge outpourings of lava occurred about 60 million years ago, forming, in part, the basalt columns shown on the cover of this book (Williamson and Bell, 2012).

Reverse faults can also develop simply as steeper parts of thrust systems in compressional regimes, as discussed in Davis and Reynolds (1996) and Price and Cosgrove (1990), and illustrated in Figure 3.7.

3.3.6 Strike-slip faults

The third of the Anderson models is for strike-slip faulting. These are also sometimes referred to as transcurrent, wrench or tear faults. Where associated with sea-floor movements, they are called transform faults. For strike-slip faulting, the maximum and minimum principal stresses are in the horizontal plane, as illustrated in Figure 3.8. The intermediate principal stress (σ_2) is vertical.

Strike-slip faults are extremely important sources of major earthquakes. Current active examples on land include the San Andreas Fault (United States), the Alpine Fault (New Zealand), the Philippine Fault and the Anatolian faults in Turkey. An ancient example is the Great Glen Fault in Scotland, which has a strike-slip displacement of about 150 km. Although the faulting is shown simply in Figure 3.8, fractures also develop in many other directions, notably Riedel shears at an acute angle to the main shearing directions. Vertically, several fractures often develop roughly parallel to the main faults, forming very complex

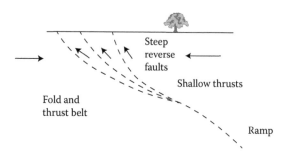

Figure 3.7 Thrusts and reverse faults in compressional zone.

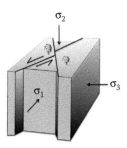

Figure 3.8 Stress regime for strike-slip faulting. (Drawing based on Ramsay, J.G. 1967. *Folding and Fracturing of Rocks*, McGraw-Hill, 568pp.)

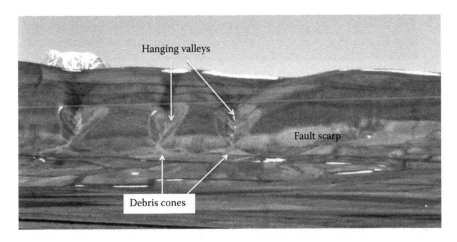

Figure 3.9 Looking north across scarp of the North Anatolian Fault near Erzincan, Turkey. A steep scarp (pale colour) and three hanging valleys with sediment fans at their mouths indicate the highly active nature of this predominantly strike-slip fault but with vertical displacement component.

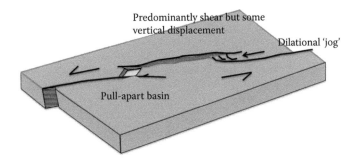

Figure 3.10 Formation of pull-apart basins and dilational jogs in strike-slip fault systems.

series of 'duplexes' and 'flower structures' (e.g., Aksoy et al., 2007), and, often, there is some vertical component of displacement, as shown in Figure 3.9.

Near-parallel branches in the fault systems result in pull-apart structures, developing as flat valleys, infilled with sediment (Figure 3.10). Elsewhere, 'jogs' occur where sections of the fault system override. These locations are often the sources for earthquakes (Scholz, 1990). An example is presented in Figure 3.11.

Figure 3.11 Damaged buildings following the 1992 Erzincan Earthquake, Turkey. The city is built on flat-lying sediments in a pull-apart basin bordered north and south by elements of the North Anatolian Fault and was severely damaged several times in historical times. (From Hencher, S.R. and Acar, I.A. 1995. *Quarterly Journal of Engineering Geology*, **28**, 313–316.)

3.3.7 Fault rocks

The fault zone of fractured, deformed and crushed rock can range from a narrow, tight plane, through to zones of fault gouge tens of metres across. As noted previously, in the upper 10 km or so of the crust, brittle mechanics processes dominate, so faulting results in fragmented breccia and non-cohesive fault gouge. At greater depths, where the rock is hot and under much higher pressure, behaviour is more plastic or 'ductile', and fault rocks are typically finer-grained (mylonite) and foliated with a parallel flow fabric (Figure 3.12).

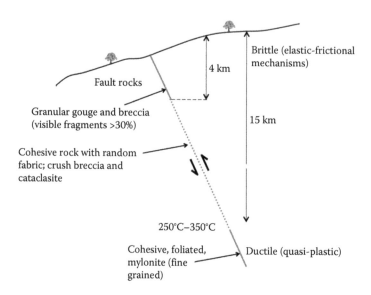

Figure 3.12 Variability of fault rock with depth of origin. (Based on Sibson, R.H. 1977. *Journal Geological Society London*, **133**, 191–213.)

Where faulting has caused severe brittle disintegration, rock quality can be extremely poor. For example, in the 'rashings' rock in the South Wales coalfield, apparently intact mudrock has been shattered so that it can be readily crumbled into silt-size sigmoidal cleavage fragments by hand. In the case of a power station in the Philippines, large concrete pad foundations had to be replaced with bored piles when the steelwork being erected settled unexpectedly because of the extremely poor quality rock associated with faulting (Chapter 7).

Faults that have developed and been active for very long periods of time can show different fault rock characteristics at different locations and, of course, the complexity is magnified where parts of the fault zone have been preferentially weathered. It can be very difficult to predict the nature of any fault zone. There are many cases where faults have been anticipated from geomorphological mapping and published map interpretation yet proved no problem for construction. Really, the only way to be sure of the nature of a fault zone is to carry out adequate ground investigation, using inclined or even horizontal boreholes as necessary, and, even then, remember that faults vary from one location to another. In the case of tunnels, where major faults are anticipated, an option is to first drive a pilot tunnel of small diameter, where support and drainage will be less of a problem than for a larger tunnel. In addition, probing ahead of the tunnel face is often essential where faults might be present, especially where they may have high water conductivity. Despite best efforts, however, faults still sometimes cause serious delays and even abandonment of projects (Chapter 9).

3.3.8 Earthquake occurrence and prediction

Large earthquakes are mostly associated with active faulting caused by the global movements of plates, although they can also result from igneous intrusion, be triggered by mining and changes in water pressure due to the construction of dams or induced by water injection associated with oil recovery and fracturing. Major damaging earthquakes in the continental areas are generally associated either with the large predominantly strike-slip faults such as the Anatolian faults or with subducting plates such as that to the west of Chile and Peru or Alaska and east of Japan, and the Philippines plate. Subduction zones are mostly responsible for the largest earthquakes historically (e.g., Alaska, 1964; Chile, 2014), although the huge 1755 earthquake that caused major damage to much of southern, coastal Portugal and north Africa, and was felt in Britain and the United States, was probably associated with a rather poorly-defined transform fault boundary 200 km offshore of Lisbon in the Atlantic Ocean. In crustal regions, most earthquakes have their origins (hypocentres) at depths of about 10–15 km, which probably is the depth marking the base of the zone of predominantly brittle behaviour (Sibson, 1977; Scholz, 1990). Earthquakes, and especially aftershocks, also occur at much shallower depths, and these can be particularly damaging. Earthquakes in subduction zones can occur at greater depths and, indeed, it was their foci, mapped geophysically, down to hundreds of kilometres, which defined the subsiding plates and helped to establish the truth of plate tectonics as a global mechanism.

Prediction of the location of future earthquakes is still, unfortunately, beyond science, despite a far better understanding of controlling mechanisms than 50 years ago. The readily understandable faulting models of Anderson (1951), whilst very illustrative and making good sense in terms of rock mechanics, are really over-simplistic for real geological fault processes. It is evident that even where the overall stresses in the Earth's crust are largely predictable from plate tectonics considerations, the actual mechanisms of faulting and stress relief can be hugely complex, with strike-slip, thrust, reverse and normal faults, all occurring essentially together within a single fault system (e.g., Aksoy et al., 2007). Furthermore, there is a major problem in understanding how major faults develop in the first place. One suggestion is that joint systems are created first, essentially as tensile patterns, and that weakness is later exploited by faulting. That might be the case for some faults but joint systems are sometimes clearly contemporaneous with faulting (Rawnsley et al., 1992; Peacock, 2001).

It can be concluded that major fault zones are not simple single-shear fractures, but complex zones of largely brittle deformation (Scholz, 1990). In civil engineering, a statistical and probabilistic approach is necessary for structural design to withstand the forces and the consequences of earthquake occurrence (Dowrick, 2009).

3.4 FOLDING

Rocks are deformed and folded by tectonic forces and often associated with major faults, especially thrusts. The mechanisms and conditions for folding are of great interest to structural geologists. Squashed and deformed fossils (Figure 3.13) and other features where the original shape is known are used as strain markers to help unravel the tectonic history at a site.

For rock engineering, the main significance of folding is the way it affects the distribution of rock strata and units at a site. Obviously, without an appreciation of fold geometries, ground investigation is unlikely to be very effective in such terrain, but that is largely a matter for engineering geologists and will be discussed later, when addressing mapping techniques and the interpretation of exposures.

Another aspect of folding is an association with weak planes and veins, such as an axial plane or fan cleavage. Similarly, when folding involves beds of different rigidity, there is often crumpling and plastic deformation of the weaker horizons (incompetent beds) that will affect mass properties such as strength and hydraulic conductivity (Figure 3.14).

During folding, inter-plane slippage can reduce shear strength along the interfaces, with the development of slickensides. Sometimes the slippage can result in high polishing, and the reduction in friction to very low values (of the order of 10° for some extensive polished horizons that are encountered in folded Coal Measures rocks). These can, of course, lead to instability in cut slopes, foundations and tunnelling.

Figure 3.13 Deformed trilobite pygidium from slate quarry near Arouca, North Portugal. Pencil for scale.

Figure 3.14 The 'Lulworth Crumple'; Stair Hole to the west of Lulworth Cove, Dorset, England. The photograph shows intense 'accommodation' folding in a relatively weak stratum – the Purbeckian – sandwiched between two more competent beds (the Portland Limestone below, the Chalk above – the white bed to the left of the photograph) that are folded into a broad monocline extending from Dorset across to the Isle of White about 50 km away in the direction of view. The folding has been largely ductile with some brittle faulting and minor thrusts resulting in repetition of beds and squeezing of some of the intermediate, more plastic, shaley horizons. The disruption is greatest towards the top of the sequence where overlain by weak horizons of the Wealden Beds (sands mainly).

Mineral veins are a common feature of fold and fault zones. Mineral-rich fluids are injected under pressure into areas of tension and low pressure (Figure 3.15). These are useful for interpreting the history of stress and strain in the rock mass, as discussed in Chapter 2.

Veins are often confined to particular brittle beds, veins and associated fractures, terminating at boundaries between relatively brittle and plastic strata, as in Figure 3.16. Such geological details can be highly significant to the engineering properties of the rock mass. They are also important because they are evidence of the geological history, including fluid

Figure 3.15 Two lines of imbricate shear gashes infilled with calcite in limestone that provide a record of tectonic strain. Picos de Europa, Spain. A5 book for scale.

Figure 3.16 Quartz veining infill to brittle fracturing in sandstone bed. Note that fractures and veins do not cut adjacent shale beds. Knife for scale (200 mm).

flow within the rock mass. The occurrence of mineral veins and their geochemistry are important data for site evaluation, with respect to the use of a site for nuclear-waste disposal, where predictions need to be made regarding likely fluid migrations over thousands of years. For example, at the Yucca Mountain investigations (for nuclear-waste disposal), it was established through careful study that calcite veining was not the result of percolation of water from the surface, but from up-welling and through-flow at elevated temperatures; for similar investigations regarding possible nuclear waste disposal at Sellafield in Britain, it was established that there were nine separate mineralisation episodes, with only one continuing at the current time (Milodowski et al., 1998).

3.5 ROCK TEXTURES, FABRICS AND EFFECT ON PROPERTIES

3.5.1 Introduction

It is important to distinguish between rock material – i.e., the intact rock that we can take as a hand sample or hit with a hammer – and the large-scale rock mass that contains discrete structures like faults and joints. At the material scale, rock 'texture' is the relationship between mineral grains that make up the rock material. The rock material may have preferential directions of weakness (discontinuities) such as intrinsic bedding and cleavage-foliation or flow-banding that can result in anisotropy of strength, deformation characteristics and permeability. Such features, combined with material texture, make up the rock 'fabric'.

The minerals that make up the rock have their own characteristics, ranging in scratch hardness on the Mohs scale of 1–10, from talc to diamond. Quartz, which is an important mineral in many rocks, has a hardness of 7, which is about the same as steel, and slightly higher than glass. Quartz lacks any pervasive fracturing, unlike feldspars (the second-most important component of many igneous rocks), which have a well-developed cleavage allowing slippage. Abrasivity stems from mineralogy and, as one might imagine, the higher the scratch hardness, generally the higher the abrasivity. Quartzite, in particular, causes severe wear to equipment such as drill bits. That said, limestone, made up largely of calcite with a low scratch index of 3, can also sometimes lead to severe cutter wear in tunnelling and dredging.

Chemistry can also be a very significant factor, not least where rocks are used in construction, or as aggregate in concrete. Alkali–silica reactivity is a major issue in aggregate selection, and the presence of minerals such as pyrite, which can easily breakdown to produce acidic groundwater, can be problematic. Hencher (2012a) addresses various hazardous issues when considering geochemistry at a material scale.

3.5.2 Cooling of igneous rock

Because igneous rocks cool from a melt, they tend to have very little pore space, although there may be a few remnant gas bubbles, generally not interconnected. Because of the way crystals grow as the rock cools, textures are typically angular, and completely interlocked, which gives considerable strength, high intact modulus, and low porosity. A typical example is shown in Figure 3.17.

The sizes of mineral grains reflect the time that the melt took to cool. If very rapid, then grains will be small, as in basalt, or even glassy, where the rock was suddenly quenched. Generally, the smaller the grains, the stronger the rock, so basalt or dolerite that cooled relatively quickly at shallow depths in or on the Earth's crust may have twice the UCS of granite. For extrusive volcanic rocks such as volcanic tuff, strength may increase markedly with time as chemical reactions take place. A fresh ash tuff extruded from a volcano as dust can be swept up as a powder, using a dustpan and brush, but given time and the right conditions, minerals will bond, and the strength can reach 300 MPa.

Figure 3.17 Thin section through granite photographed under cross polars which allows birefringence colours to be seen that help identify each mineral. Note the cleavage in the feldspar crystals that contribute to the relatively low intact strength compared to other igneous and even some sedimentary rocks. F: Feldspar. Grey with well-defined cleavage planes. Hardness 6. B: Biotite mica. Brown and flakey, easily scratched in hand sample. Q: Quartz. White/grey. No cleavage planes. Hardness 7. Note tight interlocking texture. Minerals have grown from melt to fill spaces as the magma cooled. Porosity low and UCS relatively high (probably about 100 to 150 MPa). Minerals are bonded electro-chemically.

Figure 3.18 Granite with xenoliths of country rock, partly altered to granite. Above Lake Tahoe, California, United States. Penknife for scale.

Igneous rocks are often massive and essentially isotropic but sometimes have other fabric that will affect their mechanical behaviour. Figure 3.18 shows broken xenoliths of country rock, partly replaced by solid-state mineralisation, into which granite has been emplaced and cooled.

3.5.3 Sedimentary rock

A simple classification of sedimentary rocks is presented in Figure 3.19. There are three main groups, detrital, volcanic (derived from air-borne debris and deposited on the Earth's surface as sediment) and chemical/biochemical. Detrital sedimentary rocks are derived from the breakdown and erosion of other rocks. The detritus is transported and then deposited and buried by more sediment at a later stage. Textures and mineralogy depend upon provenance, transportation, burial and cementation, as well as subsequent weathering.

3.5.3.1 Sandstone

Sandstone is rock made up of grains that are visible to the naked eye, up to a grain size of 2 mm. Clastic rocks made up of fragments larger than 2 mm (generally comprising gravel, cobbles and boulders of other, ancient rocks) are termed breccia (angular) or conglomerate (rounded) fragments. During weathering and transportation, mineral grains that are relatively unstable chemically, such as feldspar, or that are weak, such as mica, are preferentially destroyed; the chemical components combine to form new clay minerals, or go into solution, so that they might end up as biochemical or chemical deposits. Detrital grains that travel a long way become rounded due to attrition as in desert sandstones and river-laid conglomerate. Because of the many factors contributing to the texture and fabric of sedimentary rocks, it is difficult to generalise or give typical properties. Figure 3.20 shows data compiled from only two sources but is enough to indicate the wide range; some of the denser sandstones are similar in strength to fresh granite.

Grain size, mm	DETRITAL ROCK (from fragments of other rocks and clay minerals)		VOLCANICLASTIC or PYROCLASTIC ROCK	CHEMICAL & BIOCHEMICAL ROCK
	CONGLOMERATE (rounded clasts) BRECCIA (angular)		PYROCLASTIC BRECCIA (angular) or AGGLOMERATE (rounded)	LIMESTONE which includes: Chalk Calcarenite (sand and gravel size) Calcilutite (mud size matrix) Oolite (made up of rounded calcareous accumulations generally formed around a sand grain or shell fragment) DOLOMITE (Mg rich)
60			LAPILLI TUFF	
2	SANDSTONE Greywacke (generally poorly sorted) Arkose (feldspathic sandstone)		COARSE ASH TUFF	
0.06	SILTSTONE	MUDSTONE as general term	FINE ASH TUFF	EVAPORITE (salts) COAL
<0.002	CLAYSTONE	SHALE if fissile		FLINT & CHERT (Cryptocrystalline silica)

Note: See Tucker (1982) for fuller and more detailed classification.

Figure 3.19 Simple classification of sedimentary rock.

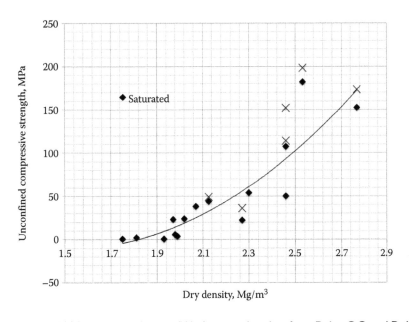

Figure 3.20 Sandstone UCS versus dry density. (Weaker samples: data from Dyke, C.G. and Dobereiner, L. 1991. *Quarterly Journal Engineering Geology*, **24**, 123–134; Stronger sample [saturated and dry]: data from Hawkins, A.B. and McConnell, B.J. 1991. *Quarterly Journal Engineering Geology*, **24**, 135–142.)

Figure 3.21 Strongly indurated upper carboniferous sandstone at Brimham Rocks, near Harrogate, Yorkshire, England. Bird for scale. The rock form was eroded in arctic conditions by wind perhaps 10,000+ years ago. Vertical water erosion features can also be seen on all three blocks to the right.

At the weaker end, the strength of sandstone is largely controlled by the packing relationships and porosity of the rock. Sandstone can be much stronger where cemented by secondary minerals such as silica, iron oxides and carbonates (e.g. calcite and siderite) as well as by chemical bonding and interpenetration of the sand grains, as illustrated in Figures 3.21 and 3.22.

3.5.3.2 Mudstone

Clay minerals are derived mostly from weathering of primary minerals, mainly from feldspars. They have a sheet structure, like mica, with the sheets connected by cations such as sodium and potassium. Young clays, derived from igneous rocks rich in iron and magnesium, such as basalt and gabbro, often comprise smectites that swell and shrink readily with moisture change, and may cause problems for construction (e.g., black cotton soils). Fortunately, by the time they become lithified, they are often changed chemically to more

Figure 3.22 Arkose in thin section under a microscope. Relatively immature sandstone, with grains of quartz (featureless, glass-like), microcline feldspar (with cross-cleavage) and lithic (rock) fragments (L). The arrows indicate several examples of deformed, concavo-convex contacts and interpenetration of grains and pressure-solution during diagenesis. There is virtually no porosity – which is an indication of deep burial. The rock will be strong and would not make a good reservoir for oil and gas, as the permeability is negligible.

Figure 3.23 The result of evaporation from smectite-rich mud washed out from rapidly degrading slopes in volcanic tuff, Turkey. Shrinkage is far in excess of 50%. The phone is placed for scale and in about 150 mm in length.

stable minerals such as illite and kaolinite. That is not always the case, however, and where mudrocks are rich in smectites, then they will erode and degrade very rapidly on exposure, as illustrated in Figure 3.23. Where the smectite content remains high in lithified mudrocks, friction angles can be very low (well below 10°), and this can be associated with deep-seated landsliding (Wentzinger et al., 2013).

During lithification and burial, the porosity of clays reduces by huge percentages, and relatively strong chemical bonds are formed as clay platelets are squeezed together. Mudstones are often massive, sometimes bedded, and often inter-bedded, with other sedimentary rocks such as sandstone. Because of their fine grain size, they act as aquicludes and as traps in oil and gas fields. Shale is the term for laminated, fissile mudstone; marl is mudstone with high calcite content.

3.5.3.3 Limestone

Limestones come in a very wide range from weak, porous chalk to massive, extremely strong massive limestone of the type found in the Pennines in Britain. The one thing they have in common is high calcite percentage, and the fact that they are soluble in weak acid. This gives rise to difficulties in rock engineering in that they may be cavernous (karstic), with consequences for construction and long-term reliability of foundations, as discussed further in Chapter 7. Despite potential susceptibility to acid attack, this is a very slow process, and strong limestone is often used as aggregate, and successfully as armourstone, in port facilities, or as riprap protection in dams (Figure 3.24). One potential problem, where used as riprap on earth dams, is that acid attack can lead to formation of carbon dioxide that, being heavier than air, may collect in tunnels, and be hazardous, as occurred at Carsington Dam in Britain (Hencher, 2012a).

3.5.4 Metamorphic rock

Metamorphic rocks are rocks that were changed by heat and/or pressure. The main types and their characteristics are listed in Table 3.1. At the small scale of intact rock, the results

Figure 3.24 Limestone blocks and concrete tetrapods in the background used in coastal protection works. Isla de Tavira, Portugal.

of pressure and temperature lead to anisotropy and sometimes flow and associated fabric, as shown in Figure 3.25.

Where rocks have been re-crystallised due to heat, the rock can be very strong, as in granulite, hornfels and quartzite, where the mineral grains form an interlocking structure; an example is given in Figure 3.26.

Table 3.1 Simplified classification of metamorphic rock

Rock name		Parent rock	Characteristics	Geotechnical characteristics
Slate	Low grade	Mudstone and more rarely sandstone	Fine grained, strong cleavage, imposed by regional stress, low temperature.	Strongly anisotropic. Material generally uniform, very strong. May weather along joints that develop parallel to cleavage.
Phyllite			Coarser than slate. Mica on foliation surfaces, lustrous. Sometimes foliation has graphite, talc or chlorite.	Weaker than slate and readily split along foliation. Presence of low friction minerals means prone to sliding.
Schist			Strongly foliated. Often micaceous but may be talc-rich, graphitic, chloritic.	Similar to phyllite but tend to be stronger though still highly anisotropic. Often made up of alternate stronger and weaker layers.
Gneiss	High grade		High grade melting of granite and other rocks. Foliations comprise segregated bands of minerals.	Tend to be stronger than schist and often massive, like granite but can have micaceous bands.
Hornfels		Any	Contact metamorphism, especially of mudrock, adjacent to igneous body. Can be extensive.	Very strong, massive.
Marble		Limestone	Heat and pressure metamorphism of limestone and dolomite.	Generally massive, completely re-crystallised.
Quartzite		Sandstone	Quartz crystals with dense interlocking texture.	Generally massive, very strong. Abrasive for drilling and excavation.

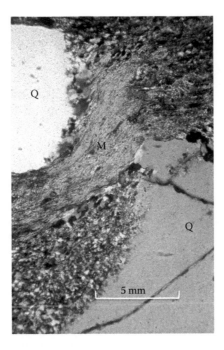

Figure 3.25 Thin section of mica schist. *Mica schist* Q: Quartz. White/grey. No cleavage planes. Hardness 7. Note re-crystallised clast material in the 'shadows' of the quartz 'porphyroclasts'. M: Muscovite. High birefringent colours under microscope with cross polars. Hardness 3. Metamorphic rock changed by heat and pressure from sedimentary rock (probably clay-rich sandstone). Tight interlocking texture but the rock is isotropic with a preferential weakness (schistosity) along the micaceous plane. Porosity low and rock probably strong at right angles to weak planes. Minerals are bonded electro-chemically.

Metamorphic rocks are often foliated pervasively (at micro and macro scales), as in slates (Figure 3.27), which have a distinct planar cleavage direction (exploited when quarrying slates for roofing materials or paving slabs).

Schist is also strongly foliated, sometimes with concentrations of low friction minerals, such as mica (even more so in phyllite) and talc along the planar fabric (Figure 3.28).

3.5.5 Hydrothermal alteration

Hydrothermal alteration is associated with igneous activity and heated or superheated water (pneumatolysis). The affected zone can be very extensive, as in the case of the Cornish granites in Britain, which are exploited for kaolin clay (Figure 3.29).

Many ore deposits owe their origin to hydrothermal processes, with mineral veins infilling fissures, or through replacement of existing rocks.

Elsewhere, the hydrothermal effects can be quite localised, as shown in Figure 3.30, where hot fluids have produced kaolin from the feldspars in the granite parent rock along an anastomosing (branching) fault.

In terms of the significance for rock engineering, hydrothermally altered rock can be very weak (in the same way as weathered profiles discussed later), and may be associated with other mineralisation, such as chlorite and talc, which may have low frictional properties. Hydrothermally altered rock is, however, often denser than weathered rock of equivalent

Figure 3.26 Thin section of granulite. Granulite is a high-grade metamorphic rock from the deep crust, related to gneiss but finer-grained. The microscope image of a low-porosity crystalline rock shows interlocking grains that grew together at high temperatures. The rock is a basic granulite from Harris, NW Scotland, and the colours are artefacts of the way in which different minerals polarise light, rather than true colours. Greys are crystals of feldspar, the brighter colours – blues, greens, reds and browns – are crystals of pyroxene or amphibole, while the central black grain is garnet. This rock texture is typical of high-grade metamorphism, and the grains have grown together in a process analogous to annealing; the field of view is about 4 mm.

Figure 3.27 Lower Palaeozoic slates, near Horton in Ribblesdale, West Yorkshire, United Kingdom.

strength, because it has not been produced by the same groundwater flow mechanisms (suffusion that washes weathering products out of a weathered rock profile, leaving a porous framework). The other vitally important aspect is that it can occur at a great depth, because it is not associated with the Earth's surface processes, and can therefore be encountered rather unexpectedly, as discussed for a tunnel in Hong Kong, and at the Pergau powerhouse

Figure 3.28 Schistocity in Carboniferous rocks, controlling the stability of cut slopes. Cachopo Road, Algarve, Portugal. The cyclist provides scale.

Figure 3.29 Water jet being used to extract 'china clay', St Austell, Cornwall.

in Malaysia, in Chapter 9. Weak hydrothermal zones may be anticipated generally near the margins of large igneous intrusions. When drilling or probing ahead for tunnels, clues indicating that you are approaching hydrothermal zones, includes general discolouration (similar to weathering), but also includes concentrations of relatively unusual minerals, including apple green epidote, quartz veining, garnets and metals.

Figure 3.30 Hydrothermal alteration with kaolin dominating in anastomosing shear zone. The pen provides the scale in the right photograph. Tunnel below Ching Cheung Road, Hong Kong.

3.5.6 Weathering

3.5.6.1 General

Weathering is the physical and chemical breakdown of rock at shallow depths (typically less than 100 m) in the Earth's crust. In extreme situations, the rock may be completely broken down to soil, with all the original minerals altered to new minerals, predominantly clay. Eventually, even the most resistant minerals, such as quartz, may be dissolved and leached away. Weathering profiles can develop to great depths, but may be very localised and complex, even with large corestones of relatively fresh rock underlain by weaker material. Most weathering occurs above the water table in oxidising conditions, but Ollier (2010) points out that other weathering processes, including reduction, hydrolysis and ionic substitution, occur below the water table, as evidenced by enriched supergene ore bodies at great depth. Thick weathering profiles dominated by chemical decomposition are associated with tropical and sub-tropical climates (e.g., East Asia, including Hong Kong, Malaysia and Singapore), but this is partly coincidental, in that most thick weathering profiles are very old (many millions of years); climates change at any particular location, not least as continents drift across the Earth's surface (evidence the presence of glacial deposits in South Africa and Coal Measures in Antarctica from previous tropical conditions).

All rocks weather, but the thickest profiles are typically in granitic rocks because a framework of resistant grains (quartz) can maintain the mass at essentially constant volume despite weathering of interstitial minerals such as mica and feldspar to clay, and washing out (suffusion) of the weathering products to leave a highly porous material. In many other rocks, weathered zones are relatively thin. That said, if the geomorphological setting is right, deep weathered profiles are found in other rock types, including basalt (Figure 3.31), sedimentary sequences (Figure 3.32) and schist (Figure 3.33).

Figure 3.31 Steep cutting through weathered basalt (developing flowing soil, rich in smectites). Centennial Highway, west of Brisbane, Australia.

Figure 3.32 Weak, weathered profile (about 20 m high) through volcaniclastic sedimentary rock and basalt (faulted). Turkey.

3.5.6.2 Disintegration

In colder climates, disintegration predominates, as in frost shattering (Figure 3.34). Excellent examples and discussion are given by Selby (1993).

Disintegration also accompanies chemical weathering in tropical conditions. Pre-existing incipient joints become full mechanical fractures (Figure 3.35), and the rock can develop

Figure 3.33 Footings being cut by hand in weathered schist profile (schistocity dipping to left). Central Harare, Zimbabwe.

Figure 3.34 Mechanical disintegration through the breakdown of mica; dominating weathering profile producing loose 'grus' sand of mainly quartz and intact feldspar, above Lake Tahoe, United States.

close micro-fractures along ground boundaries, opening up along cleavage. Sometimes this develops with a distinct parallel orientation, probably as a result of stress relief, on a par with sheeting joints (Hencher et al., 2011) and Figure 3.36.

In rock with pervasive fabric like bedding and cleavage, distinct open discontinuities begin to open up with weathering, as discussed by Hencher (2014) and illustrated in Figure 3.37.

Figure 3.35 Relict joint (planar and clay coated) through highly decomposed granite (readily broken down by hammer and trimmed with knife), Hong Kong.

Figure 3.36 Parallel microfractures developed through unloading in moderately decomposed granite. Near Lion Rock Tunnel portal, Hong Kong.

3.5.6.3 Mass weathering features

Figure 3.38 shows an idealised progressive weathering of a profile in granite. This hypothetical concept is the basis for simplistic weathering classifications for rock masses used in geotechnical engineering, as discussed in Chapter 5. Weathering profiles are often much more irregular, as illustrated in Figures 3.39 and 3.40, and this poses difficulties for investigation, characterisation and construction, with some zones rippable, whilst other zones will need to be blasted. Similarly, decisions on geotechnical parameters, to use for analysis in such profiles, are problematic.

(a)   (b)

Figure 3.37 View North, Dryrigg Quarry, near Horton in Ribblesdale, West Yorkshire, England. (a) The viewed height is approximately 70 m. The Silurian greywacke is dipping away from the viewer. Note that towards the base of the quarry, bedding is relatively indistinct with distinct bedding plane discontinuity traces separated by a few metres. In the upper part of the quarry the spacing is much closer, of the order of 1 m. (b) Exposure at the top of the quarry. The visible bedding traces are here dipping steeply with centimetre spacing (hammer for scale). The discontinuities are incipient, still with tensile strength that is gradually being reduced through weathering, allowing the slabs to spall off.

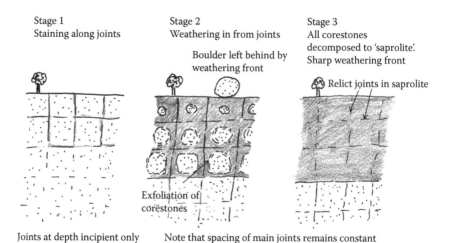

Figure 3.38 Progressive weathering in an idealised granite profile with regularly spaced incipient joints. (From Hencher, S.R. 2012a. *Practical Engineering Geology. Applied Geotechnics,* Vol. 4, Spon Press, 450pp.)

Figure 3.39 Irregular weathering profile in granite, Harare, Zimbabwe.

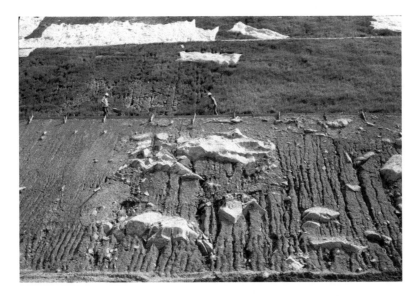

Figure 3.40 Irregular weathering profile in granodiorite, Tai Po Bypass, Hong Kong.

3.6 ROCK JOINTS AND OTHER DISCONTINUITIES

3.6.1 Introduction

Rock joints are fractures in rock that show no evidence of shear displacement. They are the most common fractures in rock masses and often govern rock mass properties and behaviour, including mass strength, deformability and permeability. They therefore have considerable significance for civil engineering, mining, water supply and the oil and gas industries.

Joints often occur as sets that can be related to the regional stresses at the time of formation, and can be characterised statistically, albeit that data should not be extrapolated from one structural domain to another where the tectonic history is different. The degree of fracturing also often depends on the degree of weathering and exposure, generally diminishing in frequency and persistence with depth.

Intrinsic rock fabric in rock, such as bedding, flow-banding, cleavage and schistosity, can open up due to weathering and unloading, as discrete fractures. These fractures are grouped, together with joints and faults, under the general term 'discontinuities'. The term 'fractures' is also used generically to mean discontinuities (either fully developed or incipient, i.e., not fully developed). Often the terms 'joints', 'discontinuities' and 'fractures' are used essentially as synonyms and I will sometimes lapse into this rather loose usage – apologies in advance. In many exposures, joint patterns can appear to be frustratingly non-systematic but, nevertheless, to get the most benefit from fracture surveys, analysis should be used to try to relate the geometrical distribution and arrangement of fractures to their likely origins, as discussed later.

3.6.2 Need for a change of approach and increased geological input in characterising fracture networks

With respect to oil and gas, Nelson (2001) reports a dramatic increase in interest in natural fracture studies, largely due to increasing percentage of oil and gas discoveries where natural fracture systems play a significant role in production. However whilst natural fractures will enhance permeability and often lead to better recovery, they can also be problematic, leading to rapid depletion and difficulties with secondary and tertiary production processes, so that the total amount of recoverable oil or gas is actually reduced.

Nelson is critical of current approaches to understanding and characterising fractured rock reservoirs, and many of these criticisms apply equally in other areas of rock engineering. He notes:

1. A general lack of in-depth *qualitative* approaches to description and characterisation of highly anisotropic reservoirs (i.e., an over-emphasis on measurement and statistical approaches).
2. Failure of geologists and engineers to recognise fractures and/or the regularity of their distribution.
3. Over-simplistic approaches in the description of fracture distributions and morphologies.
4. The need for a deterministic solution to modelling fluid flow in fractured porous media.

Nelson argues that geologists should learn to generate the fracture distribution data essential to characterise the fluid-flow properties of reservoir units. He says:

> For too long geologists have merely defined geological units and left the index quantification to the reservoir engineers who are generally less aware of the subtleties of reservoir variation and anisotropy. Quantification of small-scale features and effects are most effectively handled by the geologist but such quantification is often foreign to most geologists.

As before, while this criticism is addressed at the hydrocarbon industry, it is equally valid for geotechnical engineering generally. In essence, there is a need for greater geological input in explaining the fracture network and integrating that with a realistic interpretation of geological history at a site (including weathering and exhumation). Current practice relies too much on a statistical measurement of those fractures that can be directly observed. It is understood that this is easier said than done, in that there is little guidance in the rock mechanics literature or in the 'standards', or even in geological treatise, as to how this should be achieved.

3.6.3 Starting point for dealing with rock discontinuities

It is convenient to attempt to consider discontinuities broadly, according to their time of formation as in Table 3.2, even if this classification is rather forced, in that fractures may have complex histories. Table 3.3 is fuller and includes some typical characteristics of the various classes of joints and other discontinuities, as may be relevant to their mechanical properties, as discussed in Chapter 6.

As argued later, many discontinuities only become open mechanical fractures following weathering and unloading. Therefore, whether or not they influence the development of other fractures during geological history will certainly vary from site to site. We might be able to characterise the fracture state as it is at a particular location, but explaining how that network developed over geological history is a much more difficult problem.

Primary discontinuities are defined here as those that develop as part of the original rock formation; for example, cooling joints due to contraction in the case of igneous rocks, and hydraulic joints that propagate in thick sediment piles during burial and lithification. Secondary discontinuities are those that result from tectonic deformation, associated with faulting and uplift. The third group comprises fractures that develop in response to stress conditions at the Earth's surface.

Most rock joints are typically of the order of metres in length, commonly less than 100 m persistence even in the case of sheeting joints, but on occasion they can be much larger. Odling (1997), who studied joints in thick sandstone beds in Norway, suggested that joints, which penetrate the whole of the sedimentary sequence, could be expected to have lengths of around 14 to 20 km, which is similar to the length of the largest joints in the region, observed in aerial photographs (1:50,000). It is also on a par with the scale of faults.

3.6.4 Primary joints

3.6.4.1 Cooling (extrusive and shallow intrusive)

Some joints are evidently formed due to contraction stresses (mode 1 – tensile) as molten rock cools. Joints formed in this way are most obvious for lava and minor igneous intrusions

Table 3.2 Classification of rock discontinuities based on primary time of fracture development

Discontinuity type	Definitions
Type 1: Primary discontinuities	Planes of weakness, the origins of which are associated with early stages of rock formation, for example, cooling, hydraulic fractures formed during burial and bedding.
Type 2: Secondary discontinuities	Secondary origin due to tectonics, general flexure and stress changes during uplift. Geometry and terminations often influenced by pre-existing primary discontinuities.
Type 3: Tertiary discontinuities	Fractures formed at or close to Earth's surface. Include sheeting joints and exfoliation fractures. May be confused with pre-existing weaknesses of Types 1 and 2 that have opened up due to unloading and weathering.

Source: Modified from Hencher, S.R. 2014. *Advances in the Study of Fractured Reservoirs*, Geological Society of London Special Publication 374, 113–123.

Table 3.3 Geotechnical classification of discontinuities

I. Primary discontinuities: Associated with original rock formation processes/intrinsic cohesion characteristic of rock type

Rock type	Discontinuity type	Physical characteristics	Geotechnical aspects	Comments
General Geology	Lithological boundaries	Any angle, shape and complexity according to geological history.	Often result in marked change in physical properties including strength, fracture network and permeability.	Mappable, interpretable with respect to geological history.
Sedimentary	Bedding planes/ bedding plane joints	Parallel to original deposition surface and marking a hiatus in deposition. Usually almost horizontal in unfolded rocks.	Often flat and persistent over tens or hundreds of metres. May mark changes in lithology, strength and permeability. Commonly incipient with considerable cohesion. Opened due to weathering and unloading.	Geologically mappable and therefore may be extrapolated providing structure understood. Other sedimentary features such as ripple marks and load marks may aid interpretation and affect shear strength.
	Shaley cleavage	Close parallel discontinuities formed in mudstone during diagenesis and resulting in fissility.		
	Fissures	Common fabric feature, especially in weak rock and overconsolidated soil (such as London Clay).	May be random or relate to stress history as in other structures. May have controlling influence on engineering properties including permeability.	
Igneous	Cooling joints (extrusive and shallow intrusions)	Joints develop orthogonal to cooling surface. Systematic, often hexagonal sets form in lava, sills and volcanic plugs (to depths of several km).	Clear definition relating to cooling history.	Columns are not all developed as full mechanical fractures during original cooling episode. Fully propagate through unloading and weathering.
	Cooling joints (larger plutons)	Joints parallel to boundary of body, flow structures and cross-joints.	Cross-joints often appear as orthogonal sets; incipient (and invisible) weakness follows same pattern.	
Metamorphic	Slaty cleavage	Closely spaced. Parallel and persistent planar, integral fabric in fine-grained, strong rock.	High cohesion where intact but can be split and develop as joints due to weathering.	Formed by regional stresses and therefore mappable over wide areas.
	Schistosity	Crenulated or wavy foliation with parallel alignment of minerals in coarser-grained rock (compared to slate).	Foliation surfaces coated with chlorite, talc and mica give low, anisotropic shear strength.	Less mappable than slatey cleavage but trends recognisable.

(Continued)

Table 3.3 (Continued) Geotechnical classification of discontinuities

Discontinuity type	Physical characteristics	Geotechnical aspects	Comments
2. Secondary discontinuities: Form later so may cut, terminate against and complement earlier sets of discontinuities			
Tectonic joints	Joints formed by tectonic stresses often accompanied by folding and faulting. Sets relate to stress conditions so often conjugate; where there are several tectonic episodes, series of joint sets are superimposed upon and influenced by pre-existing discontinuities. Other 'tectonic' joints probably form due to broad stress relief and doming on unloading (e.g., Price 1966).	Joints at a microscopic scale form in tension but tensile cracks may coalesce to follow conjugate shear directions. Others are simply tensile. 'Shear' joints are often less rough than tensile fractures which may have marked distinct features such as hackle marks (see text).	Can be consistent at a regional scale especially where due to doming and unloading. Elsewhere structure can change laterally very rapidly, in response to local stress conditions and 'stress shadows' especially near major and discrete folds and faults.
Faults	Fractures on which there has been discernable shear displacement. Any scale from mm to km.	Extensive weak planes very significant for many projects. Often associated with sheared and broken rock (brittle fracture). Faults at depth may be sealed with mylonite or cemented. Weathering concentration due to fractured rock and high permeability. Association with earthquakes.	Often mappable because of displaced geology. Major faults often recognised as topographic lows geographically and as photo-lineations on aerial photos.

Discontinuity type	Physical characteristics	Geotechnical aspects	Comments
3. Tertiary discontinuities: Form close to earth's surface due to local overstressing			
Sheeting joints	Extensive joints (up to tens of metres) that form parallel to Earth's surface due to unloading and reorientation of gravitational stresses encouraging buckling. Can be generated from residual tectonic stresses, even horizontally. Mostly develop in massive, un-jointed rock masses, especially exposed igneous plutons, where there are few older joints that could accommodate stress change. Frequency diminishes with depth (few tens of metres).	Persistent, without rock bridges. Generally rough and wavy. Because they form parallel to steep natural slopes they are inherently adverse and source of landslides and rock falls. Sets of micro-fractures may develop rather than discrete fractures.	Tend to dominate landscape – fairly obvious because they are extensive and generally un-weathered and lacking much soil or vegetation. Some primary and secondary joints (such as bedding or doming fractures) may open up due to unloading and therefore take on the appearance of sheet joints.
Exfoliation fractures	Smaller scale fractures than sheet joints, due to stress relief, chemical and thermal changes. Onion-skin fractures, part of the process of rounding of relict corestones in weathering profiles.	Generally not extensive. Significant in general weakening of rock mass.	

Source: Amended from Hencher, S.R. 1987. Slope Stability – Geotechnical Engineering and Geomorphology, UK: John Wiley & Sons Ltd, 145–186.

such as sills and volcanic plugs that have cooled fairly close to the Earth's surface. The rate of advance of the isotherm downwards, from the upper cooling surface, marking the *solidus* boundary (liquid to solid transition), has been measured in boreholes in lava flows, as about 6 mm per day (Hardee, 1980), and this figure matches calculations based on thermal cooling theory (Goehring and Morris, 2008). Fractures, or at least proto-fractures, begin to form at the solidus boundary (about 900°C for basalt cooling from liquid temperatures exceeding 1000°C). Joints apparently form incrementally as tensile sections of height about 50–200 mm, sometimes displaying plume structures and hackle fringes that can be used to determine the direction of overall fracture propagation (DeGraff and Aydin, 1987). Whether the initial fracturing at such high temperatures actually forms a continuous mechanical fracture with zero tensile strength, and the fractography features recognised by DeGraff and Aydin, is, however, unlikely, as discussed later.

The tensile stresses lead to the formation of columnar structures having between three and eight sides perpendicular to the upper and lower cooling surfaces. Where the columns are well-defined and ordered, the unit is termed a 'colonnade'. If two colonnade sets develop independently, one from the top and one from the base of a lava flow, they meet, not surprisingly, in an irregular boundary with the two sets of joints mismatched, as illustrated by Budkewitsch and Robin (1994). In some situations, the upper part of the lava cools more rapidly, and the upper columns are disordered and distorted, as illustrated in Figure 3.41. This distorted layer is known as the 'entablature', and the leading hypothesis is that the disruption stems from relatively rapid cooling and inundation by water – perhaps by a diverted river or just intense rainfall, as discussed by Lyle (2000). It is then quite remarkable how distinct is the boundary between the columns that have developed from the base, as shown in Figure 3.41, and the disrupted layer above, and this whole subject area is still a major research area for geoscientists and physicists.

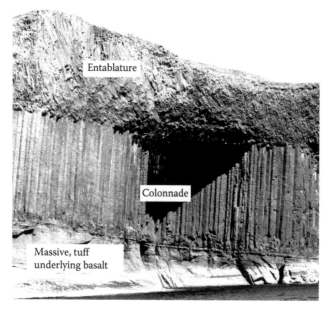

Figure 3.41 View of Staffa off the west coast of Scotland. Note the essentially planar but tilted surface at the top of the fairly massive tuff with faint bedding. The colonnaded basalt columns rise from the flat boundary essentially with no change in width until they intersect the boundary with the entablature, which has zones that are slab like and others with thin, distorted columns.

3.6.4.1.1 Other features of column jointing

With the regularity of these joints, the intuitively correct hypothesis of a series of cooling centres for the formation of polygonal columns of fractures, and the extensive research literature (more than any other kind of joints), it is surprising that there are still fundamental unknowns. Some joints are distinctly wavy, as at the Devil's Postpile in the United States, with wavelengths much larger than the incremental fracture growth. These features must be caused by some kind of oscillatory instability at fracture tips, but the full cause remains unexplained. Interestingly, I recall seeing similar undulations over 40 years ago – almost ripple like – on joints in the dolerite Whin Sill in Northumbria, England; they were a puzzle then to myself and my lecturers and, unfortunately, remain so.

At some locations the columns are bent into curves as illustrated in Figure 3.42.

Some authors attribute the shape of bent columns to complex isothermal gradients, but most choose not to discuss them. As MacCulloch (1815) wrote: *'but it will be time enough to speculate on the formation of a curved basalt column when we have something rational to offer on that of a straight one'*. I suspect that some of the deformation is a matter of plastic deformation in the hot lava flow after primary formation of proto-joint patterns at very high temperatures. That said, dramatic variations in joint patterns can certainly be attributed to local cooling phenomena and vagaries of the isotherms. For example, where a lava flow runs into the side of a glacier, then cooling columns may spread out from the ice-rock contact (Spörli and Rowland, 2006). On Mull in Scotland, a tree (John MacCulloch's tree: he found it in 1811) was engulfed by lava and, surprisingly, maintained its shape and carbonaceous remains. This and other trees in the lava flow provided the focal points for the formation of radiating columnar joints (e.g., Williamson and Bell, 2012). Elsewhere on the shoreline of Mull, rosettes of joints can be seen, also probably originally emanating from tree trunks. Similar features can be seen elsewhere in the world, as illustrated in Figure 3.43.

Figure 3.42 Curved columns at the Devil's Postpile, eastern California, United States. Note that the curved columns to the left of the exposure would definitely collapse if all the fracture traces we can see were full mechanical fractures. The debris in the foreground is evidence of the gradual process of crack propagation following the lines of the original 'proto-joints'.

Figure 3.43 Columnar joints radiating from some original cooling focal point – now eroded away. Kwangju, South Korea.

It is a surprise that even for such interesting and much-studied structures there are so many remaining unknowns, but that is the situation; for other jointing, it is even more speculative. It is to be noted that lava flows, dykes and sills, do not always show such spectacular examples of cooling joints. Quite often they are massive, with widely spaced but orthogonal jointing patterns (Figure 3.44).

3.6.4.2 Cooling and emplacement joints (plutonic)

Exposed igneous plutons (coarse-grained rocks that cooled at great depth in the Earth's crust) are often jointed. The same rock at depth, however, often shows far fewer or no visible traces of joints (Martin, 1994; Lanaro et al., 2009), which is evidence that supports the idea that visible traces of joints at shallow depths are often the result of the propagation of pre-existing joint systems, due to unloading and weathering. Analysis shows that shrinkage due to cooling of a pluton at depth could exert tensile stresses of the order of tens of MPa, certainly of the same order as tensile strength of the intact rock.

Primary cooling joints run parallel to the cooling surface, and are called 'doming joints' (Figure 3.45). Where there is an *in situ* regional stress imbalance, then tensile joints will form in line with the maximum and intermediate principal stresses. This is the interpretation made for the origin of extensive joints, striking as a set, roughly NE in the Lake Edison Granodiorite in California (Bergbauer and Martel, 1999). These authors observed joints up to 50 m in length in exposures, but individual lineations from this joint set, mapped from air photos, extend several kilometres.

A feature of some igneous plutons is the presence of orthogonal weakness directions, termed 'rift and grain'. The weakness fabric can stem from microfractures and fluid inclusions oriented in principal planes opposing the intermediate and minor principal stress directions during cooling. These weakness directions may also be developed as primary joint

Figure 3.44 Basalt from same sequence as that on Staffa, a few miles north in the Treshnish Isles. Note lack of entablature and also that joint traces, whilst predominantly vertical are wavy and much less regular than on Staffa, and that they are widely spaced. Note also that the fallen debris on the shore is of smaller dimension than the observable traces *in situ* but still sharp-edged. This implies that there are incipient fractures roughly parallel to the visible joint traces allowing the smaller, orthogonally sided debris to be produced with time.

Figure 3.45 Doming joints in granite, Yosemite National Park, United States. There is some question over whether these are actually primary cooling joints opening up by stress relief or whether they are just sheeting joints (see later in text). Professor Dick Goodman explaining the situation in foreground.

orientations, and later opened up as joints during weathering and uplift (e.g., Fleischmann, 1990). Similar orthogonal rift and grain fabrics have sometimes been interpreted, not as primary cooling features, but secondary, due to tectonics and uplift, where again they follow what were probably then principal stress directions (Wise, 1964).

Primary joints may also be associated with the stresses involved in intrusion (here, there is a little bit of juggling to be done regarding whether such joints are regarded to be 'primary' or 'secondary', i.e., tectonic – but no classification is perfect, is it?). Intrusion of large bodies of magma will inevitably be associated with massive stress changes in the surrounding rock. A good example of where the currently observed fracture system can be interpreted as contemporaneous with igneous pluton emplacement, is shown in Box 3.1 for the famous Sea Point Contact close to Cape Town, South Africa. The main joint set has the same orientation as dykes and mineralisation, as well as slithers of xenoliths of the country rock intruded by the granite. The quotation from Charles Darwin, who visited the site, is a lesson in the importance of careful observation to the interpretation of geological conditions.

3.6.4.3 Sedimentary

Master joints form during burial compaction in thick sedimentary sequences, especially sands, as they are lithified into sandstone. The primary mechanism is thought to be the development of high water pressures (see earlier discussion on thrust faulting) that reduce effective stress and allow the formation of hydraulic fractures (Engelder, 1985). These are typically orthogonal to bedding and to each other, and vertical with σ_1 due to gravitational loading. The interpretation is that after stress is relieved by the formation of one joint set in the plane of σ_1/σ_2, the horizontal stresses flip direction so that the next joints form again in the new plane of σ_1/σ_2. Examples are shown in Figure 3.46. Hydraulic fractures can also form during burial in mudstones, and these can prove particularly important for the exploitation of shale gas (Engelder et al., 2009).

3.6.5 Secondary, tectonic joints

3.6.5.1 General

Tectonic joints form as the result of mountain-building, compression and extension, general uplift, and down-warping. In principal, they are stress-release fractures resulting from some external stress field. Following Griffith's (1921) analysis showing how the presence of a crack in a solid can lead to concentrations of tensile stress at the end tips, even in a compressive stress field, joints probably are initiated in one of three ways:

1. Conversion of remote, regional compressive stress into local tensile stress as outlined by Griffith.
2. Amplification of broad remote tensional stress, so that local tensile stress exceeds intact strength of rock.
3. Local tension because of excess pore pressure as for hydraulic fractures formed during burial of sediments (Engelder, 1985).

Joints are often propagated from minor aberrations and flaws such as voids, concretions, and clasts such as pebbles or fossils, which serve to concentrate the tensile stress. Generally, once fractures are initiated, they will propagate in a straight, planar direction (rather than curve away), governed by the regional stress pattern, as explained by Mandl (2005) and Hoek and Martin (2014).

BOX 3.1 SEA POINT CONTACT

At Sea Point, west of Cape Town, South Africa, there is very good exposure of Precambrian sedimentary rocks intruded by granite. Details of the petrology are described by Walker and Mathias (1946).

To the north of the contact, in the Precambrian rock, the fractures appear relatively non-systematic.

Within the granite itself, jointing essentially follows an orthogonal grid pattern.

In the contact zone, where the granite and sediments (metamorphosed to hornfels) intermingle as 'migmatite', the dominant joint set in the granite can be seen to be followed by included slices of hornfels, and is also exploited by late-stage fluid intrusion and mineralisation. This is all evidence that the joint system in the granite and contact aureole is essentially primary, probably associated with the emplacement stresses and possibly cooling.

Charles Darwin visited the site in 1836 on the voyage of HMS Beagle and observed:

The actual junction between the granitic and clay-slate districts, extends over a width of about 200 yards, and consists of irregular masses and of numerous dikes of granite, entangled and surrounded by the clay-slate: most of the dikes range in a N.W. and S.E. line, parallel to the cleavage of the slate. As we leave the junction, thin beds, and lastly, mere films of the altered clay-slate are seen, quite isolated, as if floating, in the coarsely-crystallized granite; but although completely detached, they all retain traces of the uniform N.W. and S.E. cleavage. This fact has been observed in other similar cases, and has been advanced by some eminent geologists, as a great difficulty on the ordinary theory, of granite having been injected whilst liquefied; but if we reflect on the probable state of the lower surface of a laminated mass, like clay-slate, after having been violently arched by a body of molten granite, we may conclude that it would

be full of fissures parallel to the planes of cleavage; and that these would be filled with granite, so that wherever the fissures were close to each other, mere parting layers or wedges of the slate would depend into the granite.

3.6.5.2 Regional joints developed as tensile fractures

Regional joint patterns, in an ideal situation, can be expected to form a set of parallel fractures in the plane of σ_1/σ_2, and this is sometimes the case, although, often, joints are nowhere near as regular in their patterns. Regional stress patterns can reflect general compressive or extensional tectonics, or broad uplift, doming and stress relief (e.g., Price, 1966).

(a)

(b)

Figure 3.46 (a) Shows two orthogonal joint sets, formed during burial, the trends of which can be traced for hundreds of kilometres across the Appalachian Basin in the United States. The photo is from Engelder et al. (2009). These joints play a major role in natural gas extraction. (b) Shows similar orthogonal cross-joints, partially developed in Carboniferous sandstone, Seaton Sluice, Northumberland, England. Barnacles for scale – of the order of 10–20 mm.

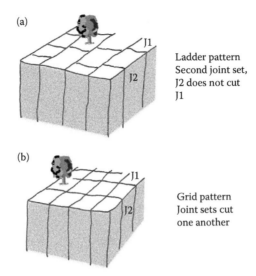

(a)

J1

Ladder pattern
Second joint set,
J2 does not cut
J1

J2

(b)

J1

Grid pattern
Joint sets cut
one another

J2

Figure 3.47 Orthogonal joint patterns. (a) Ladder pattern where joint set 1 (J1) is extensive and parallel. Second set (J2) forms orthogonally but these terminate at J1 boundary. Clearly the presence of the J1 fractures influenced the terminations of the J2 joints. In (b) the second set cross J1 without apparent influence – possibly the J1 fractures were not fully developed as mechanical fractures at the time of formation of J2.

As discussed earlier, for joints formed in sediment piles, σ_2 and σ_3 may flip in direction over time, because of stress relief or regional tectonic changes, and a new set of joints may then develop orthogonally to the first set. These are sometimes truncated at the first set, where the first set of fractures are real, mechanical discontinuities, leading to a ladder pattern (Figure 3.47a). Where the first set is not fully developed, the second set can pass straight across the first without apparent influence (Figure 3.47b). Rives et al. (1992, 1994) showed that such simple natural joint systems could be modelled using brittle varnish finishes that were stretched, compressed and twisted – giving some experimental support to the simple mathematical concepts for such fracture formation.

Tensile joint sets, often developed orthogonally, and resulting from tectonic forces or stress relief, also often dominate the jointing pattern in igneous and metamorphic rocks, although similar patterns can sometimes be attributed to cooling. An example is given in Figure 3.48, where the joint pattern probably reflects tensile fracturing during emplacement of a granite body (see also Box 3.1 earlier).

Figure 3.49 illustrates a variety of grid and ladder patterns in the joints mapped out in the Jurassic rocks at Robin Hoods Bay, Yorkshire, and reflects the stress conditions at times of fracturing. Locally, the stress field was disrupted by a major fault, as illustrated in Figure 3.50, where it is apparent that movements on the Peak Fault affected the stress trajectories, causing curvature of the joints towards the fault (and a bifurcation on the fault).

3.6.5.3 Hybrid joints

In addition to these tensile groups of fractures, joints can form under shear or hybrid stress conditions at dihedral angles (from σ_1) of 0° to about 30°, as illustrated in Figures 3.51 and 3.52, and discussed briefly in Chapter 1. Some authors dismiss shear joints because

Figure 3.48 Predominantly vertical joints in pink/brown granite with vertically aligned corestones in the lower face. In the upper face dykes of the granite have intruded into the blue/grey volcanic country rock, following a fairly regular pattern that probably reflects tensile fracturing accompanying the granite emplacement. Anderson Road Quarry, Kowloon, Hong Kong.

they should theoretically develop as 'faults', with measurable shear displacement. However, 'shear' joint set orientations do occur in nature, sometimes only developed as partial fractures – incipient, because of stress relief and stress transfer to the adjacent rock as they developed – and before they became full mechanical fractures. For this reason, though developed in a shear direction, there has been no actual measurable shear movement, so they remain joints, not faults.

3.6.5.4 Cleavage

As noted previously, cleavage is a fabric developed, mainly in sedimentary rocks, and reflects regional and local stress conditions (Figure 3.53). In slate, minerals rotate, and are re-crystallised to orientate normal to the maximum principal stress. The slatey cleavage can open up due to weathering and unloading, to form discrete discontinuities.

Locally, cleavage develops between competent (relatively strong) beds in the softer material sandwiched between, as illustrated in Figures 3.54 and 3.55.

Figure 3.49 Examples of ladder and grid sets of joints reflecting regional stress conditions in the Jurassic rocks of Robin Hood's Bay, North Yorkshire, UK. (a) and (b) show regional location and sampling points respectively. (After Rawnsley, K.D. et al. 1992. *Journal of Structural Geology*, **14**, 939–951.)

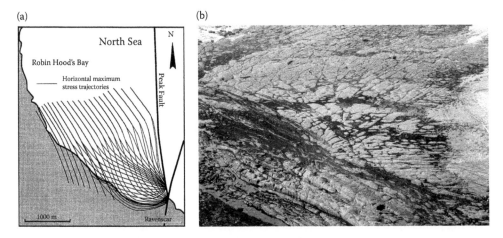

Figure 3.50 (a) Stress concentration model to explain curvature of joints towards the Peak Fault. (From Rawnsley, K.D. et al. 1992. *Journal of Structural Geology*, **14**, 939–951.) (b) In the photograph, the fault runs from the lower right corner obliquely to the left middle. Joints can be seen curving towards the line of the fault. Field of view is about 150 m.

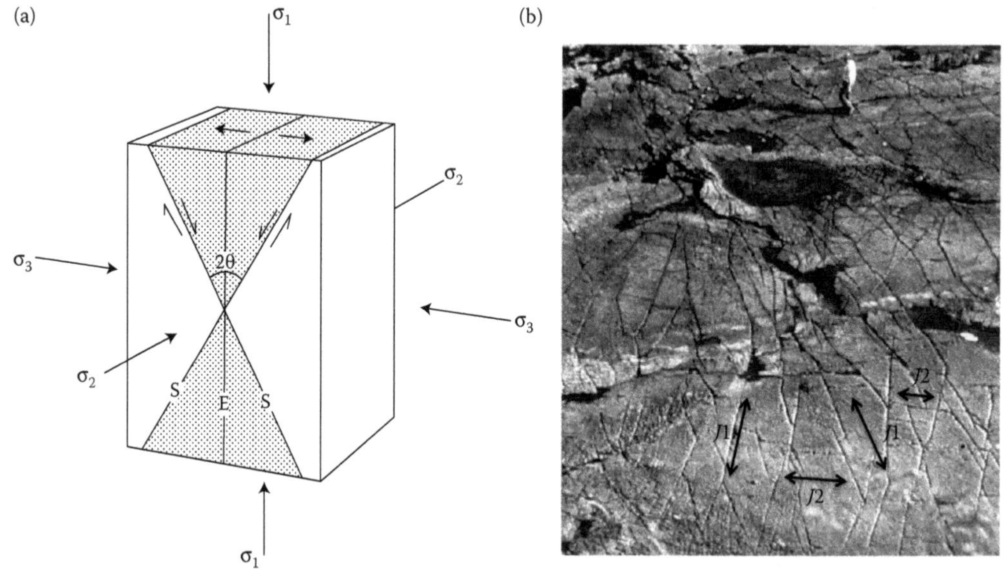

(a)

(b)

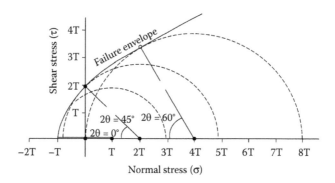

Figure 3.51 (a) Theoretical development of hybrid joints somewhere between extensional joints (E) and shear joints (S). (After Hancock, P.L. 1985. *Journal of Structural Geology*, **7**, 437–457.) (b) The photographic example is from Kimmeridge Bay, UK. Rawnsley (1990) interprets this joint network in that J1 is a conjugate set of shear and hybrid joints. At the time of formation, σ_1 was horizontal, bisecting the dihedral angle of the conjugate set. The other horizontal stress was σ_3. The second set J2 that terminates against J1 was formed with σ_1 vertical and σ_3 normal to the developing joint set.

Figure 3.52 Mohr's circle representation of 2D stress state for hybrid joint development for values of 2θ between $0°$ and $60°$.

3.6.6 Tertiary joints

Pre-existing joints open up and new joints form at the Earth's surface, due to stress changes. The most important of these are sheeting joints that develop parallel to hillsides, and are often extensive. They reflect the reorientation of gravity stresses, so that the hillside is a principal plane, with zero confining stress, and maximum principal stress parallel to the hillside, as illustrated in Figure 3.56. They are commonly seen in massive igneous rocks,

Figure 3.53 Original bedding dips shallowly to the left, cleavage is the steeply dipping fabric. Near Horton-in-Ribblesdale, Yorkshire, England.

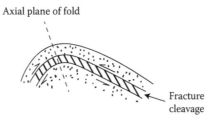

Figure 3.54 Illustration showing fracture cleavage developing parallel to axial plane of a fold.

Figure 3.55 Fracture cleavage between competent sandstone beds, Aberystwyth grits (Wales). Penknife and car fob placed on lowest sandstone bed provide the scale.

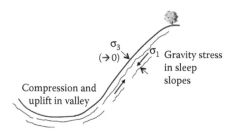

Figure 3.56 Stress conditions for the formation of sheeting joints.

Figure 3.57 Set of sheeting joints in granite, Hong Kong. The upper slabs are about 2 m thick.

such as granite, but are also found in other massive rocks, such as the sandstone of Ayers Rock in Australia (Hencher et al., 2011).

Sheeting joints, such as those in Figure 3.57, are typically wavy and rough, which gives them relatively high shear strength, but their steepness makes them prone to failure in cuttings, especially as they become weathered and the roughness component diminishes with time.

3.6.7 Joint development in geological and engineering time

It is evident that joints have a lifetime of development. When developed initially as a crack, relieving local and regional stress, they may do so without fully developing as an open, mechanical fracture. The pattern of cracks provides a blueprint for the development of visible traces of joints – and, eventually, open fractures – generally in response to unloading and weathering (Hencher, 2014). The concept is set out in Figure 3.58.

As discussed in Chapter 5, there is need to describe and characterise the tensile and shear strength of incipient fractures, and this is best done through reference to the strength of the parent rock. The current concept, implicit in rock mass classifications, and ISRM definitions, that all joints have no or negligible tensile strength (or are 'healed'), is over-simplistic, and misleading. It might tend towards the conservative for slope and tunnel stability, but is not a correct representation of reality.

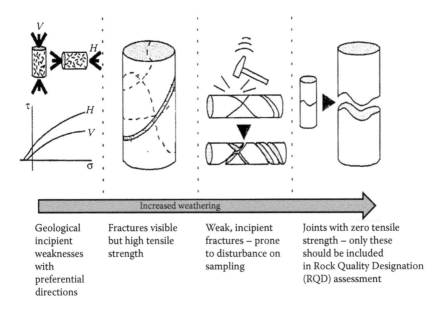

Figure 3.58 Development of joints from invisible incipient weakness through to open, mechanical fracture.

3.6.8 Shape and extent of joints

The shape and extent of joints is fundamentally important to rock engineering projects, but extremely difficult to investigate or establish. One can measure trace lengths in exposures, but extrapolation of observations from one location to different structural regimes or weathering zones will be prone to major errors. This is particularly true of characteristics such as persistence and aperture. Again, for many engineering projects, assuming full persistence and openness errs on the conservative side, but such assumptions would not be helpful in establishing a safety case for a nuclear waste repository. It is also clear from the practical world of petroleum production that, while fractures are often important, not all fractures are open and transmissive. This is a major area of uncertainty, and one that is currently being researched by investigating the characteristics of incipient fractures at Leeds University in the UK.

In the meantime, models have to be prepared, however much they might be largely guesswork, to be calibrated, perhaps against well-pressure tests in the oil and gas industry. Generally, the assumptions are that joints are essentially rectangular, limited by adjacent beds and discontinuities, or developed as ellipses within massive rock, as illustrated in Figure 3.59. There is very little evidence for the elliptical concept, but it seems reasonable for an isolated fracture development in massive rock. Of course, while the fracture 'blueprint' might well be elliptical, the properties along that disk shape may be extremely variable in terms of strength and aperture, and then, transmissivity.

3.7 MAJOR GEOLOGICAL STRUCTURES

3.7.1 Evidence from the past

People, including geologists, find it very difficult to comprehend the age of the Earth, which is currently thought to be about 4.6 billion years. The first simple life on Earth probably

Joint confined to
hard layer (and
mostly rectangular
in shape)

Joint within isotropic
rock (no boundary
intersection) – ellipse
seems reasonable

Figure 3.59 Conceptual shapes for rock joints.

dates from one billion years later, and hard-bodied animals (with shells that could be regularly preserved as fossils in the sedimented rocks) from about 600 million years ago.

Carter (2011) rates difficulties in tunnelling through mountain chains according to geological age, with the Himalayas (the youngest at 40–50 million years ago and still rising rapidly) rated as most difficult, followed by the Andes, and then the Alps; many older mountain chains are *'almost totally benign stress-wise'*. These ages are still difficult to comprehend, but important.

The evidence for huge upheavals in the Earth's history is most spectacularly given by angular unconformities, where one group of rocks overlies another, clearly separated by huge amounts of time. A classic example is in Yorkshire, Britain, as illustrated in Figure 3.60. At this location, horizontally bedded limestone, deposited in warm sea about 330 million years

Figure 3.60 Angular unconformity between underlying Silurian sandstone and overlying Carboniferous limestone, Foredale Quarry, near Horton in Ribblesdale, Yorkshire, United Kingdom.

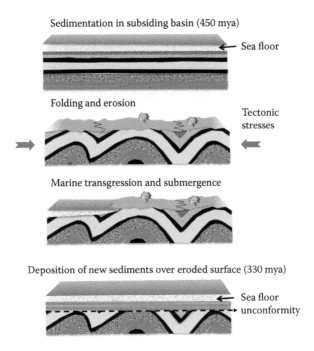

Sedimentation in subsiding basin (450 mya)

Sea floor

Folding and erosion

Tectonic stresses

Marine transgression and submergence

Deposition of new sediments over eroded surface (330 mya)

Sea floor
unconformity

Figure 3.61 Interpretation of the development of an angular unconformity.

ago, overlies folded and faulted sedimentary rocks of Silurian age (about 450 million years old). At Thornton Force waterfall nearby, a cobble beach deposit is preserved underlying the limestone – evidence of the way that the sea cut across the landscape. If you do not find this concept awe inspiring – let alone the forces and mechanics involved in the whole process – then there is no hope!

Figure 3.61 provides a very simplified representation of the process leading to the arrangement of rocks in Figure 3.60. Soper and Dunning (2005) give a fuller account, especially of the mountain-building events following deposition of the Silurian sediments, and before the mountain chain was eroded flat so that the Carboniferous sea could transgress the land mass and the overlying limestone could be deposited (Figure 3.62).

3.7.2 Evidence in the present

Some readers, having read the previous section, might be twiddling with electronic pencils, thinking 'So what! Interesting, but what is the relevance of this to practising engineers, other than something to keep in mind when drawing cross-sections through boreholes?' The important point is that we live on a dynamic Earth that is constantly changing, and engineers need to be aware of the levels of natural hazard; the obvious common natural hazards are earthquakes and volcanoes, but engineers can design against these, can't they? Earthquakes, yes, to a degree, especially so in a rich country – see the example of the design of Izmit Bridge in Chapter 7. However, if the event-tree approach to risk, commonly adopted by engineers, is flawed, then there may be unexpected and severe consequences, as at the Fukushima nuclear power plant in Japan in 2011. The structure survived the shaking well, but back-up generators designed to power the cooling systems were flooded by a tsunami, and it proved impossible to install alternative power systems in time. At the time of writing

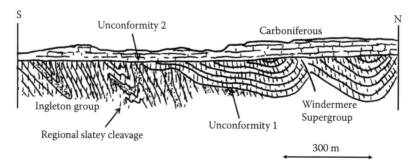

Figure 3.62 Structural relationships between the Ingletonian (currently thought to be Lower Ordovician), the Windermere Supergroup (Upper Ordovician to Silurian) and the overlying Carboniferous limestone. Note the intense folding in the Ingletonian and severe erosion that took place prior to the Windermere strata deposition. After deposition and lithification of the Windermere rocks, there has been another major tectonic event and this has imposed the regional cleavage fabric (maybe more than one). Then further erosion has taken place before deposition of the limestone. (Section is modified from Soper, N.J. and Dunning, F.W. 2005. *Proceedings of the Yorkshire Geological Society*, **55**, 241–261.)

the severe problems have still not been resolved, three years on. Volcanoes are another common hazard and localised, but the influence can be much broader. In 2010, there was a relatively minor volcanic eruption in Iceland that caused massive disruption in air traffic over Europe for about a week. It is of concern that in 1783 a much larger eruption from the same location resulted in considerable pollution and the deaths of 20% of the Icelandic population (Jones, 1999) and even 20,000 deaths in Britain; it is difficult to imagine the sociological and economic effects of a similar eruption to Europe today.

3.7.2.1 Slow processes

Many geological processes are extremely slow.

There is a remarkable site at a meander along the middle reaches of Nae Gok Cheon Stream, which is one of the branches of the Tae Hwa Gang River and not far from Ulsan City in South Korea. At this site, there is a persistent rock discontinuity in sandstone, inclined steeply, and covered with carvings (petroglyphs). Across the stream, facing the petroglyphs, is a shallowly dipping bedding plane of fine sandstone and mudstone. The bedding surface has deep indented footprints from dinosaurs, strolling across what was then a beach or mud plain, perhaps 100 million years ago. Figure 3.63 shows the bedding surfaces.

It is probable that people gathered here to be told stories depicted in the petroglyph rock face on the other side of the meandering stream. Perhaps the people who carved the petroglyphs, and told the stories, recognised that the footprints were from a huge animal set in stone.

The abstract images of the Cheon Jeon-Ri rock art are presumed to date back to the middle Bronze Age (perhaps 4000 years ago), and include overlapping lozenge patterns, concentric circles, spirals and zigzags. An example is shown in Figure 3.63.

Consider the rock mechanics and geological implications: Figure 3.64 shows the rock art carved onto two parallel discontinuities. The rearmost discontinuity was clearly as exposed then as it is today. It can be concluded that there has been little change in 4000 years. It also appears that the carvings are quite fresh – the rock has not weathered very much over the same period, despite exposure to the weather in South Korea, with temperatures annually ranging from –17°C to +35°C, snow and ice each winter, and torrential rainfall in the summer.

Figure 3.63 Engineering geologists, Professors Su Gon Lee and Robert Hack contemplating the beautiful dinosaur footprints exposed at Cheon Jeon-Ri, near Ulsan in South Korea.

Similar prehistoric cut marks are found on prominent rocks on the moor above Ilkley, West Yorkshire, Britain, where they are known as 'cup and ring marks' (Figure 3.65). The petroglyphs at Ilkley are not as varied as those near Ulsan but some are quite similar, and the date is probably about the same (4000 years ago). As at Ulsan, these are carved into sandstone, in this case the Millstone Grit – arkosic sandstone, laid down in an extensive river flood plain and delta, about 300 million years ago. Again, they show little signs of weathering (perhaps a mm or two), despite exposure to the British weather. A hundred

Figure 3.64 Examples of the abstract carvings at Cheon Jeon-Ri. Height of photo about 300 mm.

Figure 3.65 Cup and ring rock carvings on Ilkley Moor, West Yorkshire, England. Rabbit droppings, a few mm in diameter, for scale.

metres or so away from the rock carvings on Ilkley Moor, there are glacial striations, at the same level, as shown in Figure 3.66.

These are slightly indistinct, so have perhaps weathered and been eroded by a few mm since formation, probably during the last Glacial Maximum. Rock carvings from about the same age (perhaps 10–22,000 years BCE) in the Agueda Valley in Spain, appear almost fresh (Figure 3.67).

These cases illustrate the extremely slow rate of weathering and erosion (in these particular rocks, under particular climatic conditions of course). If we were to argue for 5 mm in 20,000 years, then it would take four million years to weather and erode away 1 m of rock, 400 million years to erode away 100 m. This rather clashes with the earlier discussion of

Figure 3.66 Glacial striations; scratches formed at the base of an ice sheet perhaps 21,000 years ago. Ilkley Moor, West Yorkshire, England. Rabbit droppings, a few mm, in diameter for scale.

Figure 3.67 Carving of a horse's head on a joint surface through schist, Siega Verde, Spain.

the angular unconformity at Horton in Ribblesdale, where sediments were lithified, folded, faulted, uplifted into mountains, the mountains eroded and sea transgressed, with the deposition of thick limestone beds – all in 80 million years. Phew!

Something else must be thrown into the equation.

3.7.2.2 Climatic change

One factor that helps explain more rapid erosion is climatic change. As noted earlier, it is clear that thick ice sheets covered all of Scandinavia and most of Britain only 20,000 years ago. Glaciers are very destructive, carving out deep valleys, as evident in Wales, the Lake District and Scotland, in the United Kingdom. Recent research has established that polar ice-cap melting has led to a rise of 11 mm in sea level over the last 20 years, and some scientists predict that all the polar ice will melt within 1000 years. Eleven mm per 20 years translates to 5.5 m in 10,000 years, which sounds a lot, but geologists know that the sea level has actually risen by more than 100 m in the last 20,000 years due to natural climatic fluctuations. Figure 3.68 shows sea level change over the last 20 thousand years. It can be calculated that between 18,000 and 9,000 years ago the rate of rise exceeded 12 mm per year. Over the past million years there have been cyclic fluctuations in sea level associated with natural climate change with periods of about 100,000 years as illustrated in Figure 3.69. Current predictions are that full melting of the polar ice caps will cause a rise in sea level by just over 1 m, but evidence from raised beaches either side of the English Channel show that the sea level was perhaps 5–10 m above Ordnance Datum within the last million years (Keen, 1995). This will have been affected by isostatic readjustment causing land masses to rise with the melting of ice over that period. It is highly likely that within another

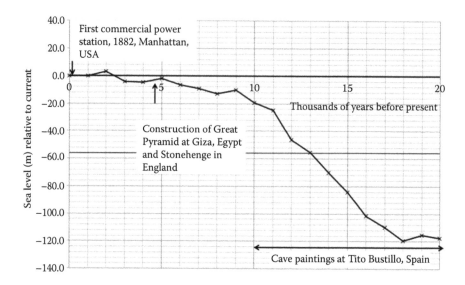

Figure 3.68 Variations in sea level against time over the last 20 thousand years. Data sources and accuracy are discussed by Miller et al. (2005 and 2011) and made available at Kenneth Miller's home page: http://geology.rutgers.edu/people/faculty/242-kenneth-g-miller.

12,000 years, Britain again will be covered in ice, and people will be able to walk to France from England – if the politics were to allow it, which would no doubt be a problem by then. Such massive fluctuations in sea level, over relatively short periods, have major consequences for landform, with rivers cutting down and the destruction of sub-sea features, and even landmasses.

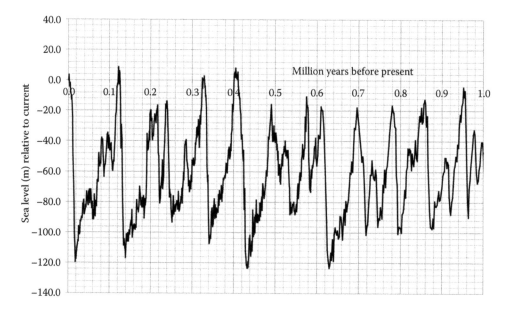

Figure 3.69 Variations in sea level against time over the last million years. Data sources and accuracy are discussed by Miller et al. (2005 and 2011) and made available at Kenneth Miller's home page: http://geology.rutgers.edu/people/faculty/242-kenneth-g-miller.

3.7.3 Faster changes

While weathering processes can be slow, it depends on the type of rock and situation. It is evident that in different parts of the world there have been regional 'weathering events', certainly involving groundwater and probably raised temperature thereby accelerating the rate of deterioration. Mudstones, especially those with high smectite content, can deteriorate very quickly when exposed in a new cutting. In the Centennial highway cutting in Queensland discussed elsewhere in this book, cuttings through weathered basalt with corestones rapidly deteriorated with the production of soil flows of purple and yellow soil rich with smectites. Weathering and internal erosion is associated with the passage of groundwater through the vadose zone. The water introduces new chemicals that encourage decomposition of the minerals making up the country rock, but it also acts to transport the decomposed and disintegrated debris out from the parent rock. Ruxton and Berry (1957) present a graphic example from Hong Kong. They describe how a tomb cut within saprolite and lined with loose brick walls was discovered at Li Cheng Uk. When excavated, it was found that the tomb had been largely infilled with material washed out from the surrounding saprolite, and deposited within the tomb. They state:

> ... for every 2 tons of original Zone 1 debris, 1 ton is now external eluviated residue and the other ton was found illuviated into the tomb.

In other words, approximately 50% of mass was internally eroded from the surrounding saprolite and washed into the tomb through the brick walls over a period of about 2000 years. This massive movement of weathering products would obviously leave the remaining *in situ* material highly porous, with a precarious structure, and with the potential to collapse.

3.7.3.1 Rapid events

Fountains Abbey is one of the largest and best-preserved ruined Cistercian monasteries in England. It is located three miles southwest of Ripon in North Yorkshire, and provides a metaphor for slow and fast geological processes. The Abbey was founded in 1132.

Figure 3.70a illustrates the relatively slow changes wrought by material weathering, over a period of about 900 years, on the sandstone blocks making up the abbey walls. Remarkably, (take note modern paint manufacturers), original paint can be found on the walls of the abbey. Figure 3.70b shows that the nave is lacking a roof. This was a rapid – rather than gradual – event that occurred soon after 1539, when Henry VIII ordered the Dissolution of the Monasteries.

3.7.3.2 Rapid natural events

We are reminded from time to time that rapid changes can occur due to geological and Earth surface processes, and rock mechanic and civil engineers, responsible for design of major structures generally, need to be aware of such hazards. I began this section by referring to Icelandic eruptions that caused extensive flight disruption in 2010 and, previously from the same location, resulted in many deaths globally in 1783. There are regular news items of massive landslides in natural slopes that cause loss of life; this has happened over the last two months, at the time of writing, in the United States and India. On May 31, 1970, a landslide from the Andean mountain Huascaran, triggered by an earthquake, buried the town of Yungay, in Peru, with an estimated loss of life of 20,000. Similar sudden changes in the landscape always accompany major earthquakes.

(a)   (b)

Figure 3.70 Slow (a) and fast (b) changes illustrated from Fountains Abbey.

As an example of how such sudden natural events can affect engineering works, major damage occurred in 2005 at a coal-handling plant site set back from the banks of the River Mahakam in Indonesia (Borneo), when a major landslide fell into the river, destroying a berthing pier for barges as well as conveyor belts and other structures. The landslide eventually extended about 400 m inland. The site was on flat land, with hills rising on each side, as seen in plan in Figure 3.71 and in the aerial photograph in Figure 3.72.

The flat area was underlain by soft clay, with un-drained shear strength below 40 kPa, and minor slumps in the river bank were to be expected from time to time adjacent to a high

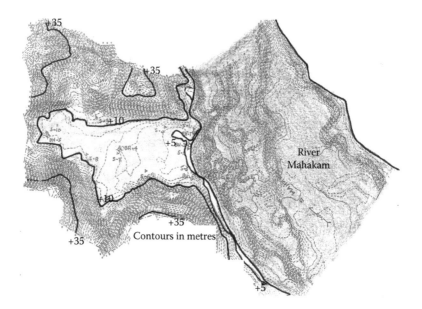

Figure 3.71 Plan of the site. River Mahakam to right, flat lying area was the site for main coal handling facilities and stockpiles.

Figure 3.72 General view of the site adjacent to the River Mahakam before the landslide.

velocity river. However, the back sapping of a landslide for 400 m over a period of a few hours was unexpected. Views of the riverbank area that failed, prior to and post-failure are given in Figure 3.73.

As part of the post-failure investigation, careful continuous sampling was conducted of the clay underlying the site. The clay was basically a valley infill with rock beneath and on each side away from the riverbank. The clay was horizontally bedded, and contained numerous wood fragments. Several of these wood fragments were dated using radioactive carbon-dating, which demonstrated a clear depth/age relationship (Figure 3.74). It was interpreted that the fairly uniform clay was deposited in a quiet, lake environment.

This provided a quandary. There are no streams or rivers leading towards the site that could have led to its silting up with fine clay in a lake environment. Away from the main Mahakam River, the flat clay-filled area is bounded by hills. The River Mahakam could not

Figure 3.73 Site prior to failure and post-failure where the river (Mahakam) has incurred into the landslide scar.

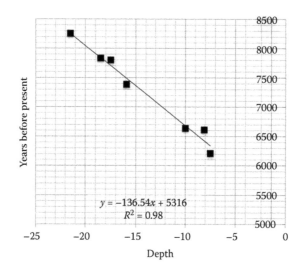

$$y = -136.54x + 5316$$
$$R^2 = 0.98$$

Figure 3.74 Dating of clay using wood fragments. Profile was deposited over about 2000 years from about 8500 years before present.

be the source of the clay as it is fast flowing. Basically, the geomorphological landscape did not make sense.

A clue came from examining bathymetric data; a chart had been prepared showing reduction in level (failure), and rise in level (deposition of debris), by comparing data preceding the failure, and with that measured a few days after the failure; this is shown in Figure 3.75.

The data showed where onshore levels had reduced in the area of the landslide scar, but also where rock promontories on either side of the coastal embayment had failed. Offshore landslide debris had raised the riverbed by up to 17 m but, intriguingly, had reduced locally by more than 7 m. It was initially presumed that this must be some kind of scour for reasons unknown. On examining the bathymetric data from each survey individually, however, it was discovered that the assumption of scour was incorrect; the local lowering (the set of close contours in line with the arrow pointing from 'A' in Figure 3.75) actually marked the erosion of a sub-river hill offshore. Figure 3.76 shows the very irregular riverbed from a few years before the landslide. To the right, there is a sub-river plateau, but this is strongly incised to depths of up to −32 m, with hillocks up to 26 m in height above the general level of the riverbed.

A few years after the landslide, the sub-river topography was very different. The profile was much smoother, with scours down to −40 m (Figure 3.77).

The data are consistent with a massive sub-river rock failure.

The point of this example is to illustrate the rapid changes that can occur even in the absence of known hazards; given the great changes that occur on the Earth's surface and that result in amazing structures such as the unconformity discussed earlier, it should perhaps come as no surprise. At the site in question, a lake has formed at the site of the landslide within the Holocene (man may well have been living in the region at the time and fishing from the lake). Water, with its mud sediment load, was fed to the lake by some river, now absent, so that the lake was gradually infilled. The Mahakam River, over the last few thousand years, has clearly migrated towards the site. The irregularity of the sub-river topography in Figure 3.75 reflects the remnants of some previous major erosional event as part of the river migration. It is interesting that studies of the Mahakam River delta, where the river

Figure 3.75 Comparison between pre- and post-failure bathymetric surveys.

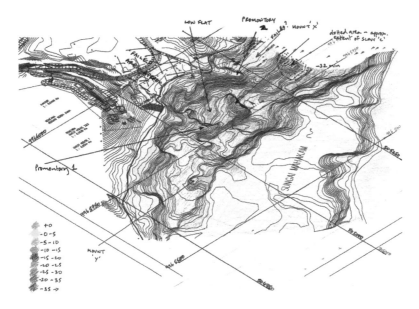

Figure 3.76 Bathymetric survey approximately four years before landslide.

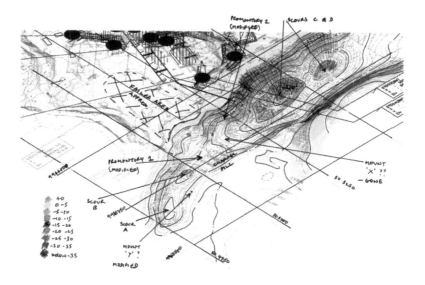

Figure 3.77 Bathymetry survey after the failure and landslide debris has been largely eroded.

meets the sea, show that the upper levels of the delta is made up of a thickness of sediments from the last 5000 years – thicker, by a factor of 6, than can be explained by the normal sediment load in the river (Storms et al., 2005). A probable contributing factor is periodic, sudden erosional events, as evidently occurred in this case, which resulted in major engineering damage running into many millions of US dollars. Such sudden erosional events are relatively common in active tectonic regions with rapidly rising landscapes, which are themselves rapidly eroded by downgrading rivers, sometimes dammed by landslides, the dams then overtopping with disastrous consequences for anyone downstream. Cook et al. (2014) report on the very rapid current erosion of a gorge on the Daan River in Taiwan, associated with uplift during earthquakes, with transformation of a deep gorge into a floodplain projected over less than 50 years. Selby (1993) reviews rates of denudation and the implications. He suggests that in one million years a mountain range is reduced in altitude by 1000 m (ignoring uplift); he also notes the importance of extreme events to these processes, as well as climate change, such as the growth and then melting of ice sheets over the relatively recent history of the Earth.

3.7.3.3 Reflections

The lessons from these examples are that change occurs at the Earth's surface gradually and slowly for the main part, but occasionally, and locally, in a sudden, dramatic manner. In civil engineering, hazards from earthquake, flooding, rainfall and wind, are dealt with in a largely statistical way – accumulating historical data and making predictions of worst case and probable case scenarios for the design of structures. Where there is a will and the resources, as in Hong Kong for natural terrain landslides, catchments are examined, and potential debris flows, and travel distances calculated, so that dangers can be mitigated in some way. However, as advocated elsewhere, the world is mostly made up of 'forgiving' sites, where there are few hazards and little need for ground investigation or complex design to make civil engineering projects safe. Rarely, there are 'unforgiving' sites – a disaster waiting to happen – and one of the real skills in engineering, and engineering geology, is in identifying and distinguishing these sites, where great caution is required, and where, sometimes, the best mitigation is to build elsewhere or even relocate existing towns and infrastructure. Building beneath a natural hillside, with a historic record of large failure, or simply the potential for large failure, is not sensible. Individuals might be prepared to accept the risks, as for example the farmers who have re-populated the slopes of Vesuvius since the destruction of Pompeii and Herculaneum, but as professional engineers and geologists, we should be looking out for the dangers, with an appreciation of the intense and sudden damaging events that can occur at the Earth's surface. It is not just a matter of using codes of practice and determining Factors of Safety. It is a cautionary statistic that one in 100 dams fail in some way, and, typically, nine out of 10 ground failures are due to geological causes.

Chapter 4

Hydrogeology of rock masses

There was nothing wrong with Southern California that a rise in the ocean level wouldn't cure.

Ross Macdonald
The Drowning Pool

4.1 INTRODUCTION

Fluid flow in rock masses is important for many reasons:

- Fundamental geological considerations. Fluids in the rock mass play a major role in hydrothermal alteration, secondary cementation of sediments and mineralisation, hydraulic fracturing, fault propagation and in the development of weathering profiles
- Civil and mining engineering projects such as prediction of water flow into tunnels and the design of drainage systems to improve slope stability
- Water supply from pumped wells
- Oil and gas reservoir exploitation
- Nuclear waste repository studies

The principle, that flow through the rock mass is driven by a difference in fluid potential (hydraulic head) between two locations, is common to all of these applications and to all fluids (oil, gas, water). The following discussion relates primarily to groundwater but the same principles generally apply to oil and gas reservoirs.

4.2 FUNDAMENTAL CONCEPTS AND DEFINITIONS

4.2.1 Porosity

Total porosity is the percentage of a rock mass that comprises voids, which includes discrete voids and fractures – either filled with fluid or gas. The different types of porosity are illustrated in Figure 4.1. In some soils, such as alluvial gravel and sand, the porosity is dominated by the matrix porosity but in many rocks, the fracture network dominates the interconnected porosity. In some masses (porous sandstone especially) both matrix and fracture porosities are important.

In many rocks only a relatively small proportion of the porosity will be interconnected so that throughflow can occur. An even smaller proportion would drain under gravity. It is a

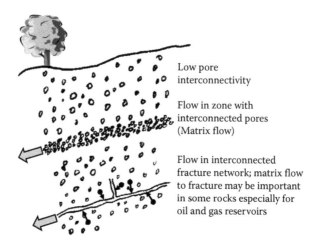

Low pore
interconnectivity

Flow in zone with
interconnected pores
(Matrix flow)

Flow in interconnected
fracture network; matrix flow
to fracture may be important
in some rocks especially for
oil and gas reservoirs

Figure 4.1 Porosity and flow in rock masses.

matter of economic importance that, as a consequence of poor connectivity the percentage of oil and natural gas recovered from reservoirs without enhanced recovery techniques is often only 10%–30%. Enhanced recovery such as injection of steam and other gases can improve recovery up to perhaps 50%–60% but the rest is left in the ground.

There will be a wide range in distribution of pore sizes in the rock matrix depending on the origin and geological history of the rock. Figure 4.2 shows the change in total porosity with degree of weathering in a series of samples of granite. Range of pore size and total porosity was measured using a mercury porosimeter.

Whereas porosity increases with weathering, the porosity of sediments decreases with burial as illustrated in Figure 4.3. This is partly due to self-weight compaction and partly due to infilling of the pores with cementing agents and finer sediment.

In terms of fracture permeability the basic model commonly used is of two parallel planes separated by an aperture but that is not realistic as discussed later. Examples of localised channel flow in rock masses are presented in Figures 4.4 and 4.5.

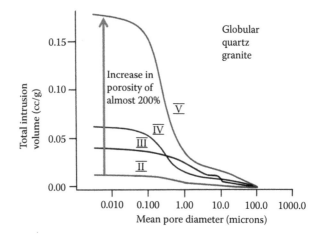

Figure 4.2 Increase in porosity with degree of weathering from Grades II to V granite. (Modified from Ebuk, E.J. et al. 1993. *The Engineering Geology of Weak Rock*, Balkema, 207–215.)

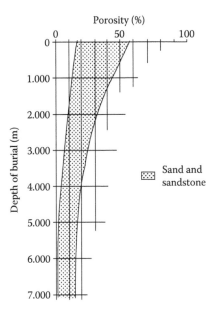

Figure 4.3 Typical variation of porosity of sand and sandstone with depth (various sources).

Figure 4.4 Local flow in face (dark areas and orange staining), Aznallcollar open pit mine, Spain. Height of view approximately 100 m.

Figure 4.5 Local water flow from fractures below 'water table', Dryrigg Quarry, Horton in Ribblesdale, Yorkshire, England.

4.3 HYDRAULIC CONDUCTIVITY AND PERMEABILITY

Fluid movement through the rock mass is driven by fluid potential differential, termed 'hydraulic head difference'. For most situations the head difference is essentially the difference between the levels to which water rises in boreholes or piezometers at different locations, although the mechanics and analysis are more complex in reality. Hydraulic conductivity is a measure of the rate at which a fluid moves through a soil or rock mass and is expressed by Darcy's Law:

$$Q = KA\frac{\Delta h}{l}\,\mathrm{m^3 s^{-1}} \tag{4.1}$$

where Q is volume of flow per second (m³/s), A is the cross-sectional area through which the flow passes, Δh is hydraulic head difference between two locations and l is the distance between them along the direction of flow. K is the constant of proportionality otherwise known as hydraulic conductivity (m/s or m/day). The rate of flow changes with the viscosity of the fluid (e.g. oil vs. water vs. gas) so there is a need to define a parameter called 'intrinsic permeability', k, so that:

$$k = \frac{Kv}{g} \tag{4.2}$$

where v is fluid kinematic viscosity (m²/s) and g is acceleration due to gravity (m/s²). Intrinsic permeability has the dimensions of area (m²). The oil industry uses a unit of intrinsic permeability called the Darcy, which is approximately equal to 10^{-12} m² although values are more commonly quoted in milliDarcies. In practice, in hydrogeology and geotechnical engineering, the term permeability is used, colloquially (incorrectly), and interchangeably with hydraulic conductivity (with the parameter of m/s rather than m²).

4.4 MEASURING HYDRAULIC CONDUCTIVITY

4.4.1 Difficulties

Because of the dominating influence of fracture permeability on large-scale hydraulic conductivity in most rock masses, this is one parameter that can certainly not be measured at the lab scale and, even in the field, tests can be extremely difficult to conduct and interpret. There have been major hydrogeological investigation campaigns in association with nuclear waste studies to establish hydraulic characteristics with an aim of predicting migration of radionuclides but results are often disappointing. Fractures identified in boreholes or excavations that are expected to conduct water do not; intersections that are expected to be important to flow paths turn out not to be. This no doubt largely reflects the very low conductivity of the rock masses being investigated for nuclear waste studies. In such masses the flow will inevitably be confined to narrow channels along fracture planes and at their intersections which are very difficult to target in investigations.

Water tests in boreholes can lead to a false impression of the hydrogeological conditions if preferential flow paths are not sampled within the test zones. An example is from high-quality ground investigations at Sellafield, UK. One of the strata, important to the hydrogeological modelling, was the Brockram, cropping out and found at depth over a large part of the study area. From early tests this stratum was judged to have low hydraulic conductivity and was modelled essentially as an impervious capping layer. Later investigations unexpectedly encountered highly conductive master joints in the Brockram and this necessitated a complete change in the way this rock had to be modelled numerically (Hencher, 1996a,b). Baecher and Christian (2003), commenting on the selection of hydraulic conductivity parameters for analysis, advise caution over the use of probability distribution function (pdf) representations of the data. The Brockram example illustrates this in that the mass hydraulic conductivity, controlled by very-widely spaced master joints, and measured in later tests, was much higher than the range of values that had been used for numerical modelling up to that time.

4.4.2 Water tests in boreholes

In fractured rock masses with higher conductivity, or where the matrix permeability contributes more strongly, testing is easier to conduct. In individual boreholes falling or rising head, 'slug' tests are conducted usually using inflated packers to constrain a test interval. Water is added or removed from a borehole and then the time measured incrementally for the hydraulic head to return to its initial ('standing') level. An alternative is to add water continuously to a borehole, maintaining constant hydraulic head and measuring the water outflow over the test zone. This is known as a constant head test. Methods of interpreting data are provided in standard references such as BS 5930:1999 but these generally assume isotropic matrix permeability. Where one is dealing with fracture permeability the interpretation is more difficult – flow from or to the borehole may be essentially one-dimensional (1D) along a channel, 2D on a plane or 3D through pervasively and closely fractured rock or in masses where the matrix dominates as in porous sandstone or some weathered rocks (Black, 2010).

4.4.3 Lugeon testing

A common test used in rock engineering empirically to judge the need for grouting and to establish the efficiency of grouting as a quality-control measure, say in dam foundations, is the Lugeon test. Water is pumped into the ground under constant pressure generally between two inflated packers. The pressure is then increased through two more stages measuring the flow of water from the test interval. Further testing is carried out at two further stages, reducing the pressure sequentially. The Lugeon value is calculated as volume of water (litres/metre of test interval/minute) multiplied by 1/test pressure (MN/m²). Assuming isotropic and homogeneous conditions one Lugeon is generally taken as equivalent to hydraulic conductivity of 1.3×10^{-7} m/s (Fell et al., 2005a).

Details of the test and its interpretation are given in Houlsby (1976) and Lancaster-Jones (1975). Different ground behaviours can be inferred by the varying hydraulic conductivity K at each stage. For example, if the calculated K increases for the higher pressure stages and then falls again as the pressure is reduced, it can be interpreted that at higher stress, the fractures are jacked open thereby increasing permeability. If the calculated K values increase through the five stages then it might be interpreted that infill is being washed out.

4.4.4 Pumping tests

On a larger scale, pumping tests are carried out, preferably using observation boreholes to measure water drawdown at various distances from the pumping well. Interpretation can be based on the steady-state extraction rate and shape of the 'drawdown curve' away from the well but there are many options and potential difficulties for interpretation especially where the geology is not uniform and if the ground is sloping (Kruseman and deRidder, 2000). Where the rock mass is of low conductivity, the drawdown curve will be very steep, and very little water can be extracted from the well; in higher conductivity rocks, the curve will be more shallow with water flowing towards the well from a greater distance.

For illustration, Figure 4.6 shows an idealised situation with a well fully penetrating an unconfined aquifer. At steady state (once drawdown has stabilised), the pumping volume per second is measured (Q m³/s) and the water table will have been drawn down to a 'cone of depression' around the pumping well. The hydraulic head, far from the well, where there is zero drawdown, is H metres and the distance from the pumping well at which there is zero drawdown is R metres. Then, assuming a steady state has been reached in the subsurface (i.e. the cone has ceased to expand) the volume flowing through any cylindrical shape at r metres away from the well must be the same so that:

$$Q = Ki(2\pi rh) \tag{4.3}$$

where K is the hydraulic conductivity in m/s.

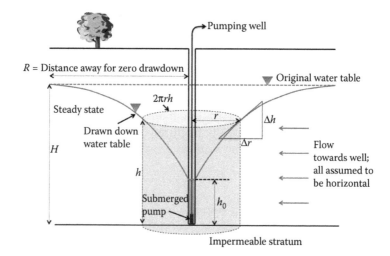

Figure 4.6 Pumping test to measure hydraulic conductivity.

The hydraulic head at various distances from the pumped well is defined as h, so that the hydraulic gradient, i, is the tangent to the drawdown curve:

$$i = \Delta h/\Delta r \tag{4.4}$$

The radius of the well is a metres.

Then:

at the well, $r = a$ and $h = h_0$
at distance, $r = R$ and $h = H$

Integrating between the limits of $r = a$, and $r = R$ yields:

$$Q = \pi K(H^2 - h_0^2)/\ln R/a \tag{4.5}$$

which allows K to be found from the pumped well drawdown if an assumption is made about the radius of the cone R (commonly it is assumed that $R = 200a$). There are difficulties with this interpretation, not least that all flow is taken to be horizontal whereas close to the well there will be considerable vertical movement. The interpretation is improved where there are at least two observation wells between the well and the far field. More sophisticated approaches to pumping test interpretation are based on the temporal development of the cone – guidance on both the steady state and transient interpretations is given in Kruseman and deRidder (2000), Todd (1976) and BS5930:1999 (British Standards Institution, 1999).

Further discussion of the options for measuring hydraulic conductivity specific to rock masses is given by Beale and Read (2014).

In the oil industry pressure tests are carried out in individual wells and these interpreted with respect to the fracture and matrix contributions to flow as discussed in more detail in Section 4.9.3.

4.5 TYPICAL PARAMETERS

Many rocks have very low permeability as materials (matrix) even where they have high porosity such as in chalk, young mudrocks and volcanic pumice. The main exception to this is sandstone, which may have interconnected high porosity, which allows flow. As noted earlier, flow in rock is often controlled by fractures as is indicated by some of the high value hydraulic conductivity data presented in Table 4.1. Bear (1972) quotes values of $k = 1$ to 10^{-3} m/s (10^8 to 10^5 milliDarcies) for highly fractured rock which is about the same as for gravel (Plumlee, 1999). These data are indicative only. Other sources give even wider ranges (e.g., Beale and Read, 2014).

4.6 UNCONFINED AND CONFINED AQUIFERS AND STORAGE

4.6.1 Unconfined conditions

An unconfined aquifer is where there is direct connection between the aquifer and the earth's surface so that water can infiltrate directly from rainfall or from a lake or river (Figure 4.7). The water level in the rock mass can rise and fall without restriction, in response to inflow or drainage. The top of the saturated zone is known as the water table where the pressure is atmospheric. Water pressure below the water table usually increases linearly with depth.

Table 4.1 Hydraulic conductivity of rock masses (indicative only)

	Hydraulic conductivity (m/s)
Sedimentary rock	
Sandstone	3×10^{-10} to 6×10^{-6}
Siltstone	1×10^{-11} to 1.4×10^{-8}
Shale	1×10^{-13} to 2×10^{-9}
Limestone/dolomite (non-karstic)	1×10^{-9} to 6×10^{-6}
Karstic limestone	1×10^{-6} to 2×10^{-2}
Salt	1×10^{-12} to 1×10^{-10}
Coal	2×10^{-8} to 5×10^{-7}
Crystalline rock	
Permeable basalt	4×10^{-7} to 2×10^{-2}
Unfractured basalt	2×10^{-11} to 4.2×10^{-7}
Fractured massive igneous and metamorphic rock	8×10^{-9} to 3×10^{-6}
Unfractured massive igneous and metamorphic rock	3×10^{-14} to 2×10^{-10}
Weathered granite	3.3×10^{-6} to 5.2×10^{-5}; 5×10^{-9} to 4×10^{-3} (according to Anon (1997))

Sources: Domenico, P.A. and Schwartz, F.W. *Physical and Chemical Hydrogeology*. 1990. Copyright Wiley-VCH Verlag GmbH & Co. KGaA; Anon. 1997. *Tropical Residual Soils*. Geological Society Engineering Group Working Party Revised Report, 184pp. for weathered granite.

4.6.2 Confined conditions

Aquifers are described as confined where there is a low permeable horizon capping the aquifer that prevents the fluid rising to the level matching its pressure within the aquifer. An example is an aquifer of sandstone, overlain by mudstone and where the catchment area (where rainfall infiltrates) is at some distance and at a higher level. If a borehole penetrates the capping layer, fluid will rise in the borehole to match the hydraulic head in the aquifer. Examples of confined aquifers include the London and Paris basins where chalk is the aquifer, confined by overlying clay beds. Artesian water has also been drawn from deeper boreholes. One such borehole at Slough, Buckinghamshire, UK, yielded almost 500,000 L/h from a depth of about 300 m in the Lower Greensand (Blyth, 1967).

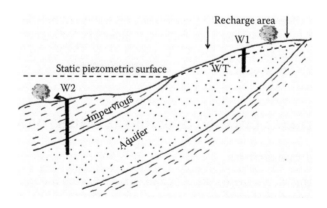

Figure 4.7 Unconfined and confined areas in the same aquifer. Well W1 will encounter water at the water table (WT) and water level will rise and fall with rainfall infiltration. Well W2 is in a confined area and will issue water under artesian conditions reflecting the pressure in the confined aquifer.

If fluid is extracted from a confined reservoir or aquifer, the rock mass will remain saturated, but the fluid pressure will drop. Reduction in pressure leads to an increase in effective stress (weight less water pressure) and therefore compaction due to self-weight as well as a small amount of fluid expansion. The fluid that is released by pumping in a unit volume of the formation, for a unit reduction in pressure, is known as the specific storage and is a function of the density of fluid, porosity and the relative compressibilities of the rock mass and the contained fluid. In the case of dual porosity formations both the rock material matrix pore space and the fractures in the rock mass will contribute to the elastic storage; the parameter is used in slope engineering to estimate the quantity of water to be drained to achieve a reduction in water pressure and the rate at which this can be achieved (Beale and Read, 2014). Specific yield, also known as the drainable porosity, is less than the effective porosity and is a measure of the fraction of the aquifer volume that will drain simply through gravity.

4.7 COMPARTMENTALISATION, AQUICLUDES AND AQUITARDS

The heterogeneity of rock masses is reflected in their hydrogeological characteristics, especially close to the earth's surface. Some zones will be readily recharged because of their exposure to rainfall or because of standing water that can infiltrate and migrate under gravity down to the water table. The way that water infiltrates and the rate at which it can do so influences the mode and timing of landsliding in a strong manner (e.g., Iverson, 2000; Hencher, 2010). Some types of shallow and deep landslide mechanisms triggered by water pressure and actions are illustrated schematically in Figure 4.8.

Low permeability strata such as massive mudstone, igneous dykes and structures such as faults or sheeting joints infilled with clay due to weathering or as sediment infill, will act as barriers to water flow and lead to compartmentalisation of the rock mass. The same is true of hydrocarbon migration at depth within the earth's crust so that oil and gas accumulates below confining horizons and such traps are the prime targets for geological and geophysical investigation in the oil and gas industries.

In geotechnical engineering the term aquiclude is used to describe an impermeable barrier; an aquitard is a low permeability stratum or horizon that restricts flow. Examples of compartmentalisation and flow concentration by geological variation at a zonal scale are illustrated in Figure 4.9.

4.8 FLOW PATHS

4.8.1 Flow paths in rock (unweathered)

Fracture flow in rock is poorly understood and extremely difficult to investigate or characterise. As noted earlier, at the simplest level rock discontinuities are modelled as two parallel plates separated by an aperture (Snow, 1968). The important finding of this analysis is that the flow through a fracture is proportional to its aperture cubed, i.e. flow through rock is highly sensitive to the openness of the fractures.

In reality flow paths are tortuous, localised and extremely difficult to identify. What matters with flow is not only the openness of the fracture network but connectivity along the full transmission path (and head along that path). The author has experience in tunnels at about 150 m depth below the sea where some zones of highly fractured rock were dry whereas other sections in seemingly intact rock were seeping water. Clearly local

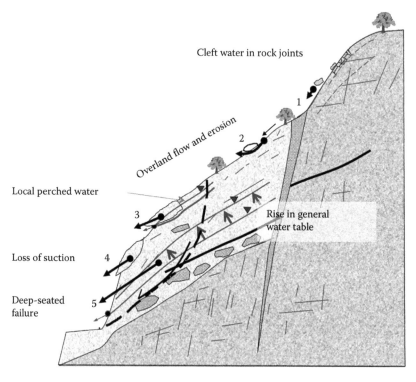

Cleft water in rock joints

Overland flow and erosion

Local perched water

Rise in general
water table

Loss of suction

Deep-seated
failure

Figure 4.8 Modes of failure in a rocky hill slope, triggered by rainfall and throughflow. Most failures (types 1–4) will be relatively small and occur due to cleft water pressures, erosion, minor perching and loss of suction in weathered profiles. Such failures tend to occur during or very shortly after rainstorms. Deeper-seated failures due to rises in deeper groundwater often occur days or weeks after a storm.

observations of fracture state can be poor indicators of hydraulic conductivity of the rock mass at a larger scale.

Representative elemental volume (REV) is the concept of a volume of rock that can be taken to have repeatable hydraulic properties.

REVs have been established for numerical models (e.g., Gnirk, 1993) but the idea that the hydrogeology of a rock mass can be represented as an REV is still problematic in reality (e.g., Davy et al., 2006).

If such an REV was to exist then the rock mass could be modelled as an equivalent continuum (see Figure 4.10), if not then the mass needs to be dealt with as a discontinuous framework and there are software packages that can do this as discussed later. The problem then is always a lack of data and it will never be possible to establish the geometry of the fracture network in any real sense (Black et al., 2007). Numerical models as illustrated in Figure 4.11 are developed based on available information and some geological interpretation (although see discussions in Chapter 3 regarding the difficulties). In practice, modelled flows seldom match those measured in test and production wells in the oil industry very accurately so models need to be adjusted iteratively, adding or removing fractures and adjusting their properties, to try to match the model to observed behaviour.

The importance of single transmissive fractures or sets of fractures, linking over great distances, must be emphasised. Dershowitz and LaPointe (1994) report how new oil wells caused large drops in production to existing wells at distances of several kilometres

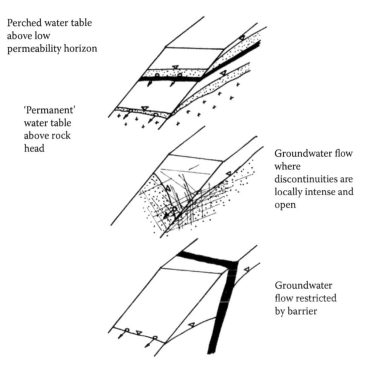

Perched water table
above low
permeability horizon

'Permanent'
water table
above rock
head

Groundwater flow
where
discontinuities are
locally intense and
open

Groundwater
flow restricted
by barrier

Figure 4.9 Flow paths and restrictions in heterogeneous rock masses. (After Hencher, S.R. *Slope Stability – Geotechnical Engineering and Geomorphology*. 145–186. 1987. Copyright Wiley-VCH Verlag GmbH & Co. KGaA.)

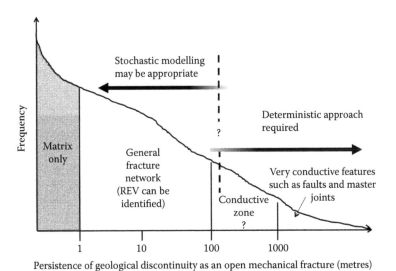

Persistence of geological discontinuity as an open mechanical fracture (metres)

Figure 4.10 Approach for measurement, analysis and numerical modelling of the hydrogeology of rock masses. (After Nirex 2007. Geosphere Characterisation Report: Status Report, October 2006, 198pp.)

Figure 4.11 Fracman model of groundmass.

within two days whilst other wells between them were unaffected. Such behaviour could not be predicted without extremely good knowledge of the fracture network and understanding of potential connectivity. There are many similar examples in tunnelling where excavation at one location has unexpectedly affected structures at great distance. Muir Wood (2004) reports the case of water flow into an adit causing unacceptable settlement to the Zeuzier arch dam in Switzerland, 1.5 km away. This and other cases are discussed by Strozzi et al. (2011). Beitnes (2005) discusses major environmental difficulties relating to inflow from above and through the invert of the Romerksporten rail tunnel in Norway. Subsequent grouting of a 2.2 km section of the tunnel cost as much as did the construction of the whole 13.8 km tunnel. The spatial intensity of occurrence of the most transmissive features may not be adequately sampled during investigation and this is a major problem for all geotechnical investigations, not least for potential nuclear waste repositories where it is recognised that large-scale conductive features need to be identified and dealt with in a deterministic manner. This is a major area of continuing research in nuclear waste studies.

4.8.2 Preferential flow paths in weathered rock

Whilst flow through fresh rock is difficult to investigate and characterise, weathered rock masses are even more complex and so consequently are their hydrogeological characteristics. Note the very wide range of values in hydraulic conductivity for weathered granite in Table 4.1. In practice and empirically, the mass might be represented by a simple set of parameters as in an REV approach for less weathered rock, but those parameters probably do not actually represent the physical and mechanical processes taking place in anything other than very simple situations.

Channelised fracture networks from the fresh parent rock often persist though the various stages of weathering. This type of preferential flow needs to be considered in investigation, hydrogeological modelling and design. Natural pipe systems develop and probably follow original structural paths (especially master joint or fault intersections), but may also be formed by seepage pressure in weak saprolite or in superficial soils such as colluvium. They also develop at permeability contrasts (e.g., colluvium overlying saprolite).

Pipe systems carry water during and following storms but may become clogged or collapse seasonally (Figure 4.12) so that, as closed-end pipes, they can lead to excess water pressures during the following wet season. Pipe systems can provide a large store of water

Figure 4.12 Well-graded alluvial sand in a Mazier sample tube (74 mm diameter), sandwiched within weathered granite at a depth of about 20 m.

that can maintain water pressure in a slope far longer than from normally infiltrated water. More details of piping and its link to landsliding processes are given in Hencher (2010).

4.8.3 Establishing hydrogeological conditions in weathered rock profiles

To allow prediction of water pressures there is no substitution for realistic hydrogeological models based on an appreciation of geology preferably backed up by data from careful instrumentation. This can prove difficult to achieve in practice partly because most ground investigation contracts are conducted in a single campaign, which allows little time for a proper assessment of the hydrogeological situation before the installation of instruments.

In an investigation of a series of major landslides in Hong Kong, several of which occurred in engineered slopes following ground investigations and instrumentation it was found that, in all cases where piezometric data were available, these did not reflect peak water pressure at the failure surface as inferred from post-failure observations or from back analysis (Hencher et al., 1985). One problem was the common inability of the installed instrumentation systems to record rapid transient rises and falls in water levels. A further problem is that many piezometers were installed at geological locations where they could not detect the critical perched water tables which developed prior to failure.

Obviously identifying hydrogeological situations dominated by natural pipe systems and discrete channels can be particularly difficult and rather 'hit or miss'. Geomorphological features such as lines of boulders, vegetation and dampness may be indicative of shallow sub-surface flow and these can be studied by remote sensing techniques including infrared photography. Trenches can be used to find shallow pipes but deeper systems are much more difficult to identify and are rarely observed in the author's experience. Subsurface investigation of large tracts of land to identify sources of infiltration and throughflow paths is not generally feasible although in a highly populated area subject to high risk, the cost might be justified. Open pipe systems will generally just be seen as poor core recovery and perhaps misinterpreted as core loss. If pipes are suspected, these can be investigated by downhole investigation techniques such as TV cameras, borehole periscopes or geophysical tools. Tracer tests using dyes or saline solutions can be used to trace sources of water and get some measure of velocities of throughflow.

Resistivity surveys can be helpful in identifying underground streams and zones of enhanced conductivity as illustrated by the case study at Yee King Road (Hencher et al., 2008).

4.9 CHARACTERISATION AND PREDICTION OF HYDROGEOLOGICAL CONDITIONS FOR PROJECTS

4.9.1 Slopes

For rock slope stability, assumptions are generally made as in soil slopes that there is an overall groundwater table with water pressures distributed linearly with depth as illustrated in Figure 4.13. The level of the water table might be measured in boreholes using standpipes or might be inferred from observed dampness in a slope face and the presence of vegetation.

Where fractures are open, allowing rapid infiltration, 'cleft' water pressure can develop quite quickly in some situations and to take that pressure fully into account can lead to onerous but rather conservative design assumptions. For a more refined analysis a better model is needed of the pressures that develop in the fracture network on a site-specific basis. One of the best examples known to the author is the investigation and instrumentation that was carried out by Golder Associates in a densely populated area of Hong Kong Island (the North Point Study). The area is underlain by granite with many adversely oriented sheeting joints. Some of these are severely weathered. The risk is high but so would be the cost of preventive measures such as anchors and drains or of rehousing as might be the outcome if a broad, conservative approach was taken to the design parameters, and this justified a more sophisticated examination of the problem.

The design engineers designed a monitoring system with a number of pneumatic piezometers installed specifically on the various sheeting joints identified in boreholes. These instruments allowed water pressure to be monitored against time during storms as illustrated in Figure 4.14. Measurements showed rises in water head above the steady-state water table, after major storms, of up to 10 m. The most important observation however was that different piezometers at different locations responded at different times. They also responded differently from storm to storm. The data indicated pressure pulses travelling down the joints following a storm. These data allowed it to be demonstrated that the conditions for stability were far less onerous than if the peak level in each borehole (adjusted up for even worse storms in the future) were combined in a single piezometric surface for design as would often be done.

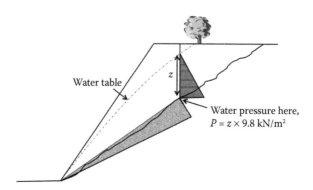

Water table

z

Water pressure here,
$P = z \times 9.8 \text{ kN/m}^2$

Figure 4.13 Typical assumption regarding water pressure in a rock slope. Water exerts a destabilising force in a vertical tension crack; there are often vertical joints in rock slopes. Water pressure also exerts an uplifting force (buoyancy effect) on the potential slip surface. (After Hoek, E. and Bray, J.W. 1981. *Rock Slope Engineering*, Institution of Mining and Metallurgy [revised 3rd edition], 358pp.)

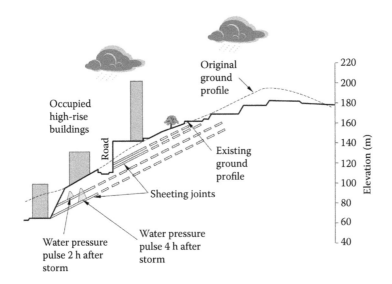

Figure 4.14 Schematic representation of instrumented slopes at North Point, Hong Kong showing localised pulses of water pressure following a storm. (Based on Richards, L.R. and Cowland, J.W. 1986. Stability evaluation of some urban rock slopes in a transient groundwater regime. *Proceedings of the Conference on Rock Engineering and Excavation in an Urban Environment*, IMM, Hong Kong, 357–363 [Discussion 501–6].)

4.9.2 Underground openings

Groundwater flow into an underground opening depends upon a number of factors, as illustrated in Figure 4.15, including

1. Construction below water table with head differential towards opening.
2. Sufficient water in storage within the rock mass to keep flow running for a significant time; in the case of subsea or sub-lake tunnels there may be unlimited recharge.
3. A connecting network of fractures, daylighting in the face, each of sufficient hydraulic conductivity to contribute to flow.

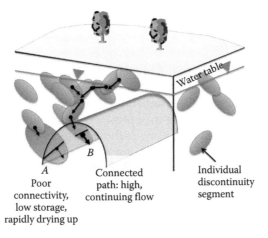

Figure 4.15 Conditions for inflow to tunnels. Representation of discontinuities as ovoid disks or rectangles is often used in numerical models and used here schematically.

It follows that inflow is highly dependent on the connectivity of the water-carrying fracture system away from the face as illustrated in Figure 4.15. At location *A* the tunnel intersects a major discontinuity but the connectivity is poor so the flow would soon dry up. At location *B* there is a smaller fracture intersection but more extensive connectivity. If that series of fractures connected to a major source of water such as a lake or even a syncline with high conductivity as in the case of the Ping Lin Tunnel in Taiwan, then the flow could be high and continuous.

Odling (1997) describes fracture network models in sandstone, in terms of their connectivity and describes fractures that do not lie on direct pathways through the rock mass as 'dead ends'; ignoring the dead ends, we are left with the 'backbone', a term she borrowed from percolation theory (Stauffer, 1985). In Figure 4.15, the fractures leading to point *B* comprise a 'backbone'.

Most problems with water inflow to tunnels arise where highly conductive and laterally extensive features, such as brittle shear zones and faults infilled with granular materials, are encountered. Where there may be highly permeable zones with high or unlimited recharge in advance of the tunnel then predrilling in advance of the tunnel, with blowout preventers, is highly advisable.

Excessive inflow of groundwater to tunnels and mines can cause delays, collapse and danger to tunnel workers during construction. These major inflows are usually associated with three different types of geological feature:

1. Fractured and crushed shear zones
2. Solution channels and caves
3. Super-conducting rock joints

Anthropogenic works such as old mine workings also can be a major hazard.

Breaching shear zones often leads to a sudden inrush of water. The Seikan railway tunnel in Japan is 53.8 km long and runs between the two most northern islands (Matsuo, 1986). Much of the tunnel is undersea (23.3 km). Despite being a minimum of 100 m below the sea floor, several inundations occurred. The most serious in May 1976 was due to the tunnel intersecting a zone of fractured tuff that had hydraulic connectivity through to the sea floor. A tunnel boring machine had to be abandoned and the tunnel advanced by drill and blast with extensive grouting necessary in advance, to restrict inflows. This led to very slow tunnelling rates and major delays.

4.9.2.1 Setting limits for inflow

At a less catastrophic level, tunnel owners and their designers will generally set a limit for what is acceptable inflow in the long term. This might well be with a view to avoiding drawdown of local water tables with potential for settlement damage to the overlying infrastructure. Where bad conditions are encountered it may prove very difficult to meet the specifications. Regarding the SSDS project under Hong Kong harbour, described in more detail in Chapter 9, the original specification limit for water ingress for one of the tunnels, tunnel *C*, was approximately 1,000 L/min for the whole 5.3 km tunnel. During tunnelling, water inflow peaked at 10,400 L/min for a single discrete fault zone. For tunnel *F*, the allowable inflow was about 200 L/min/km compared to actual of 1,400 L/min/km even for the completed tunnel.

Pells (2004) quotes an example from Australia where the permitted inflow contractually was 33 L/min/km, which equates to a mass hydraulic conductivity of 7×10^{-9} m/s. He notes that this is very optimistic (unachievable) for a tunnel in sandstone and shale, below the water table, even with pervasive grouting.

4.9.2.2 Predicting inflow into an underground opening

Predictions of inflow to tunnels are generally made using analytical solutions such as that of Goodman et al. (1965). For a situation with infinite recharge (say below a lake or the sea), the only parameters are the radius of tunnel and the depth below the water table together with hydraulic conductivity. Such solutions can give reasonable predictions for 'general country rock' but of course the key variable is hydraulic conductivity, which is difficult to measure or predict especially for a tunnel at depth where there will be little geotechnical data.

Kong (2011) reviews the various methods for predicting water inflow with reference to measurements taken from directionally drilled boreholes as part of the investigation for the planned drainage tunnels for stabilising the Po Shan area of Hong Kong – an upgrade from the previous scheme of long, inclined drains. Kong uses an equation proposed by Raymer (2001), which is based on that of Goodman et al., but with a correction of one of the parameters and an empirical adjustment from experience of actual flows. The revised equation is as follows:

$$q_s = (2\pi K(z + h_1)/\ln(2z/r) \times 1/8 \tag{4.6}$$

where
  q_s = steady-state inflow per unit length of tunnel (m³/s)
  K = equivalent hydraulic conductivity of rock mass (m/s)
  z = thickness of ground cover above tunnel (m)
  h_1 = depth of standing water above ground surface, if present (m) (e.g., lake or sea)
  r = tunnel radius (m)

Kong recommends the use of this analytical approach combined with an empirical method of McFeat-Smith et al. (1998) that provides estimation of inflow based on IMS rock class (weathering grade and degree of fracturing) linked to the experience of mainly land tunnels in Hong Kong.

4.9.2.3 Experience of inflow

Nilsen (2014) presents a review of inflow case studies to Norwegian subsea tunnels. One of his findings is the lack of correlation between large water inflows during construction and pre-tunnel predictions both in terms of hydraulic conductivity and geophysical rock mass quality. In his examples, Lugeon tests were carried out in boreholes at some distance from the tunnels and he makes the observation that extrapolating from those locations, where there were high Lugeon values, to tunnel level did not correlate with locations of high inflow. There are two main lessons that stem from this observation, reconfirming earlier discussion in this chapter:

1. Extrapolating observations of open fractures from one location to another, even only tens of metres away, is highly dubious without a realistic understanding of geological structure.
2. The Lugeon test only measures water outflow to the rock mass locally – not ingress from a large volume of the rock mass. Locally high measured conductivity is no evidence of extensive connectivity. Conversely, a low local Lugeon value is no guarantee against nearby systems of other connective fractures, that were not sampled over the test length.

Nilsen notes regarding prediction from geophysical tests that, in the Karmsund gas pipe tunnel, inflow occurred at one location with low seismic velocity as might be expected (poor quality ground) but that there were many other features with similar or even lower seismic velocity where there was no water ingress. Again this serves as an illustration of the importance of connectivity.

Holmøy and Nilsen (2014) reviewed the occurrence of inflows to hard rock tunnels and how they related to various 'geological parameters' on the basis of six reasonably well-documented case histories. Of the various correlations they attempted the most interesting findings were

1. Almost all water ingress was from fractures that were sub-parallel to the current major principal stress and making an angle with nearby major faults of 45° ± 15°. It was interpreted that such joints were probably influenced by tectonic stresses of relatively recent geological age (and therefore might be relatively open and not infilled).
2. Water inflow did not decrease with depth of rock cover. This is surprising and contrary to a common assumption that with depth, fracture frequency will be less and aperture tighter and therefore water flow reduced.
3. There is no linear correlation between Q value (rock quality – see Chapter 5) and water inflow.

The difficulties in understanding or predicting connectivity are well illustrated from tests for nuclear waste disposal research activities. Even where extensive investigation has been carried out as at the Underground Research Laboratory, Manitoba, prediction of flow is often inaccurate (Martin et al., 1990). One particular fracture through the rock mass was encountered which had characteristics considered ideal for testing methods of instrumentation and analysis. Lang (1988) described the situation as a 'unique opportunity'. In the event, despite such favourable circumstances, whilst the mechanical response of the fracture was predictable, *'predictions of the permeability and hydraulic pressure changes in the fracture, and the water flows into the tunnel, were poor'*. Similar difficulties and poor results are reported by Rouleau and Raven (1995) for tracer tests in Ontario and by Nirex (1996) at Sellafield, UK where high quality and well-constrained cross hole-testing between two boreholes showed that zones previously identified as flowing showed no observable response in the cross hole tests. These examples simply demonstrate the complexity of the problem of characterising the rock mass and making predictions about fluid flow and the need to take a cautious, managed risk approach, generally involving due consideration of all potential hazards and being prepared to deal with these. Often this will require predrilling in advance of the excavation face.

4.9.2.4 Mining

Morton et al. (1988) developed a hybrid aquifer-inflow finite element model to predict inflow to a proposed gold mine in South Africa, over a period of 6 years. The model extended over an area of about 100 km^2 and was calibrated using pumping test data. An account was taken of various development stages due to pumping from the existing mine shafts. Comparisons between predicted pumping rates and actual inflows were very good although it was noted that the model would not account for sudden influxes associated with major faults.

Beale and Read (2014) give a detailed and up-to-date review of hydrogeology associated with large open-pit mining including investigation techniques, analysis and mitigation of adverse effects on slope stability.

4.9.2.5 Nuclear waste repositories

The proposed development of repositories and disposal sites for nuclear waste has driven considerable research in rock engineering over the last 30 years or so. Many of the issues for design are essentially the same as for other underground openings but there are particular issues regarding thermal loading, and fluid flow that are especially important to nuclear waste studies. This is because of the extremely long engineering life of such proposed sites – many thousands of years compared to perhaps a nominal 100 years for many engineering structures.

Probably the greatest issue is establishing hydrogeological conditions at a site and ensuring through experimentation and analysis that risk targets regarding the escape of polluted fluids to the biosphere will be achieved. Eventually this will need to be demonstrated to a sceptical public. This is no mean task as predictions from water and other fluid experiments to date have been at best, marginally successful (Olsson and Gale, 1995). This no doubt can be put down largely to the geological unknowns in many geological scenarios. The one perhaps most promising observation has been at the Canadian underground research facility at Manitoba where it was found that there are no joints in the massive granite at depth (Martin, 1994). If there are no joints then flow will be extremely slow though a low permeability matrix. It is the lack of joints that also makes salt an attractive potential host rock for repositories.

By comparison the Borrowdale Volcanics near Sellafield, targeted in the UK for nuclear waste disposal, have had a very long geological history with burial, tectonic uplift, joint development and faulting. The prospect of proving a safety case for such rock seems bleak, no matter how intensive the investigations, testing and analysis. Probably the only way forward for such rock masses is through better structural geological characterisation methods (establishing hypotheses and testable rules for fracture network geometries including apertures) although it seems unlikely this will be achieved within many decades without extensive research.

In the meantime, empirical testing and back analysis will allow some advance though it is unlikely to be convincing. Numerical modelling provides a way forward in data-limited situations where there is some tolerance to risk, as in open-pit mining. However the lack of data, quantifying hydrogeological conditions in rock masses, means that the results and predictions from numerical predictions will always be questionable.

4.9.3 Oil and gas

4.9.3.1 Dual porosity and well testing

Oil and groundwater reservoirs can be considered, conceptually, to range from essentially homogeneous porous media to rock masses in which fractures dominate both storage and flow whilst the intervening intact rock is essentially impermeable. Intermediate to these extreme cases are reservoirs where transport and storage are shared between the matrix blocks and fractures. Such 'dual porosity' models are important to oil and gas production. The relative contributions vary with changes in effective stress and distance from the production well during extraction of fluids.

Such fractured reservoirs show somewhat complex behaviour during well tests. The pressure variations are interpreted as reflecting the relative contributions of the fracture network and matrix blocks to the flow system at different times. If such interpretations of the mechanisms controlling flow are correct and can be extrapolated throughout the reservoir then extraction might be optimised.

Warren and Root (1963) provided a solution for the double porosity model, which accounted for pressure changes during production. It is assumed that matrix blocks and fractures are uniformly distributed within their model. It is also assumed that the fractures have a relatively high permeability and a low storage capacity. The fractures carry the fluid to the wellbore. The matrix blocks have a relatively low permeability but a high storage capacity. The role of the matrix is to feed fluid to the fractures. It is also assumed that any reservoir volume contains large numbers of fractures and matrix blocks so that the REV is small.

The solution of the double porosity model by Warren and Root showed that the pressure behaviour within a wellbore is controlled by two parameters, λ and ω.

The flow coefficient, λ, relates to the ease with which fluid can seep from the matrix blocks into the fractures and ω is the ratio of the storativity of the fracture system to that of the rock mass as a whole (fractures plus matrix blocks). These parameters are a function of compressibility, 'porosity' and permeability of both matrix and fractures.

These generalised parameters are inevitably gross simplifications for real rock masses on the scale of an oil reservoir but provide a starting point for the interpretation of well tests.

4.10 GROUTING

4.10.1 Purpose of grouting

Grouting is used to reduce the permeability of rock masses and therefore flows. Examples of its use are:

1. Beneath dams, to reduce water loss from reservoirs.
2. In tunnelling, to reduce water flow both during construction and for the permanent works. Grouting might also improve ground conditions through compaction and by providing some cohesion to otherwise open fractures.
3. To form an isolation barrier.

4.10.2 Options and methods

Usually the need for grouting and results of grouting are assessed using the Lugeon test as discussed earlier. Rubber packers are inflated in boreholes to isolate a test section and then water injected using three increments of pressure, ascending and then descending whilst measuring water loss. Once the Lugeon value has been measured then steps can be taken to reduce this by injecting grout into the fracture network from boreholes under pressure. Further Lugeon tests are conducted post-grouting to check the reduction in permeability. In tunnels the need for advance grouting may be judged from inflow measurements from boreholes drilled in advance of the tunnel face.

In practice it often possible to reduce permeability by a factor of one or two orders of magnitude for example from 10^{-5} to 10^{-7} m/s, but it depends upon the nature and connectivity of the fracture network. Grouting is usually done using normal cement, sometimes combined with other material such as fly ash. If the rock mass is tight but permeability unacceptable, then ultrafine cement, silica gel or other chemical grouts might be used.

To inject grout into a tight rock mass requires very high pressure and this needs the rock confining pressure also to be high otherwise grouting might actually open up joints and make matters worse. In dam construction the grout curtain below the dam is usually completed after the dam is constructed to allow higher pressures to be used. In a tunnel, grouting needs to be done in advance of the tunnel construction. Post-grouting a leaking tunnel is often ineffective (Pells, 2004).

4.11 HYDROGEOLOGICAL MODELLING

4.11.1 Modelling geology as isotropic

Geological profiles are generally depicted for groundwater modelling as made up of discrete, homogeneous and often isotropic units of given hydraulic conductivity. Refinements in models might be to incorporate local aquitards and barriers – such as fault seals or more variable geological conditions with local litho-facies represented in some detail (e.g., Fogg et al., 1998). Generally, however geological profiles are modelled as simple confined or unconfined aquifers (Todd, 1980).

4.11.2 Anisotropic flow models

In rock, flow is predominantly through joints and other fractures and the situation can be anticipated to be more complex than for soil. Nevertheless, for practical purposes there is a need to try to compute an equivalent hydraulic conductivity, perhaps by idealising the discontinuity network as a regular series of parallel fractures (Hoek and Bray, 1974). Available sophisticated software can model the fracture network fairly realistically. For example, FracMan (Golder Associates) is a discrete fracture simulator (Dershowitz et al., 1996) that allows the geometry of discrete features, including faults, fractures and stratigraphic contacts to be modelled deterministically and/or to meet predefined statistical criteria. The models can then be used to model flow through a unit of the rock mass using the associated software Mafic (Golder Associates). As discussed earlier verification tests (predictive) to predict flow through rock masses have met with mixed success (Olsson and Gale, 1995). Attempts to represent rock masses as uniform units for a site at Sellafield, which is probably the best investigated, analysed and modelled site in the United Kingdom, are discussed in detail by Heathcote et al. (1996).

Chapter 5

Characterising rock for engineering purposes

'*When I use a word*', Humpty Dumpty said, in rather a scornful tone, '*it means just what I choose it to mean – neither more nor less*'.

'The question is', said Alice, '*whether you can make words mean so many different things*'.

'The question is', said Humpty Dumpty, '*which is to be master – that's all*'.

Lewis Carroll
Through the Looking Glass

5.1 INTRODUCTION

Rock masses need to be characterised for engineering projects. This involves firstly defining a geological model which represents the relationships between the various rock units at a site. Boundaries between rocks of different ages might be conformable, unconformable, faulted or intrusive as discussed in Chapter 3. The parent rock may be weathered locally and the extent and nature of weathering is often an important consideration.

The amount of effort and cost of establishing an adequate ground model can vary enormously. It is not a simple matter of more drill-holes means better information as one might believe from some of the literature. Many sites are 'forgiving' meaning that even minimal or poor quality ground investigation will not lead to any difficulties for construction or performance. In other situations it may take considerable work to understand the ground conditions. For example, Figure 5.1 shows a section of new road under construction in Queensland, Australia. The original ground investigation was typical for such work with boreholes to a depth of 20 m or so every 100–200 m along the line of the road together with a few trial pits. As the road was about to be completed, heave occurred in the freshly laid blacktop together with a few small landslides. Eventually to understand and remediate the on-going movements required about 5 km of rock coring, a huge array of instruments, pumping tests, radioactive dating, geophysical surveys and three separate geological investigations together with other specialist testing and numerical modelling. This was despite almost 100% rock exposure. The site was 'unforgiving'. One of the main skills of an engineering geologist is recognising the potential complexities and difficulties at a site and how risks can be mitigated.

5.2 INITIAL STAGES OF SITE INVESTIGATION

The importance of desk study and preliminary reconnaissance mapping is emphasised for producing a preliminary ground model that makes sense geologically and incorporates the

Figure 5.1 Difficult ground conditions: Centennial Highway, west of Brisbane, Australia. (Some details are given in Starr, D. et al. 2010. *Queensland Roads*, **8**, 57–73; Wentzinger, B. et al. 2013. Stability analyses for a large landslide with complex geology and failure mechanism using numerical modelling. *Proceedings International Symposium on Slope Stability in Open Pit Mining and Civil Engineering*, Brisbane, Australia, 733–746.)

probable geological history at the site. The requirements and techniques for desk study are dealt with in detail by Clayton et al. (1995) and Hunt (2005) and discussed by Hencher (2012a). For rock engineering specifically, we are generally particularly interested in fracture networks. Providing there is reasonable exposure, considerable geological insight can sometimes be gained using air photos to establish general joint patterns and the locations of major faults that often (but not always) appear as topographical lineaments because of the more readily erodible nature of faulted rock and gouge. Following desk study, through consideration of the nature of the project, any ground investigation should be targeted at the important 'residual unknowns' specific to the project. An investigation for slope stability along a highway will (should) be very different to that for a tunnel. A checklist technique is recommended whereby each element: geology, environmental factors and project considerations, is considered in turn and in combination – the 'Three Verbal Equations' of Knill (1976, 2002) as discussed in detail in Hencher (2012a). This approach will usually be cost-effective making sure that there are no 'surprises' other than those that truly could not be foreseen and that money is not wasted providing irrelevant data.

5.3 FIELD MAPPING

Before specifying any sub-surface ground investigation, full use should be made of existing exposures to help develop the preliminary ground model. Geologists are trained to produce 3D ground models by mapping alone. They examine the natural topography, any exposures, even soil, to interpret the likely geological conditions at a site. In many parts of the world this exercise will be greatly facilitated by existing maps published by geological surveys or in journals. Mapping existing exposures prior to conducting expensive sub-surface investigation will be cost-effective in allowing any ground investigation to focus on the important unknowns for the project. The degree to which this is possible of course varies from site to site and the amount and quality of exposure. A good geological map will show the relationships between different rock units, allow the geological history at a site to be understood, and aid interpretation of data obtained from drilling or other investigation methods. It should be remembered however that geological maps, even those published by prestigious geological surveys, are always an interpretation of an area, based on interpolation and analysis of available data. There will always be errors and certainly a lack of detail with respect to an individual project site. This is especially true for thematic maps such as those on mining, landslide hazard or settlement. It is important that assumptions and predictions are checked at a site-specific scale and as a project proceeds – excavations may well reveal unanticipated ground conditions that might affect the design assumptions in an important way.

Guidelines on geological mapping, specifically for engineering projects, are provided by many authors, including Dearman and Fookes (1974), the Engineering Geology Working Party on the preparation of maps and plans in Anon (1972) and the California Division of Mines and Geology (1981). Figure 5.2 illustrates a stage in the mapping process. The geologist has identified a major fault boundary in the footprint of a dam under construction. He has marked numbered locations on the fault for surveyors to pick up accurately and mark on a plan; records of conditions and measurements are kept in the geologist's field notebook.

Figure 5.2 Field record numbers painted on major fault boundary. Foundations for Queen's Valley Reservoir Dam, Jersey, UK.

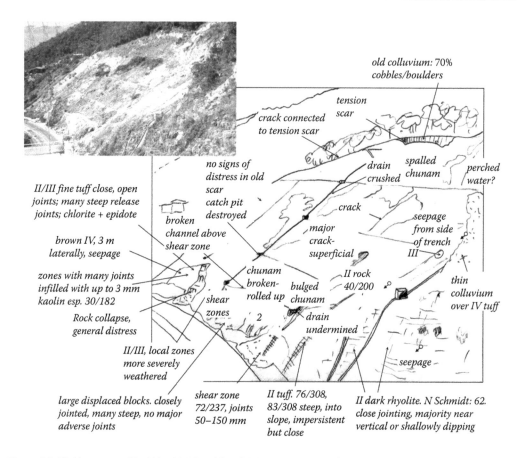

The following labels appear on the figure:

old colluvium: 70% cobbles/boulders

tension scar

crack connected to tension scar

no signs of distress in old scar catch pit destroyed

II/III fine tuff close, open joints; many steep release joints; chlorite + epidote

broken channel above shear zone

drain crushed

spalled chunam

perched water?

crack

brown IV, 3 m laterally, seepage

major crack- superficial

seepage from side of trench

III

zones with many joints infilled with up to 3 mm kaolin esp. 30/182

chunam broken- rolled up

II rock 40/200

thin colluvium over IV tuff

Rock collapse, general distress

shear zones

bulged chunam

2

drain undermined

II/III, local zones more severely weathered

seepage

large displaced blocks. closely jointed, many steep, no major adverse joints

shear zone 72/237, joints 50–150 mm

II tuff. 76/308, 83/308 steep, into slope, impersistent but close

II dark rhyolite. N Schmidt: 62. close jointing, majority near vertical or shallowly dipping

Figure 5.3 Field notes on Tin Wan Hill Road Landslide, Hong Kong, 1983.

Figures 5.3 shows notes from a preliminary reconnaissance survey of a distressed slope. From this preliminary mapping it was established that the slope was primarily in highly fractured volcanic tuff of reasonable rock quality (predominantly material weathering grades II and III) and seemed to be a generalised asymmetrical failure triggered by high ground water pressure that had dissipated as the rock mass dilated. There was no evidence in field exposure that the failure involved sliding on adverse discontinuities. Sub-surface drilling was carried out using triple tubes with foam flush and impression packers to measure discontinuity directions. The GI confirmed, as anticipated from the field mapping, that the landslide was essentially through fractured rock that could be analysed as a generalised Hoek–Brown failure.

5.4 TRIAL EXCAVATIONS

Where there is little or no relevant, existing exposure sometimes it is very useful to construct trial trenches or adits. These have advantages over borehole investigations for rock in that the extent of discontinuities and large-scale structural features can be investigated, mapped and recorded in better detail. Relationships between fracture sets can be examined and analysed. An example is shown in Figure 5.4.

(a)   (b)

Figure 5.4 Trial trenches to examine major faults and rock structure for North Anchorage of Izmit
Suspension Bridge, Turkey (see Chapter 7). (a) Photograph looking across Izmit Bay and a major
active fault with trial trench extending for about 100 m in foreground. (b) Describing and logging
fault in detail. Hammer for scale.

5.5 DISCONTINUITY SURVEYS

Measurement in field exposures is a very important technique for characterising the nature
of rock fracture networks albeit that relevance might be limited because of distance from
project, different structural domains or degree of weathering. These difficulties and limitations however are often outweighed by the advantage in seeing the extent and large-scale
roughness and continuity of discontinuities and their terminations – which contributes to
a better understanding of geological history as outlined in Chapter 3. Of course where the
exposure is directly relevant to the engineering project as in assessing stability of an existing
rock slope or bearing capacity in a foundation pad or raft footprint then such surveys are
vital (Figure 5.5).

Some of the features important to engineering including strength and conductivity, to be
observed and recorded if possible, are illustrated in Figure 5.6. Many of these characteristics are extremely difficult to see or quantify – especially persistence and the existence or
otherwise of rock bridges. This is an area where additional geological research is necessary

Figure 5.5 Clar compass clinometer attached to aluminium base plate being used to measure dip and dip direction of discontinuity. Glensanda Quarry, Scotland. Note: in standards it is suggested to record dip direction followed by dip (as in 290/57) but when using a Clar, you read dip first, before direction, so it makes sense to record as 57/290.

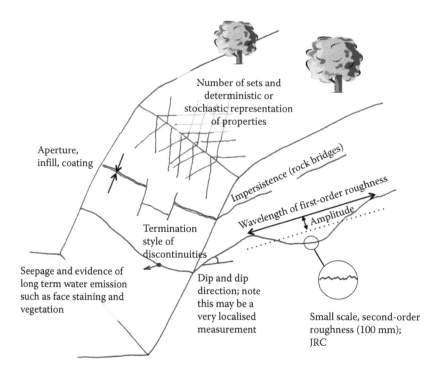

Number of sets and deterministic or stochastic representation of properties

Aperture, infill, coating

Impersistence (rock bridges)

Wavelength of first-order roughness

Amplitude

Termination style of discontinuities

Dip and dip direction; note this may be a very localised measurement

Seepage and evidence of long term water emission such as face staining and vegetation

Small scale, second-order roughness (100 mm); JRC

Figure 5.6 Features to be measured/observed in field exposures. (From Hencher, S.R. *Slope Stability—Geotechnical Engineering and Geomorphology.* 145–186. 1987. Copyright Wiley-VCH Verlag GmbH & Co. KGaA.)

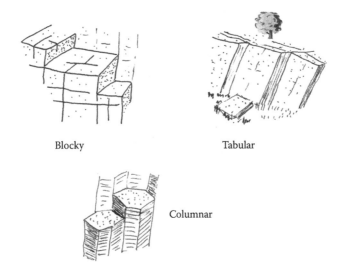

Blocky   Tabular

Columnar

Figure 5.7 Typical joint patterns and descriptive terms. (Modified from International Society for Rock Mechanics. 1978. *International Journal of Rock Mechanics and Mining Sciences and Geomechanics Abstracts*, **15**, 319–368.)

if rock mechanics modelling is to be improved. Persistence and extent of rock bridges will be a function of original discontinuity formation and subsequent geological history including unloading and weathering but there is very little guidance on how to measure or predict these in the literature (see discussion in Chapter 3). Roughness – large- and small-scale can be measured on exposed surfaces and then extrapolated to the hidden geology, linking borehole observations where available, based on a geological appreciation of the origin of the discontinuities. Some guidance on describing field exposures is given by the ISRM (1981) although the terminology recommended there is not generally used to this author's knowledge. Some terms to use in describing the general nature of rock joint systems is presented in Figure 5.7. Various terminologies for block size are given in national standards (such as BS 5930:1999 in the UK) but the author has never found these useful; it is recommended instead to record measured dimensions and ranges. The same applies to joint spacing, bedding and aperture.

The recommended procedure (after Hencher and Knipe, 2007) for assessing rock masses at exposures is

1. Carry out a reconnaissance of the exposure. View it from different directions.
2. Identify geological relationships by eye supplemented by close up examination of features; split exposure into mappable structural and weathering zones/elements. Sketch model.
3. Broadly identify the discontinuity sets that are present, where they occur and what their main characteristics are.
4. Measure data to characterise each set geologically and geotechnically. This should include data on fractography (hackle marks etc.) and coatings. Record locations on plans and on photographs. Record variations in degree of weathering and with structural regime.
5. Plot data and look at geometrical relationships. Consider geo-history and how the various sets relate to one another and to geological structures such as faults, folds and intrusions.

6. Considering geo-history, decide whether all joints that might be expected have actually been identified. Search for missing sets.
7. Analyse and reassess whether additional data are required to characterise the discontinuities that are particularly important for the engineering problem.
8. Where data collection is remote from the actual project, consider how data are likely to vary spatially.
9. Consider sensitivity of the fracture systems and the mechanical behaviour to stress changes that might be brought about by the proposed works.
10. Create a scorecard to highlight critical components of the discontinuities present and their potential impact.

As the recommendations from Hencher and Knipe imply, it is often best to first establish a fracture model by eye and then to characterise it through measurements. However, in some situations a more statistical approach is taken using line or window surveys in which a tape is laid along a face and then each discontinuity that is intersected, is recorded (Figure 5.8). An example sheet is provided in Figure 5.9. Shorthand terms are given in Figure 5.10. It is to be emphasised that rock mass description and joint network characterisation is not a simple mechanical exercise that can be delegated to some junior member of a geotechnical team armed with a blank scanline sheet. There is always judgement involved and interpretation to be made. There are also dangers in this approach that discontinuity sets striking roughly parallel to a scan line will be under-represented so the normal recommendation is to take three orthogonal line scans. Actually most geological exposures are not so simple and will frustrate any statistical approach, however well intentioned. There is also a common problem with accessibility and safety.

Figure 5.8 Postgraduate students from Leeds University being trained in characterising a rock mass in a disused granite quarry, Malvern Hills, UK.

DISCONTINUITY SURVEY	Date	Location	Scanline/Window	Name	Sheet
	23/07/14	Erewhon Quarry	West facing, D-D' (see plan and photo)	SH	4 of 10

Ch./No.	Type	Dip	Az.	Observations
0.5	J	44	351	Mo, St FeO, >5m, term. bedding
0.7	J	42	228	We, St FeO, 2m, term. bedding
1.2	J	85	010	Hi, Pl, 5m, term. rock
1.5	J	58	102	Op, infill IV 25mm, slow seepage
2.3	B	25	272	Mo, Cn, Pl, Sl. (mstn)
3.4	J	44	350	Hi, Un, Chlorite
4	J	85	190	Mo, Pl/rough.
4.5	J	57	104	Mo, Cn, JRC10
6	F	60	260	Clay gouge, 100mm, >10 m hidden
6.3	J	50	270	Op, clay 10mm, damp, >5 m hidden
7	J	89	008	Hi, Pl
7.2	J	60	100	Hi, Cl, Un, JRC8
8.5	B/C	34	288	mstn/sstn, Op, Sn, Pl, slow seepage
9	J	46	348	Hi, Un, Sm, Chlorite
9.1	V	82	185	Hi, infill Qtz 25 mm
9.3	J	55	105	Mo, Cn, Pl, JRC4
10	J	45	350	Mo, Un,
10.4	J	43	352	We, Un, Sm, FeO
10.7	J	62	098	We, Cn, Pl, Sm
11	J	65	104	Mo, term. bedding
11.3	J	90	012	We, term. bedding
11.5	J	44	230	Op, Pl, seepage
11.9	B	40	265	We, Cn, (false/ discontinuity?), Sm, sstn
12.6	F	89	340	Pl, Sl, 200mm soft clay breccia
12.9	B/C	35	295	sstn/mstn, Op, Sn, Pl, damp, veg
13.2	B	38	298	shaley mstn, We, Pl, damp, seepage
13.5	J	70	190	Mo, (tight, set 150 mm, term. at bedding)
14	J	75	355	Same as above - blocky
14.1	J	30	090	Forming blocks in mudstone, term. at beds

Ch./No.	Type	Dip	Az.	Observations
14.3	B	29	279	Mo, Cn, Pl, Ripples (sstn) > 15m
15	J	55	100	Hi, Ro, > 5m (hidden)
15.2	B	28	275	Op, 10mm soft clay, Pl, Sm/Sl, Dry
15.5	J	62	275	Op, Pl, Sl, veg. sl. seepage
16	J	85	188	Mo, Pl, < 2m, T bedding
16.4	J	45	235	Mo, Ro, JRC 12
17.8	J	74	185	Hi, Qtz >8m (crosses bedding)
18	J	42	249	Mo, St, JRC8-14
20	J	65	099	Mo, Cn, Pl, JRC4
21.2	J	38	225	Mo, Cn, Sm, > 4m hidden
25	J	44	252	Op, Pl, damp > 4m hidden
25.6	F	80	320	Shear zone, 200mm, >25m term. hidden, strong seepage
28	J	86	010	We, Ct thin chlorite
28.4	J	85	005	Hi, Qtz partial infill
28.9	J	70	100	We, Cn, Pl, Sm
32.1	J	88	195	Hi, Ro, >5m hidden
33.5	J	80	194	Hi, Ro, >5m hidden
33.8	J	42	235	We, thin clay, weathered
35	B	32	281	Hi pebbly band 100mm dry
35.2	J	55	104	Mo, Ro, 1.5 m term. bedding
38	J	88	187	Mo, Sm, > 4m hidden
38.5	J	44	342	We, FeO, weathered
38.7	J	40	230	We, FeO, weathered
40.1	B	35	260	We, Pl, damp

See field notebook and photo overlays for sketch of sets and additional geological descriptions.

Client: BB Quarries   File: BB/Rock face stability/field data. For notation see accompanying Figure 5.10

Figure 5.9 Scanline survey.

1. Defect type	2. Strength relative to rock	3. Coating or infilling
B Bedding X Foliation C Contact L Cleavage J Joint S Schistosity V Vein F Fault	Hi High Mo Moderate We Weak Op Open fracture Above terms are especially appropriate for core logging. In exposure if cannot tell record as U = unknown	Cn Clean Sn Stain Ct Coating or Infill Record nature of mineral coating or infill Where weathered, use material weathering grades as per Figure 5.48

4. Planarity	5. Roughness	6. Aperture
Pl Planar Un Undulating St Stepped Estimate 1st order waviness	Sl Slickensided Sm Smooth Ro Rough Estimate JRC over 100 mm	Record where significant in exposures using terminology such as recommended by ISRM.

7. Termination (visible persistence)	8. Water (and any evidence of seepage)
Measure or estimate visible persistence in metres and nature of terminations (e.g. in rock, hidden, terminates against other discontinuities)	Dry Stained/vegetation Seepage (estimate flow and nature)

For borehole logs, record depth and dip relative to plane at right angle to core axis
In exposures record dip and dip direction (azimuth) as well as location.

Examples from borehole log (depth of feature given):

4.15m: B, We, Pl, Sl, 35°
Interpretation: Distinct bedding-parallel incipient defect with low tensile strength, planar, slickensided and dipping at 35 degrees

5.1m: J, Mo, Ct FeO, Pl, Sm, 25°
Interpretation: Incipient joint significantly weaker than parent rock but still relatively strong, coated with iron oxides, planar and smooth and dipping at 25 degrees.

Figure 5.10 Notation for logging discontinuities in borehole logs and scanline surveys.

5.6 REMOTE MEASUREMENT

Remote survey using ground-based laser scanners is being commonly used, especially to access difficult locations (Sturzenegger et al., 2011). The success of laser and photogrammetric techniques depends upon access, available viewpoints, exposure and the visibility/reflective nature of discontinuities. In laser surveys triangulation of reflection points from individual discontinuity segments is necessary and for a rough, variable surface it may be difficult to resolve the point cloud. A geologist in the same situation would use a clipboard or view from a distance to get an average or otherwise representative orientation measurement.

Figure 5.11 shows data recorded from a manual window survey plotted using a stereonet as discussed later. Figure 5.12 shows the results from a terrestrial laser survey and stereonet prepared automatically at the same location. The comparison is impressive.

Laser surveys can also be used underground to provide 3D images and to record the orientation of discontinuities. Seers and Hodgetts (2014) present a case example from an old mine in Cheshire (Figure 5.13).

(a)

(b)

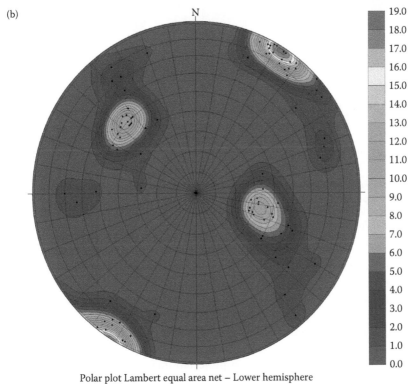

Polar plot Lambert equal area net – Lower hemisphere

Figure 5.11 (a) Photograph of blocky rock slope in northeast Spain and (b) stereonet representation of discontinuity measurements taken by hand. (From Slob, S. 2010. *Automated Rock Mass Characterisation Using 3-D Terrestrial Laser Scanner*. Unpublished PhD thesis, Technical University of Delft, 287pp.)

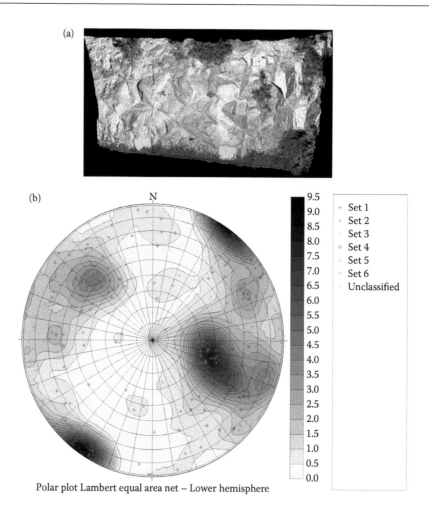

(a)

(b)

Polar plot Lambert equal area net – Lower hemisphere

Figure 5.12 (a) Point cloud image from laser survey at same site as Figure 5.11. (b) Stereographic representation of interpreted discontinuities from point cloud data. (From Slob, S. 2010. *Automated Rock Mass Characterisation Using 3-D Terrestrial Laser Scanner.* Unpublished PhD thesis, Technical University of Delft, 287pp.)

The model produced by Seers and Hodgetts can then be manipulated and interrogated to take measurements. The data collected remotely can then be input directly into a discrete fracture numerical model. Thomas Seers advises that the image in Figure 5.13 is a small part of a model created using 13 scans which were merged together (the whole model was about 120 m long). Photos are not needed to reconstruct the surface, though the colouring on the model is created using calibrated images taken from a camera on top of the scanner (each vertex is coloured with the nearest projected image pixel). It is possible to project photo textures directly onto the surface (photo-realistic model) if the camera calibration is known, which may be helpful for interpreting structures such as fracture traces which are not manifest on the surface reconstruction itself.

With regards to applications to tunnelling, laser scanning is now being used to record as-built geometries but can also be used during construction to obtain detailed rock mass and excavation information without costly delays or disruption with a simple tripod setup (Fekete et al., 2010). Examples of application of the technique are calculation of shotcrete

Figure 5.13 High-resolution digital outcrop model of the West Mine main chamber, Alderley Edge, Cheshire, UK. The model is a 3D reconstructed triangular mesh surface generated using terrestrial laser-scanner-derived 'point cloud' data (groups of vertices defined by X, Y, Z co-ordinates representing an exposure surface). The high degree of realism in the model is achieved by mapping referenced digital imagery to the triangular mesh surface. The lateral extent of the image is approximately 12 m. (After Seers, T. and Hodgetts, D. 2014. Comparison of digital outcrop and conventional data collection approaches for the characterization of naturally fractured reservoir analogues. Geological Society of London Special Publication 374, *Advances in the Study of Fractured Reservoirs*, 51–77.)

thickness, installed bolt spacing and mapping of exposed discontinuities as per the earlier examples in Figures 5.12 and 5.13.

Laser scanners are also used for monitoring movements in natural slopes and in open-pit mines together with 'total stations' that are used to measure distances between 'prisms' set on the target being monitored and the measuring station as discussed later.

Photogrammetric techniques using stereo pairs of photographs have been employed for many years to characterise rock slopes, and measure characteristics of landslide scars and similar. Accuracy is improved where there are key points and markers surveyed on the slope. An example of a 3D model prepared from uncontrolled digital photographs taken by the author at about 10 paces apart on the opposite side of a valley is presented in Figures 5.14 and 5.15.

Figure 5.14 3D visual model of slope (partially stripped and failed) to be stabilised above the spillway for Kishanganga Dam in Indian Kashmir. Produced from photographs taken using hand-held digital camera.

Figure 5.15 Sectional view through slope shown in Figure 5.14 from model prepared from photographs taken at right angles to slope. Access was difficult and dangerous to allow improved measurement, but the technique and model allowed cross sections to be prepared that were sufficiently accurate for design (as discussed in Chapter 8).

5.7 INTERPRETATION

The best place to interpret geological structure is in the field and the best person to carry out this interpretation is a structural geologist. Usually at sites with a long geological history the interpretation takes time to examine the field relationships of the various lithological units, their contacts, faulting relationships and joint patterns. There will be an iterative process whereby the geological history is hypothesised and then tested by collecting further data and confirming or otherwise the theories of how that particular rock mass took on its characteristics. It is not a simple mechanical process. The end product may comprise maps and certainly sketches in notebooks of the various important features. It is often useful to take photographs at an early visit and then use these to draw overlays of structures. For a large site, photos taken at a large distance as well as air photos will often help interpret the local geology in a broader regional context as well as help to establish the locations of observations and sampling points. Figure 5.16 shows a geological block diagram prepared by a junior geologist based only on field mapping in an area of sparse exposure.

5.8 ROSE DIAGRAMS

Once one has collected a series of discontinuity measurements by face mapping, line surveys or down boreholes, this then needs to be interpreted, first in terms of structural geology and secondly regarding engineering significance. One tool that can be revealing is a rose diagram. In rock engineering we usually record orientation data as dip and dip direction (azimuth) and this is enough to fully describe a local measurement of orientation of a dipping plane. Another measure that is important for structural interpretation is 'strike'. This is

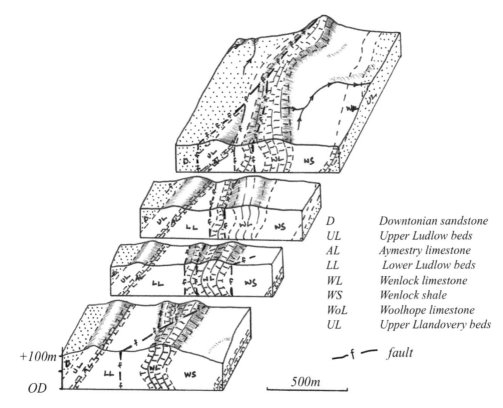

D	*Downtonian sandstone*
UL	*Upper Ludlow beds*
AL	*Aymestry limestone*
LL	*Lower Ludlow beds*
WL	*Wenlock limestone*
WS	*Wenlock shale*
WoL	*Woolhope limestone*
UL	*Upper Llandovery beds*

Figure 5.16 Hand-drawn block model of folded strata near Knightwick, Worcestershire, England. This model has been drawn and extrapolated to a depth of about 100 m based solely on measurements and interpretation of less than 20 rock exposures and interpretation of topographic features together with desk study. This would serve as an adequate preliminary ground model as the basis for planning further ground investigation. Length of view is approximately 2 km.

the orientation of a line drawn horizontally along a dipping plane. It is at right angles to the dip direction. Strike lines, drawn on a map, show regional geological trends for structures so are very helpful.

The geometrical data in the scanline example in Figure 5.9 are recorded as dip and dip direction. Inputting this data into the program DIPS (Rocscience) one can produce the 'rose' diagram shown in Figure 5.17. This shows the distribution of strikes for the reported data. One could refine this by distinguishing between the strike orientations of bedding, joints, mineral veins and faults. This plot is helpful in showing that the majority of data recorded in Figure 5.9 falls into two strike directions, essentially orthogonal to one another, and this will assist in the interpretation.

5.9 STEREOGRAPHIC INTERPRETATION

5.9.1 Introduction

Stereographic projections are used for many applications in geology including crystallography and structural geology as a means for demonstrating geometrical relationships between dipping planes. The technique has been adopted in rock engineering as a way of

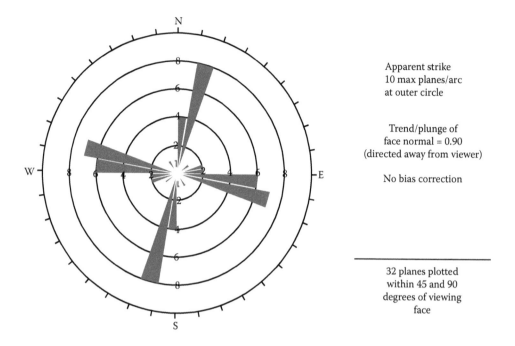

Apparent strike
10 max planes/arc
at outer circle

Trend/plunge of
face normal = 0.90
(directed away from viewer)

No bias correction

32 planes plotted
within 45 and 90
degrees of viewing
face

Figure 5.17 Rose diagram of strike data from data in Figure 5.9. (Using DIPS by Rocscience.)

demonstrating structural relationships and also for preliminary analysis of rock slope stability and the potential for key blocks to fall out in tunnel roofs and sidewalls as discussed in Chapters 8 and 9.

The basic concept is quite easy. One imagines a hemisphere, like the bottom half of a football,* that is fixed in space. We mark a North point on the horizontal circumference. Then a dipping plane, say dipping at 54° towards the 240-degree direction (SW), will cut the hemisphere as illustrated in Figure 5.18a. It will intersect the horizontal upper surface of the lower hemisphere as a straight line (actually representing the strike of the plane) and cut the lower surface of the hemisphere as a curve. All the points along the intersection can be projected back up to the zenith of the full sphere as illustrated in Figure 5.18b and this will appear as a curve (a 'great circle') in the horizontal plane. This may sound difficult but a little practice will soon make things clear. Before we leave Figure 5.18b, note that if we take a line normal to the dipping plane and intersecting the lower hemisphere, this defines a single point known as a 'pole'. We will find that we can represent each dipping plane as a complete great circle or as a unique pole – both representations have their uses for analysis. Also before leaving Figure 5.18, imagine making the dip of the plane illustrated shallower. The projection will move towards the circumference and in the extreme, a horizontal plane will plot as the circumference. The pole to a horizontal plane will plot as the centre. If the plane were dipping vertically (with the same strike), it would plot as a straight line running between 150 and 330 on the circumference. There would be two poles to the vertical plane – one plotting on the circumference at 240, the other at 060 (note that azimuths should always be recorded with three digits so that they cannot be confused with dip angle).

---

* Usually the convention is to use the lower half for analysis – we could use the upper half equally well.

(a) Discontinuity dipping 54/240

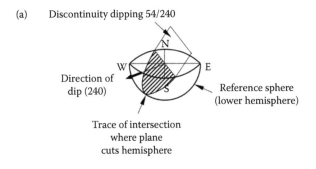

Direction of
dip (240)

Reference sphere
(lower hemisphere)

Trace of intersection
where plane
cuts hemisphere

(b) Projection of
trace into
horizontal plane

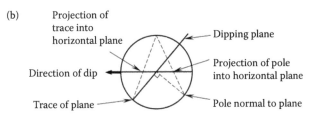

Dipping plane

Direction of dip

Projection of pole
into horizontal plane

Trace of plane

Pole normal to plane

Figure 5.18 Principles of stereographic projections. (a) Plane dipping at 54° in a direction 240°, intersecting a hemisphere fixed in space. (b) View at right angles to dipping plane. Plane trace plots as a 'great circle' in the horizontal. A 'pole', normal to the plane plots as a single point. (From Hencher, S.R. *Slope Stability—Geotechnical Engineering and Geomorphology.* 145–186. 1987. Copyright Wiley-VCH Verlag GmbH & Co. KGaA.)

5.9.2 Stereonets

A stereonet is like a 3D protractor. It is a stereographic representation of a series of reference planes and lines within the hemisphere (lower or upper). The stereonets can be 'equal angle' or 'equal area' – the first better for preserving angular relationships, the latter for statistical analysis but in practice both produce similar results (Hoek and Bray, 1981). Generally, we use equal angle projections.

Two projections are used as shown in Figure 5.19. These projections can be downloaded from www.cambridge.org/gb/download_file/202611. Download these and print them out on good quality paper or card. Make sure that they are printed out to the same diameter because they are used interchangeably. Use a drawing pin (thumb tack) to make a hole in each one, exactly in the centre. Push the pin through from the back so that a piece of tracing paper can be placed over each net (maybe reinforced with sellotape). The 'equatorial' stereonet has a series of great circles spaced at 2°, running from top (North) to bottom (South) resembling the lines of longitude on a map of the earth. The circumference is also marked in 2° intervals for the full 360°. The projection has four radii, each of 90° running from the centre to the four poles (N, E, S, W – 0°/360°, 090°, 180°, 270°). The 'polar' net has a full set of radii at 2° intervals.

5.9.3 Plotting data

One might be tempted to jump straight to software such as DIPS that can be used to plot and analyse discontinuity data but it is very useful and instructive to carry out an exercise by hand in order to understand the concepts properly.

(a) (b)

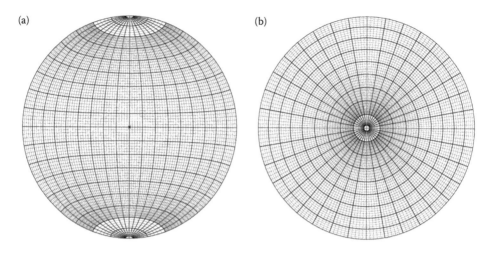

Figure 5.19 (a) Equatorial equal-area net. (b) Polar equal-area net. Equatorial and polar stereonets downloaded from www.cambridge.org.

5.9.3.1 Step 1: Plot a plane

Take the first data line from Figure 5.9 – a joint (moderately strong, stained with iron
oxide, with persistence greater than 5 m, terminating against bedding planes and dipping
at 44/351).

Plot the plane as illustrated in Figure 5.20 and as follows:

1. Mark N, E, S and W locations on the reference stereonet (downloaded).
2. Place tracing paper over equatorial net (pierced by drawing pin), draw circumference
   with pencil and mark N.

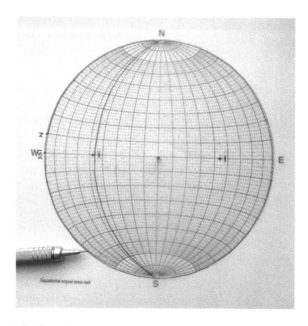

Figure 5.20 Plotting a single plane as a great circle.

3. Mark a tick on the circumference at 351° (nine degrees West of N).
4. Rotate the tracing paper so that the 351 tick lies on the E–W diameter of the underlying stereonet (Figure 5.20).
5. Measure 44° in from the circumference and then copy the corresponding great circle onto the tracing paper.
6. Measure 44° out from the centre towards the East (along the same diameter) and mark a cross which is the pole to the great circle for the plane 44/351 (here marked as 1).
7. If you wish, rotate the tracing paper back until the Norths of the tracing and underlying net again match and admire your work!

5.9.3.2 Stage 2: Plotting a second plane and measuring the intersecting wedge

Any two planes, if physically adjacent, will intersect to form a wedge as illustrated in Figure 5.21.

Take the second data line from Figure 5.9 – a joint dipping 42/228. Plot this plane in the same way as for the first plane. Mark a tick at 228 on the circumference; rotate the tracing paper until the 228 mark coincides with the W or E on the underlying stereonet and trace on the great circle and pole for Plane 2. The great circles intersect and the point of intersection defines the wedge geometry. Rotate the tracing until the point of intersection is on the E–W diameter and measure the dip (plunge) of the line of intersection in from the circumference (24°) (Figure 5.22). Rotate the tracing until the Norths match and measure the direction that the wedge is pointing – 289° – as illustrated in Figure 5.23.

This same analysis can be conducted using DIPS and the corresponding plot is shown in Figure 5.24.

5.9.3.3 Plotting large amounts of data

One could plot every plane as a great circle and determine all potential wedges but this would soon look confusing as illustrated in Figure 5.25a. By hand it would also take quite a lot of work with lots of rotating of the tracing paper. A quicker way is to work with the poles alone (Figure 5.25b). It is then relatively easy to identify groups of discontinuities that

Figure 5.21 Two sets of joints intersect to define a wedge structure. The joints meet along a 'line of intersection' that has a unique geometry, in this case dipping steeply out of a rock slope. Black Hills Quarry, Devon, England.

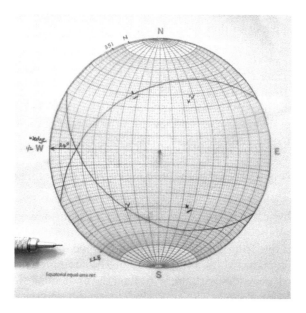

Figure 5.22 Determining dip of wedge intersection.

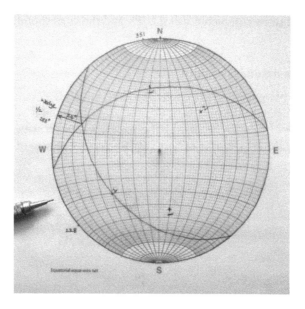

Figure 5.23 Determining direction of potential sliding for a wedge.

form 'sets' of similar geometry (Figure 5.25c). This can be done using software but can also be carried out very quickly by hand as follows:

1. Place tracing paper on the polar net with N, E, W and S marked up (Figure 5.19b).
2. Mark North on the tracing paper then rotate so that N on the tracing paper overlies S on the polar net (this is the trick!).

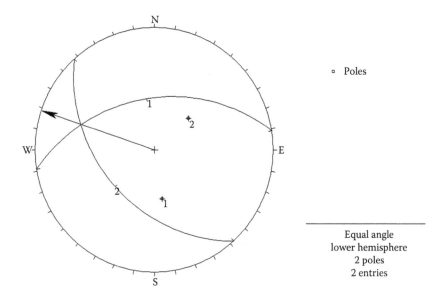

o   Poles

Equal angle
lower hemisphere
2 poles
2 entries

Figure 5.24 Wedge formed by Planes 1 and 2 (from Figure 5.9), plotted using DIPS.

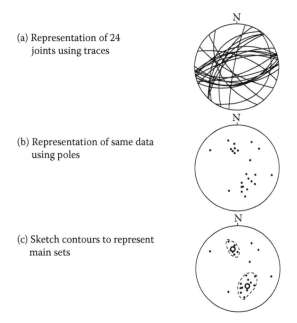

(a) Representation of 24
    joints using traces

(b) Representation of same data
    using poles

(c) Sketch contours to represent
    main sets

Figure 5.25 Representation of large amounts of data as great circles (a) and as poles (b) that can be examined statistically or manually to define clusters of data (sets) illustrated in (c).

3. Using the underlying net, plot each pole measuring dip away from the centre along the radii as illustrated in Figure 5.23 for all the discontinuities in Figure 5.9. For information this took about 10 min (Figure 5.26).

Now place the tracing, with poles marked, onto the equatorial net (of the same diameter as the polar net), now with the Norths corresponding. The poles are in the same

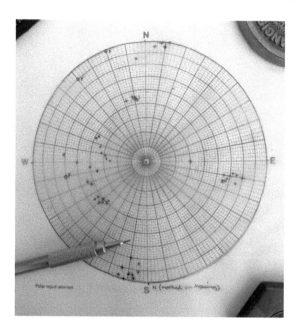

Figure 5.26 Plotting discontinuity poles using a polar net (inverted with N on South). Crosses are joints, dots are bedding, faults identified as F, one vein as V.

locations as if they had been plotted directly onto the equatorial net as illustrated for Figures 5.20 and 5.22. Several joint sets have now been identified in Figure 5.27 by their clustered geometries (J1 to J5) as well as bedding. Faults and one mineral vein are identified individually.

It can be seen that most joint sets and bedding fall in a single great circle (a girdle); the pole to that great circle falls with the joint set here labelled as J5. All poles could be rotated by about 35° in direction 284 so that bedding would be at the centre of the stereoplot. Set J5 would then be a vertical set clustered at 284 and 104. An interpretation of this fracture network is therefore that bedding was originally horizontal with maximum principal stress vertical and orthogonal joints developed in directions indicative of the minor principal stresses at the time of joint formation (J3, J4 and J5). J1 and J2 have formed in a shear direction as discussed in Chapter 3. At some later stage the rock mass has been tilted through about 35° towards the 104 direction. It appears that the faults are unrelated to and probably post-date the other discontinuities. This plotting by hand (plus analysis) took perhaps 20 min. An advantage of doing it this way is that each pole has been individually plotted (and thought about). For further analysis one would really need to consider the set characteristics (persistence, roughness and so on).

Plotting the same data from Figure 5.9 using DIPS is a matter of cutting and pasting the lists into the software. The equivalent plot to Figure 5.27 is illustrated in Figure 5.28.

The use of stereographic projections in preliminary analysis of rock slopes is discussed in Chapter 8.

5.10 ROUGHNESS MEASUREMENT

Geological discontinuities are never truly planar at any scale and deviation from that planarity affects many properties, especially shear strength and transmissivity. Many are wavy

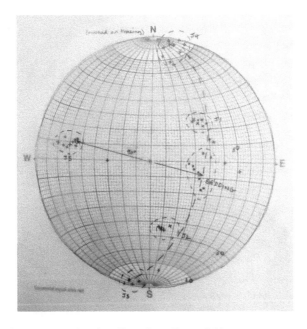

Figure 5.27 Sets identified using a polar plot. (Data from Figure 5.9.)

and variably dipping at large scale, reflecting their origins and/or later deformation. This large-scale roughness, which is expressed relative to mean or average dip of a discontinuity, is termed 1st order and can be characterised by field measurement as illustrated in Figure 5.29. Dip measurements are taken at grid intersections (perhaps 0.5 m spacing) using a range of sizes of plates. The plates might just be circular pieces of wood or aluminium plates

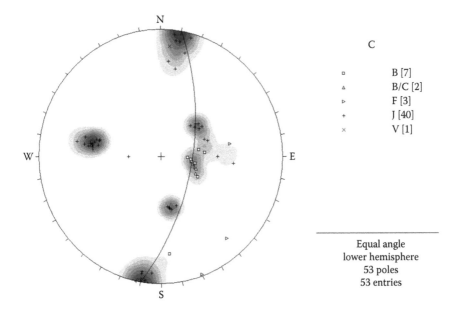

Figure 5.28 Data from Figure 5.9 plotted using DIPS software (Rocscience).

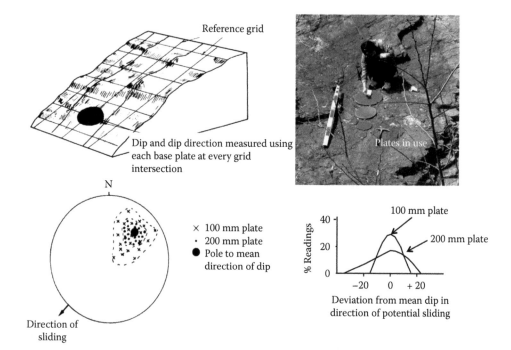

Figure 5.29 Recommended method for characterising large scale roughness in the field. (After Fecker, E. and Rengers, N. 1971. Measurement of large scale roughness of rock planes by means of pro-filograph and geological compass. *Proceedings Symposium on Rock Fracture*, Nancy, France, Paper I–18; International Society for Rock Mechanics. 1978. *International Journal of Rock Mechanics and Mining Sciences and Geomechanics Abstracts*, **15**, 319–368; Richards, L.R. and Cowland, J.W. 1982. *Hong Kong Engineer*, **10**, 39–43; Hencher, S.R. *Slope Stability—Geotechnical Engineering and Geomorphology*. 145–186. 1987. Copyright Wiley-VCH Verlag GmbH & Co. KGaA.)

that will adopt orientations reflecting local roughness features. Generally the smaller the plate, the more scatter in orientation, because smaller roughness features – such as cross-joint steps and sole markings or ripple marks on bedding surfaces – will be sampled. Larger plates will bridge across minor roughness features and represent more closely the broad orientation of the discontinuity. Figure 5.30 shows the scar of a large rock slope failure where sliding took place on a single, undulating plane. Figure 5.31 shows orientation measurements taken over an area of about 35 m² of the failure surface on a grid of 0.5 m centres. As usually found, scatter in orientation decreases with increasing size of plate but in this case is still high for the largest plate. Considering the scatter in measurements for the single discontinuity in Figure 5.30, it is a cautionary observation that in Figure 5.9, listed discontinuities are represented by single dip and dip directions. No record is made of the variability across the discontinuity and this is typical of the way that discontinuity data are recorded. The person taking measurements should try to record representative orientations but clearly this is (generally) a gross simplification of real rock mass geometry and it is worth bearing this in mind when considering the validity of any analysis and the potential for errors.

The way roughness data are used in shear strength determination is addressed in Chapters 6 and 8. In the case illustrated in Figure 5.30, whilst the dip was very variable and wavy, the contribution to shear strength of that roughness was largely negated by about 700 mm infill of clay-bound gravel breccia that prevented the discontinuity walls coming into contact (Hencher, 1983).

Figure 5.30 Scar from rock failure. Sliding occurred on single daylighting plane (fault). Man for scale, lower right. South Bay Close, Hong Kong.

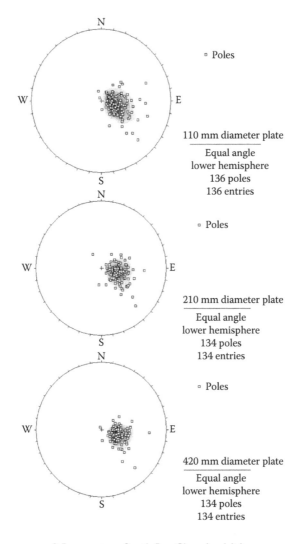

Figure 5.31 Roughness survey at 0.5 m centres, South Bay Close landslide.

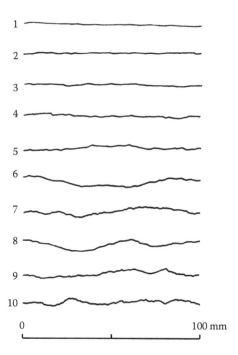

Figure 5.32 Roughness profiles. (From International Society for Rock Mechanics. 1978. *International Journal of Rock Mechanics and Mining Sciences and Geomechanics Abstracts*, **15**, 319–368.) These profiles are linked to Joint Roughness Coefficients (JRC). (After Barton, N.R. and Choubey, V. 1977. *Rock Mechanics*, 12(1), 1–54.)

Small-scale roughness is sometimes characterised through reference to a series of profiles included in ISRM (1978) and reproduced here as Figure 5.32. The profiles are not included as part of description for the BS 5930 (BSI, 1999), where recommendation is made instead to just use descriptive terms such as smooth, stepped and undulating and without any quantification. The ISRM profiles are increasingly rough from 1 to 10, but not in a systematic, progressive way and it is often a difficult matter to assign a value to a discontinuity (Beer et al., 2002). Roughness is also evidently different in different directions. There are for example, sharp small steps in profiles 4, 5, 7, 9 and 10, which might play a dominant role in peak shear strength depending on wall strength and direction of shear. They are also of course only 2D cross sections of parts of larger 3D surfaces so are only indicative of general conditions. The profiles 1–10 are correlated to JRC by multiplying by 2 (profile 1 equates to JRC 0–2, profile 2 to JRC 2–4 and so on). JRC is an important component of the empirical criterion of Barton and co-workers for estimating shear strength (discussed in Chapter 6) but has no role in other methods at determining shear strength (Hencher and Richards, 2015) other than as a broad classifier of relative roughness.

Roughness profiles can be measured using pin profilers (Figure 5.33) or other specialised profilometers such as laser scanners but these are mainly lab-based. When it comes to judging rock shear strength and the contribution from field roughness, the contribution from small scale (<100 mm) and textural roughness is largely a matter of characterisation (partly using plate data), and judgement, preferably in conjunction with good quality shear testing as discussed in Chapter 6. The chief consideration is whether the minor roughness features are generally and similarly distributed across the joint surface in question (pervasive) or just localised. If generally distributed then they can be included in shear strength considerations.

Figure 5.33 Characterising small scale roughness using pin device. Kishanganga Dam, Kashmir, India.

5.11 GROUND INVESTIGATION TECHNIQUES

5.11.1 Introduction

Once full use has been made of existing information, maps and photographs interpreted and field mapping carried out, it may be appropriate to carry out sub-surface investigation. Again, a checklist approach is recommended targeting important residual unknowns.

5.11.1.1 Geophysics

It is beyond the scope of this book to describe geophysical methods of investigation in any detail but some of the main methods are summarised in Table 5.1.

Seismic surveying is particularly important for characterising major structures within large volumes of rock.

Reflection seismology is used extensively to understand the distribution and geometry of rock volumes particularly to predict and map the location and volume of oil and gas but also in nuclear waste studies and for large engineering projects. Sound waves are generated by a source and reflect from boundaries of contrasting acoustic impedance. The waves recorded in geophones are processed to produce a seismic reflection image of the sub-surface. A typical output is shown in Figure 5.34a. This forms the basis for interpretation by geologists to define the various structures (Figure 5.34b).

Extensive geophysical surveys were conducted during investigations of a potential nuclear waste disposal site at Sellafield in NW England and the British Geological Survey (1997) presents summary interpretations of results. Geophysics, particularly seismic methods, is also used to characterise quality of rock mass, largely as reflected by change in P or S-wave velocity (Simons et al., 2001). Cross-hole tomography has been used with some degree of success in characterising sites as part of studies for nuclear waste disposal research. At the major research site of Stripa in Sweden it was concluded that the zones of low velocity on the tomographic maps correlated 'quite well' with the fracture zones that had been identified on the basis of core logs and borehole photography (Gnirk, 1993).

Magnetic and gravity surveys (airborne or by sea) can also help in preparing broad geological models as well as locating particular features of interest such as old mine workings and underground cavities. Magnetic surveys carried out in Hong Kong waters were particularly useful for identifying the extent and distribution of faults that were encountered during sub-sea tunnelling as illustrated in Figure 5.35.

Table 5.1 Geophysical methods

Method and operations	Measurements	Outcome	Application
Seismic reflection and refraction on land or on sea. Source of energy may be explosive, a 'boomer' or a sledgehammer on land Used in tunnels to predict ground conditions in advance of face	Time for seismic shock waves to travel to receivers either placed on land or towed behind seismic source at sea	Depths to rock units with different seismic velocities (related to strength and elastic properties)	Regional geological studies, identification of field relationships and major faults. Depth of infilled valleys
Cross-hole seismic and seismic tomography	Measurements between boreholes	Identify rock of different quality	Used in major investigations, for example, nuclear waste studies
Resistivity and self-potential	Potential drop between electrodes	Anomaly maps; conductivity profiles	Depth to groundwater. Underground natural pipes and cavities; change in strata
Ground-probing radar	Reflections from surfaces below ground	Depth profile	Detailed, shallow geological studies; location of pipes and artifacts
Gravity, land or marine	Variations in gravity field	Anomaly maps	Detailed geological studies. Cavities and mine shafts
Air and marine magnetic surveys	Variation in Earth's magnetic field (e.g. presence of basic rocks with high proportion of magnetite)	Magnetic maps showing anomalies	Regional geological studies; they can be helpful to find buried mine shafts
Radiometric ground and air surveys using equipment such as Geiger counters	Natural radioactivity	Anomalies. Ore bodies	Exploration for metals
Down-hole logging Many different tools (see Schlumberger web pages for example)	Many types	Bed thickness, clay mineralogy, resistivity, salinity	Mostly in oil and gas explorations but also in mining and civil engineering especially for discontinuities
Acoustic televiewer		Discontinuity logging	

Resistivity surveys can prove very useful for identifying voids. Figure 5.36 shows results from a survey to trace the lines of underground streams in pseudokarstic conditions in weathered granite.

5.11.1.2 Rock drilling

There are numerous techniques used for drilling and sampling rock and some of these are listed in Table 5.2. I have not included here techniques used in deep oil and mining applications where either rock roller bits are used to advance the hole quickly, sometimes with water jet assistance. In such investigations geology is logged from the chippings produced and using down-hole geophysical and optical tools to measure data and to record discontinuity characteristics.

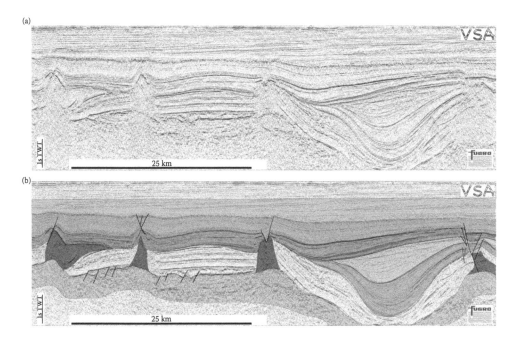

Figure 5.34 This seismic reflection image is from the eastern part of the Dutch Graben, southern North Sea with the processed image in (a) and the geological interpretation in (b). In (a) dark lines mark strong impedance contrasts between rock units. Note that as the section is derived from sound waves, the vertical axis represents time. The interpretation in (b) reveals four dome-like structures (dark grey), which are probably salt bodies; the more continuous reflections indicate boundaries between major sedimentary rock units. (Data courtesy of Fugro and kindly released through the Virtual Seismic Atlas. http://www.seismicatlas.org/)

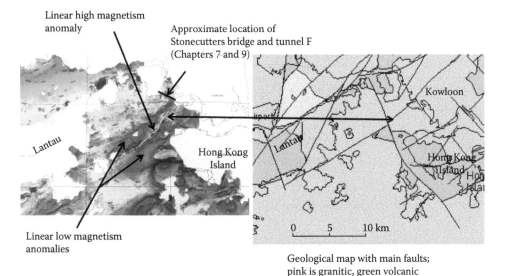

Figure 5.35 Sea-borne magnetic survey of area between Hong Kong Island and Lantau and corresponding area of geological map. Linear features (relatively low and high magnetism) can be clearly made out that probably depict the major Tolo Channel Fault. This fault caused considerable difficulties for the construction of Tunnel F for the SSDS Project in Hong Kong and a similarly aligned fault caused difficulties for piling for Stonecutter's Bridge as discussed in Chapters 9 and 7 respectively.

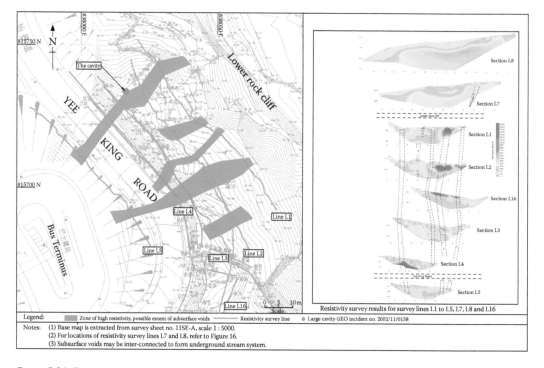

Figure 5.36 Resistivity survey data to depths of about 15 m. Lines of high resistance shown (red) are locations of underground streams. (From Halcrow China Ltd. 2003. *Interim Report on Detailed Hydrogeological Study of the Hillside near Yee King Road, Tai Hang.* Landslide Study Report LSR 4/2003: 123pp; Further details are given in Hencher, S.R., Sun, H.W. and Ho, K.K.S. 2008. *Geotechnical and Geophysical Site Characterization* [eds., Huang, A.B. and Mayn, P.W.], London: Taylor & Francis, 601–607.)

Table 5.2 Main drilling methods for rock coring

Purpose	Method	Comments
General rock use	Double tube barrel	Core enters inner barrel that does not rotate with outer barrel and bit. Less disturbance than single barrel, well-boring rig Can be difficult to extract core from inner barrel without disturbance
	Triple tube barrel	Split tube sampler container inside non-rotating core barrel. Easy extraction and least disturbance
Deep rock	Wire-line drilling	Sample barrel can be separated from drill bit and drill line and withdrawn from hole without disassembly. Saves considerable time on deep holes
Directional drilling	Drill rods with wedges or gyroscopic set up to control drilling direction	Hole can be deviated in a specific direction, say to follow a proposed tunnel line or to install cables
Weak, weathered and mixed ground	Mazier barrel	Soft ground cutter spring-loaded. In rock, cutter retracts and drill bit takes over. Sample enters plastic container to be cut open later

Figure 5.37 Rotary rig drilling inclined hole, Hong Kong.

Figure 5.38 Triple tube barrel assembly. Diamond bit with integral holes for flush fluid, lower right. To its left, core catcher assembly that prevents core being dropped when retrieving full core barrel. At the top, split metal tube barrel to hold core.

A typical rotary drilling rig is shown in Figure 5.37, drilling a hole inclined at 45°.

In civil engineering, one of the best tools for rock sampling is the triple tube core barrel illustrated in Figure 5.38. The core sample enters the core barrel through the bit annulus and is restrained by a core catcher. When the barrel is retrieved (Figure 5.39) – generally after drilling about 1.5 m the central split barrel is removed and opened so that the core samples can be examined and placed in a core box.

The Mazier sampling tube is commonly used in Hong Kong and elsewhere in the Far East for high quality sampling in mixed and weathered rock. The device has a sharp cutting shoe for soft ground that retracts when encountering harder material so that drilling continues using a rotary bit. A Mazier tube sample is illustrated in Figure 5.40.

Figure 5.39 Removal of core barrel to extract core from inclined hole. Izmit Bridge North Anchorage investigation, Turkey.

Figure 5.40 Plastic sample holder (full of completely decomposed granite) extruded from Mazier barrel. Cutting shoe assembly to right of barrel. (More details are given in Hencher, S.R. 2012a. *Practical Engineering Geology. Applied Geotechnics.* Vol. 4, Spon Press, 450pp.)

Table 5.3 Flush mediums used to cool drilling bits and remove cutting debris

Flush type	Nature	Comments
Air	Used mostly in percussive holes	Used in percussive/rotary holes for blasting holes/anchorages etc.
Water	Cheap	Can soften sample and cause core loss.
Mud	Used to stabilise open holes (without casing)	Especially important for deep drilling as in oil and gas investigation and production wells.
Foam	Use polymer foam products, mixed on site	Reduced core loss and less sample disturbance compared to water. Preferred flush for high quality drilling using triple tube and Mazier barrels.

Figure 5.41 Rotary drilling rig using foam flush to cool bit and remove cuttings, being used here with Mazier core barrel, Yee King Road landslide, Hong Kong.

When drilling, some flushing medium is used to cool the bit and to remove cutting debris. Types of flush are listed in Table 5.3. Generally, in weak and weathered rock or where there may be weak zones that might be washed out by water flush (e.g., when investigating a landslide), then foam flushing generally produces the best recovery. An example of a rig, using polymer foam flush is shown in Figure 5.41.

5.11.2 Sampling and storage

Recovered samples need to be handled carefully and placed into core boxes with depths marked and spacers used to represent any lost core or samples taken for testing. Core should be photographed with a scale and colour chart and logged as soon as possible,

Figure 5.42 Logging core in the field immediately on recovery from core barrel. Note use of plastic sheet
to help retain moisture content. Sebuku mine, Indonesia, Borneo.

particularly where the core might deteriorate due to changes in moisture content as illustrated in Figure 5.42.

Great care should be taken when transporting core to avoid creating new fractures or opening up weak incipient discontinuities. For any ground investigation, thought should be given to how core is to be stored so that it is accessible and can be examined by project engineers throughout design and construction, preferably in reasonable comfort with boxes laid out and adequate lighting and air conditioning if appropriate (Figure 5.43).

Core drilling is expensive so it is worth spending time and money to ensure that core is kept well and can be examined. One of the greatest failings of a geotechnical team is to

Figure 5.43 Core shed with core laid out for logging. Batoka Gorge Dam site, Zimbabwe.

order a lot of sampling and then not even to examine the samples (actually common practice from experience).

5.12 DESCRIPTION AND CLASSIFICATION OF ROCKS

5.12.1 Introduction

Systematic description is generally applicable to ground investigation – logging boreholes and exposures. Descriptions should be thorough and unambiguous so that the end user, perhaps in a design office, will know what has been observed on site. The scope and style of routine description of soil and rock for engineering purposes in logging are well established and suitable for many projects. For more difficult sites then much fuller geological characterisation may be needed following the geological literature.

There are numerous national standards for the description and classification of rocks and these vary in definitions; many use the same terms but with different meanings. One of the earliest attempts to provide fairly comprehensive guidance for logging rotary core through rock was the Geological Society Engineering Group Working Party Report, Anon (1970). This report and another on Maps and Plans (Anon, 1972) set many standards that were then followed, particularly in the United Kingdom. Other bodies (ISRM and IAEG in particular) also established their own working groups and came up with their own sets of terms to describe rock features. The ISRM publication on Suggested Methods for the Quantitative Description of Discontinuities (ISRM, 1978) is a particularly well-illustrated and useful guide. This was followed in the United Kingdom by the preparation of fuller guidance on the description and classification of both soil and rock in the BS 5930: 1981 Code of Practice for Site Investigations. The Geotechnical Control Office in Hong Kong (1988) published Geoguide 3 on Soil and Rock Descriptions, which largely followed British practice, but with some distinct differences (improvements) especially regarding the description and classification of weathered rock. The British Standard, BS 5930 was revised and republished in 1999 and was a better document. Most recently amendments have been made as part of introduction of Eurocode 7 and some of the changes are poorly thought through and retrograde as discussed by Hencher (2008). Meanwhile other countries have adopted their own schemes (e.g. Australia, New Zealand and China) and whilst there are common aspects, often the same terms (and certainly the same properties) are redefined in different ways, which is confusing. US practice is set out in Hunt (2005) and CALTRANS (2010).

In the following discussion, I will follow European practice as far as possible and will point out any particular difficulties and where alternatives should be considered.

5.12.2 Order of description

No preferred order is given in EN ISO 14688 (BSI, 2002) or 14689 (BSI, 2003) so, for the UK, the BS 5930: 1999 recommendations should be adopted in logging.

For rock the BS recommended order is as follows:

A. Material characteristics:
1. Strength
2. Structure
3. Colour
4. Texture

    5. Grain size
    6. Rock name (in capitals, e.g., GRANITE)
  B. General information:
    1. Additional information and minor constituents
    2. Geological formation
  C. Mass characteristics:
    1. State of weathering
    2. Discontinuities

BS 5930: (BSI, 1999) should be consulted for all terms and definitions as used in British practice. Other countries have their own terms and practice guidance so the engineering geologist needs to be aware of local usage wherever he is working.

For rock, key issues are intact rock strength, nature of discontinuities, weathering and rock mass classification.

5.12.3 Strength

Intact strength of rock material is very different from mass strength as addressed in Chapter 6. Uniaxial compressive strength can be estimated quite readily in absolute terms (MPa) by using index tests as set out in Figure 5.44. Often rock breaks in tension or shear rather than compression but nevertheless strength is classified by compressive strength as in a UCS test (Chapter 6).

There are major differences in terminology between 'old' BS 5930 and amended BS 5930 (which now ties in to ISRM guidance). Of importance is that 'weak' rock now ranges from UCS 5 to 25. The old boundary at 12.5 MPa has been omitted. This is actually considered a retrograde step because the 12.5 MPa level marks where rock core can be just about broken by hand which is an important and practical index. This test is used to differentiate between highly and moderately weathered rocks in many weathering classifications and is used in foundation design. Deere and Deere (1989) use the test to specify what is 'sound' rock to be included in RQD (which has knock on implications for rock mass classifications as discussed below).

5.12.4 Joints and discontinuities

Joints, faults and other fractures are generally termed discontinuities and of major importance to rock engineering but rather poorly defined as discussed by Hencher (2014).

The ISRM (1978) defines discontinuities as having zero or low tensile strength.

For Europe BS EN ISO 14689-1:2003 defines a discontinuity as a surface, which breaks the rock material continuity within the rock mass and that is open or may become open under the stress applied by the engineering work; the tensile or shear strength across or along the surface is lower than that of the intact rock material.

In other words, whereas a discontinuity is defined by ISRM as a mechanical fracture, the European standard broadens this to include incipient fabric (such as cleavage or bedding planes) that might open up under stress. The implication is that a feature might be described as a discontinuity in some logs but not in others.

The distinctions might appear trivial but where it matters is when characterising the rock mass. Rock mass classifications discussed later mostly incorporate discontinuity or joint spacing as a fundamental parameter.

It is best to follow the ISRM definition that a discontinuity is a mechanical break with no or very little tensile strength. Other fabrics and discontinuities that retain tensile strength

BS5930: 1999; GCO (1988);		BS EN ISO 14689-1:2003; ISRM (1981); amended BS 5930: 1999 (2010)	
Term & UCS (MPa)	Identification	Identification	Term & UCS (MPa)
Extremely weak <0.5	Easily crumbled by hand; indented deeply by thumbnail.	Indented by thumbnail	Extremely weak <1.0
Very weak 0.5–1.25	Crumbled with difficulty; scratched easily by thumbnail; peeled easily by pocket knife.		
Weak 1.25–5	Broken into pieces by hand; scratched by thumbnail; peeled by pocket knife; deep indentations (to 5 mm) by point of geological pick; hand-held specimen easily broken by single light hammer blow.	Crumbles under firm blows with point of geological hammer, can be peeled with pocketknife	Very weak 1–5
Moderately weak 5–12.5	Broken with difficulty in two hands; scratched with difficulty by thumbnail; difficult to peel but easily scratched by pocket knife; shallow indentations easily made by point of pick; hand-held specimen usually broken by single light hammer blow. Point load strength (PLS) 0.2 – 0.5 MPa.	Can be peeled by a pocket knife with difficulty, shallow indentations made by a firm blow of geological hammer	Weak 5–25
Moderately strong 12.5–50	Scratched by pocket knife; shallow indentations made by firm blow with point of pick; hand-held specimen usually broken by single firm hammer blow. Point load strength (PLS) 0.5 – 2 MPa.	Cannot be scraped or peeled with a pocket knife, specimen can be fractured with single firm blow of geological hammer	Medium strong 25–50
Strong 50–100	Firm blows with point of pick cause only superficial surface damage; hand-held specimen requires more than one firm hammer blow to break. PLS 2 – 4 MPa.	Specimen requires more than one blow of geological hammer to fracture it	Strong 50–100
Very strong 100–200	Many hammer blows required to break specimen. PLS 4 – 8 MPa.	Specimen requires many blows of geological hammer to fracture it	Very strong 100–250
Extremely strong >200	Specimen only chipped by hammer blows. PLS > 8 MPa.	Specimen can only be chipped with geological hammer	Extremely strong >250

Figure 5.44 Terminology for intact rock compressive strength.

(including bedding and cleavage) should be described as incipient. It is recommended that incipient discontinuities should be logged and characterised as 'high' tensile strength (close to that of intact rock), 'intermediate' or 'low' (readily split) as in Figures 5.45 and 5.46. In geological terminology incipient fabric such as cleavage, where it has little effect on intact strength, should be called 'faint' as in faintly-defined cleavage.

Geometrical description and measurement of discontinuities is addressed in the ISRM guidelines. Extrapolation should not be made from one exposure or sampling point to another without careful consideration of possible geological change including structural regime and degree of weathering.

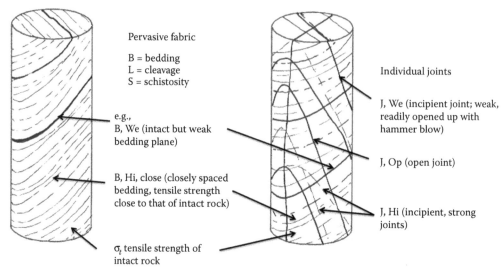

Pervasive fabric

B = bedding
L = cleavage
S = schistosity

Individual joints

J, We (incipient joint; weak,
readily opened up with
hammer blow)

e.g.,
B, We (intact but weak
bedding plane)

J, Op (open joint)

B, Hi, close (closely spaced
bedding, tensile strength
close to that of intact rock)

J, Hi (incipient, strong
joints)

σ_t tensile strength of
intact rock

Relative strength of discontinuities
Hi = High strength; similar to parent rock
Mo = Moderate strength (judged to be intermediate between Hi and We)
We = Weak; may be broken during transport, during drilling or by tap with hammer
Op = Open fracture

Figure 5.45 Suggested terminology to describe mechanical fractures and incipient discontinuities and rock fabric. (After Hencher, S.R. 2014. Characterising discontinuities in naturally fractured outcrop analogues and rock core: the need to consider fracture development over geological time. Geological Society of London Special Publication 374, *Advances in the Study of Fractured Reservoirs*, 113–123.)

5.12.5 Rock quality designation

Rock quality designation (RQD) is the basis for most commonly used rock mass classifications. It is a simple scheme employed in logging core, devised by Dr. Don Deere in the early 1960s as a practical way of differentiating between different qualities of rock mass. Prior to that date quality of the rock mass was largely inferred from core recovery but, as methods of drilling improved, good recovery could be obtained in poor rock masses so a new approach was necessary (Deere and Deere, 1989). Pieces of sound rock core ('*sound*' is defined by Deere and Deere (op. cit.) as slightly or moderately weathered rock, i.e., rock that could not be broken by hand*) that are greater than 100 mm in length, measured along the core axis are summed and expressed as a percentage of the core run. The methodology is illustrated in Figure 5.47. Only natural, open rock discontinuities (with <u>zero</u> tensile strength) are to be part of the RQD assessment.

The method is a useful index (on a par with the empirical SPT test in soils and weak rocks), essentially a broad indicator of rock quality in core, but it has limitations:

1. The definition is per core run – not lithology; clearly where there are mixed rock types these may have very different fracture characteristics so the RQD assessment is rather a blunt instrument (and was never intended otherwise). Deere and Deere do suggest

---

* See later discussion on weathering classification.

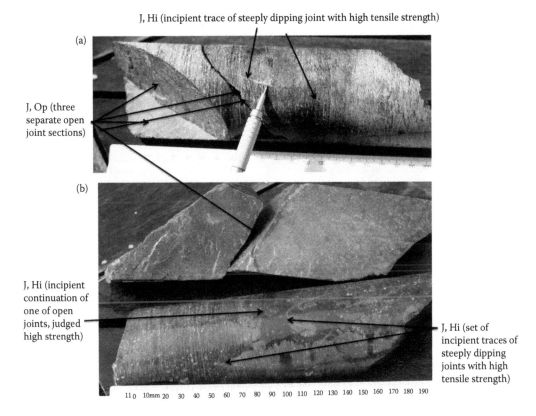

(a)

J, Hi (incipient trace of steeply dipping joint with high tensile strength)

J, Op (three separate open joint sections)

(b)

J, Hi (incipient continuation of one of open joints, judged high strength)

J, Hi (set of incipient traces of steeply dipping joints with high tensile strength)

11 0   10mm 20   30   40   50   60   70   80   90   100   110   120   130   140   150   160   170   180   190

Figure 5.46 Example of rock core with incipient and open discontinuities. (a) Section of core, andesitic tuff, Island road, Hong Kong. (b) Same core, disassembled. (After Hencher, S.R. 2014. Characterising discontinuities in naturally fractured outcrop analogues and rock core: The need to consider fracture development over geological time. Geological Society of London Special Publication 374, *Advances in the Study of Fractured Reservoirs*, 113–123.)

that the RQD might be determined for different lithologies in a sequence rather than sticking to the prescribed core run approach.

2. There will be directional bias. For example, in vertical boreholes where the geology is dominated by vertical joints those fractures will be under-sampled.

3. RQD is insensitive where discontinuities are generally >100 mm in spacing yet 100 mm spacing is actually 'very close' for natural rock exposures. In other words, a rock mass containing mechanical fractures with zero tensile strength at >100 mm has the same quality definition as massive rock with no discontinuities at all.

4. There are difficulties in definition with particular rock types with pervasive weak planes like shales.

5. There are problems with recognising natural discontinuities.

Appended to Deere and Deere (1989) there are more than 20 pages of discussion from practicing geotechnical professionals of the US Corps of Engineers at that time; the main comments can be summarised as follows (some quotes):

1. Over-simplistic. Over-reliance on RQD in logging.

2. 'The fact that RQD is correlated with such terms as 'excellent', 'good' and 'poor' and is sometimes even correlated with allowable bearing capacities (Peck et al., 1974),

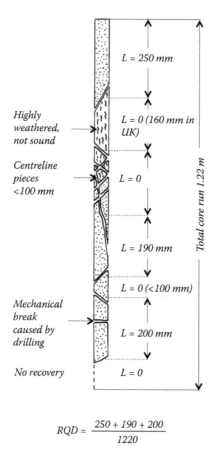

Figure 5.47 RQD determination. (Redrawn from Deere, D.U. and Deere, D.W. 1989. *Rock Quality Designation [RQD] after Twenty Years.* Contract Report GL-89-1, US Army Corps of Engineers, 67pp plus Appendix.)

affords much opportunity for its misuse. Designers must not rely on RQD alone as a basis for foundation design decisions. At best, it can only serve as a tool of limited use in the engineering qualities of bedrock'.

3. 'Anyone who uses RQD as a tool to understanding properties of bedrock must also understand the limitations inherent in such a simplistic approach to assessing the engineering properties of bedrock. When considered by itself, outside the context of the local geology, that is the lithology of the rock, the geologic structure of the bedrock and the potential influence of bedrock weathering, RQD becomes a meaningless number'. ... 'Therefore, to be properly used, RQD must be considered as only one small part of the overall geologic evaluation and cannot be used as the sole basis for determining the engineering qualities of bedrock'.

As illustrated in Figure 5.47, European practice (certainly the UK) has shifted from the original Deere specification for RQD (and therefore by default for RMR and Q) in failing to specify that the intact pieces of core need to be 'sound'. Following this practice, any solid core of 'rock' counts towards RQD which includes any material with UCS >0.6 MPa according to EC7 which is very different to the original concept of Deere. Historically, RQD has been applied in the UK to weak rock such as chalk for more than 40 years. Mortimore (2012) gives examples of applying the Q rock mass classification to rock of UCS only 6 MPa although this must be questionable practice where the Q-system empirical base is generally from much stronger rocks.

5.12.5.1 RQD in three dimensions

The adoption of RQD as a simple way of differentiating rock core of different fracturing (and soundness) has led to many derivatives.

Palmström (1982) suggests that RQD can be estimated from the number of discontinuities per unit volume based on visible discontinuity traces surface exposures using the following relationship:

$$RQD = 115 - 3.3\,Jv$$

where Jv is the sum of the number of joints per unit length for all discontinuity sets and known as the 'volumetric joint count'. The fact that this really needs to be done in '*each rock type encountered, each structural domain, and for the upper weathered zone*' (Deere and Deere, 1989) seems to have been rather overlooked.

There is a practical difficulty here also in that traces of discontinuities will seldom equate with mechanical fractures in core (the incipient joint problem as discussed earlier). The engineering geologist must beware of counting all visible traces in defining RQD otherwise the ground will be assigned a much lower quality rating than is really justified. This can have major consequences in assessing the potential of tunnel boring machines or road headers to make progress when cutting rock.

RQD is sometimes measured from photographs (e.g., Priest and Hudson, 1981); but again it is impossible to ascertain that joint traces visible in photos have no tensile strength which has knock on effects for assessing rock mass quality.

5.13 ROCK MATERIAL AND MASS CLASSIFICATION

5.13.1 Introduction

At some stage in the design process, rock units need to be defined within a ground model on the basis of similar anticipated mechanical behaviour. The criteria might be strength, deformability, permeability or excavatability or based on some other intrinsic quality such as chemistry or degree of fracturing. As an example, at the simplest level there is a distinction made in tunnelling between 'soft' ground or 'hard' ground which helps to define the approach and methodology that will be needed; the former is generally excavated manually using an excavator of some kind, the latter probably requiring blasting. Broad descriptive or numerically derived classifications are also used in defining types of materials or mass zones either for primary descriptive characterisation or for anticipating performance based on measured parameters.

5.13.2 Weathering classification

As discussed in Chapter 3, many rocks are weathered to great depths, especially in tropical and sub-tropical areas of the world. Weathering effects should be described and recorded and may be interpreted directly from changes in colour, discontinuity spacing, degree of infill on joints and intact strength. In some circumstances, weathering classifications are useful to characterise rock at the scale of an intact, uniform sample or at the mass scale. The classifications provide a shorthand description, which is often treated synonymously with material or mass strength. Weathering classifications are really used just for preliminary grouping of materials and masses into ground model units but there is no direct link to defining design parameters such as strength (unlike for GSI) or need for tunnel support measures (for classifications such as Q) as discussed later and in Chapter 6.

5.13.2.1 Material-weathering classifications

Having used weathering classifications a great deal in practice, the author is convinced that material-weathering classifications such as that used in Hong Kong and presented in Figure 5.48 are the most useful for characterising rock that weathers from a strong condition progressively to a soil. There is no guidance given by ISRM or in the relevant European standards on weathering classification at the material scale but actually, when most workers talk in terms of 'highly' or 'completely' weathered rock etc., they are commonly thinking in terms of samples that might be tested in the laboratory, i.e. degree of weathering at the material scale. The classification is essentially one of strength but includes a number of other index tests such as colour change, Schmidt Hammer rebound and whether or not a sample disaggregates in water (slakes) that reduce subjectivity (Martin and Hencher, 1986). In many parts of the world profiles of saprolite (soil-like material but with the recognisable fabric and texture inherited from the parent rock) are found that can be tens of metres thick and need to be differentiated into zones that include quite a wide range of strength and other properties. The scheme in Figure 5.48 (and similar) have been used for logging many different tropically weathered igneous and sedimentary rocks and even in temperate climates and is essentially a shorthand strength rating. Examples of the use of weathering grade classification are given in Figure 5.49. In practice, the boundary between grades III and IV is often taken as distinguishing between rock-like and soil-like behaviour for slope analysis and for foundation design and also as the break between 'sound' and 'unsound' rock in RQD. Classifications have been developed for weaker specific rock types like chalk and mudstones and these are reviewed by Norbury (2010) and in Anon (1995).

5.13.2.2 Mass weathering classifications

Larger scale exposures or ground model units can be described and characterised according to the volumetric proportions of the various material classification grades as illustrated in Figure 5.50. The current BS EN ISO 14689-1:2003 and in various ISRM documents suggest weathering classifications at the mass scale based on proportions of 'rock' and 'soil'. These classifications are coarse, ill-defined and generally found to be unworkable from the author's experience; the arguments are set out in Martin and Hencher (1986) and Hencher (2008). The scheme recommended by the Engineering Group of the Geological Society of London (Anon, 1995), referred to below as the EG Scheme, and which is very similar to that adopted in Hong Kong at the mass scale (GEO, 1988) is illustrated to the left of Figure 5.50. The scheme, which works reasonably well in practice is based on perceived geotechnical performance and thereby serves as a good starting point for characterising units in a ground model. The reasoning for defining the boundaries is discussed in Martin and Hencher (1986). In this scheme 'rock' is defined where large pieces cannot be broken by hand and this is the

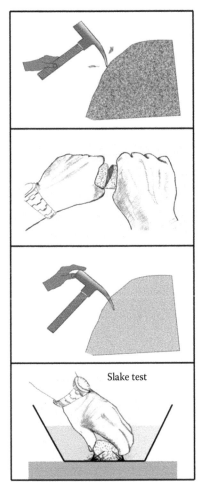

Grade I: Fresh

Grade II: Slightly
weathered
- Discoloured

Grade III: Moderately
weathered
- Considerably weakened but
  cannot break by hand

'*Sound*' rock
generally counting
towards RQD

Grade IV: Highly
weathered
- Broken by hand
- Doesn't slake
- 0–25 Schmidt hammer
  rebound value

Grade V: Completely
weathered
- Slakes (disaggregates) in
  water
- Zero rebound on Schmidt
  hammer
- Geological pick can be
  pushed into material
- Original rock fabric is
  still present

'*Unsound*' according
to Deere and Deere
(1989) and should not
be counted in RQD

Slake test

Grade VI: Residual soil
- Soil derived by
  weathering or rock *in situ*
  but lacking original
  rock texture or fabric

Figure 5.48 Weathering classification for intact rock that is strong in the fresh unweathered condition. (Based on Moye, D.G. 1955. *Journal of the Institution of Engineers*, Australia, 27, 287–298; Geotechnical Control Office 1988. *Guide to Rock and Soil Descriptions* (Geoguide 3). Geotechnical Control Office, Hong Kong, 189pp.)

boundary between highly and moderately weathered rocks in the material classification (Figure 5.48). It also corresponds to the concept of what comprises 'sound' rock for RQD determination as discussed earlier. In Figure 5.51, some examples are given of weathering profiles. The upper profile could be characterised using the ISRM or EC7 schemes by separating zones of greater or less than 50% rock although as 'rock' in EC7 includes any material that has UCS > 0.6 MPa perhaps the whole profile would have to be described as 'slightly weathered', without further differentiation. The EG scheme would allow more detailed differentiation into different zones.

In the lower example, following the EG scheme the profile would be described as Zone 6 (100% soil) further quantified as grade IV material. Using the EC7 or ISRM approaches, given the lack of corestones, the observer would have to describe the exposure as either 100% rock, 'slightly weathered' or 100% soil 'completely weathered' neither of which conveys a reasonable description back to the design office.

Weathering
grade ranging
from Grade II at
centre of
granodiorite
corestone (with
exfoliation
fractures) to
Grade IV
between
corestones.
Very thin Grade III

Uniform Grade IV
tuff being
prepared for
direct shear test

Figure 5.49 Examples of the use of weathering classification at the material scale.

In practice, from experience, it is often best to describe geotechnical mass weathering units using material classification as the building blocks rather than the mass classification. This is a flexible approach. Weathering profiles can be very variable and trying to force a pre-defined mass classification on the situation is often very difficult.

5.13.3 Other rock mass classifications

5.13.3.1 Introduction

Various rock mass classifications have been developed largely to estimate support requirements for underground excavations linked to case histories. Several of these, aimed at describing how the ground mass might behave during tunnelling, are discussed in Chapter 9. An exception is the Geological Strength Index (GSI), which is aimed instead at predicting

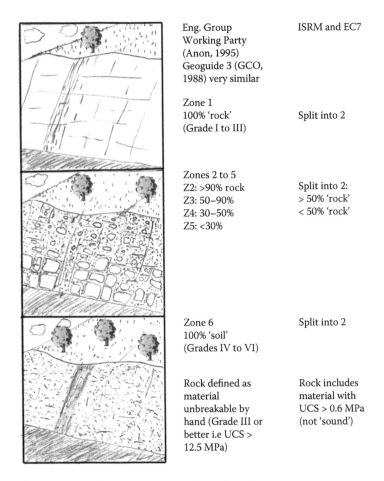

	Eng. Group Working Party (Anon, 1995) Geoguide 3 (GCO, 1988) very similar	ISRM and EC7
	Zone 1 100% 'rock' (Grade I to III)	Split into 2
	Zones 2 to 5 Z2: >90% rock Z3: 50–90% Z4: 30–50% Z5: <30%	Split into 2: > 50% 'rock' < 50% 'rock'
	Zone 6 100% 'soil' (Grades IV to VI)	Split into 2
	Rock defined as material unbreakable by hand (Grade III or better i.e UCS > 12.5 MPa)	Rock includes material with UCS > 0.6 MPa (not 'sound')

Figure 5.50 Mass classification for heterogeneous weathered rock.

engineering parameters. These classifications really come into their own whilst tunnelling and where a decision has to be made quickly as to the level of support that is required to stabilise the rock mass.

Rock mass classifications are attempts to represent sometimes complex geological conditions as single numbers. Hoek (1999) described the process as putting numbers to geology. It is significant that these have predominantly been prepared by non-geologists. Exceptions are those produced for slope stability assessments by Selby (1980) and Hack (1998).

5.13.3.2 Rock mass rating

The rock mass rating (RMR) of Bieniawski (1976, 1989) has stood the test of time as a useful system despite many question marks over definitions. The five parameters for the RMR system and their ranges of assigned points are as follows:

1. Uniaxial compressive strength of rock material (0–15)
2. RQD (3–20)

(a)

(b)

Figure 5.51 Examples of application of weathering classification at the large, mass scale. (a) This Korean heterogeneous rock mass could be differentiated using either approach from Figure 5.50 but there is more scope for addressing the detailed variability using the scheme from the Eng. Group Working Party. (b) This severely weathered, weak granite would be defined as 'Zone 2: Slightly Weathered' followed ISRM or Eurocode 7; according to Anon (1995) it is Zone 6, highly weathered.

3. Spacing of discontinuities (5–20)
4. Condition of discontinuities (0–30)
5. Groundwater conditions (0–15)

Points are assigned for each parameter and then summed to rate the rock as very good to very poor. It is to be noted that degree of fracturing and condition of those fractures covers 70% of the rock mass rating and that there is a high level of double counting between factors 2 and 3. There is also somewhat of a conundrum in that rock with RQD of 90%–100% (which might include blocky masses with discontinuities spaced at

100 mm) is allocated the full 20 points but rock with joint spacing 60–200 mm is only allocated 8 out of 20 points.

Orientation of discontinuities is used to adjust the summed rating according to whether discontinuities are adverse relative to the engineering project.

RMR is used (as is RQD by itself) to correlate with rock mass parameters including rock mass strength and deformability as discussed in Chapter 6.

The use of RMR in tunnel design is outlined in Chapter 9.

5.13.3.3 Q System

The Q system of Barton et al. (1974) can also be used to classify the quality of rock mass – as a predictive tool in estimating tunnel support requirements for a planned tunnel or cavern, for judging whether ground conditions during tunnelling were as expected or not (contractual issues) and for making decisions on temporary and permanent support requirements during tunnelling. Barton (2000, 2005) also discusses ways of using the Q system in predicting TBM performance. Q has a value range from 0.001 to 1000.

$$Q(\text{quality}) = RQD/Jn \times Jr/Ja \times Jw/SRF$$

where
RQD is rock quality designation
Jn is joint set number
Jr is joint roughness number
Ja is joint alteration number
Jw is joint water reduction factor

and SRF is a stress reduction factor (relating to loosening of rock, rock stress in competent rock and squeezing conditions in incompetent rock).

Ranges and descriptions for each parameter are given in the original publications by Barton but also in Hoek and Brown (1980) and at Hoek's Corner at the Rocscience web page.

5.13.3.4 RMi

RMi has been proposed as an alternative to RMR and Q and is perhaps less prescriptive than the other classifications, ties into geology rather more readily, avoids RQD which is rather a contentious classification and can also be used to make predictions of the need for support in underground excavations.

5.13.3.5 Geological strength index

The GSI introduced by Hoek et al. (1995) provides a means of estimating rock mass strength and deformability through broad classification based on rock type and quality of rock mass and is rather more 'geologically friendly' and applicable than are most such classifications. A chart for estimating GSI is presented in Figure 5.52. It is linked to the Hoek–Brown criterion which is very useful for estimating the shear strength and deformability of rock masses as discussed in Chapter 6, but the application of this approach is limited to situations where the rock is highly fractured and where there is no structural control of shear displacement which is relatively rare in nature.

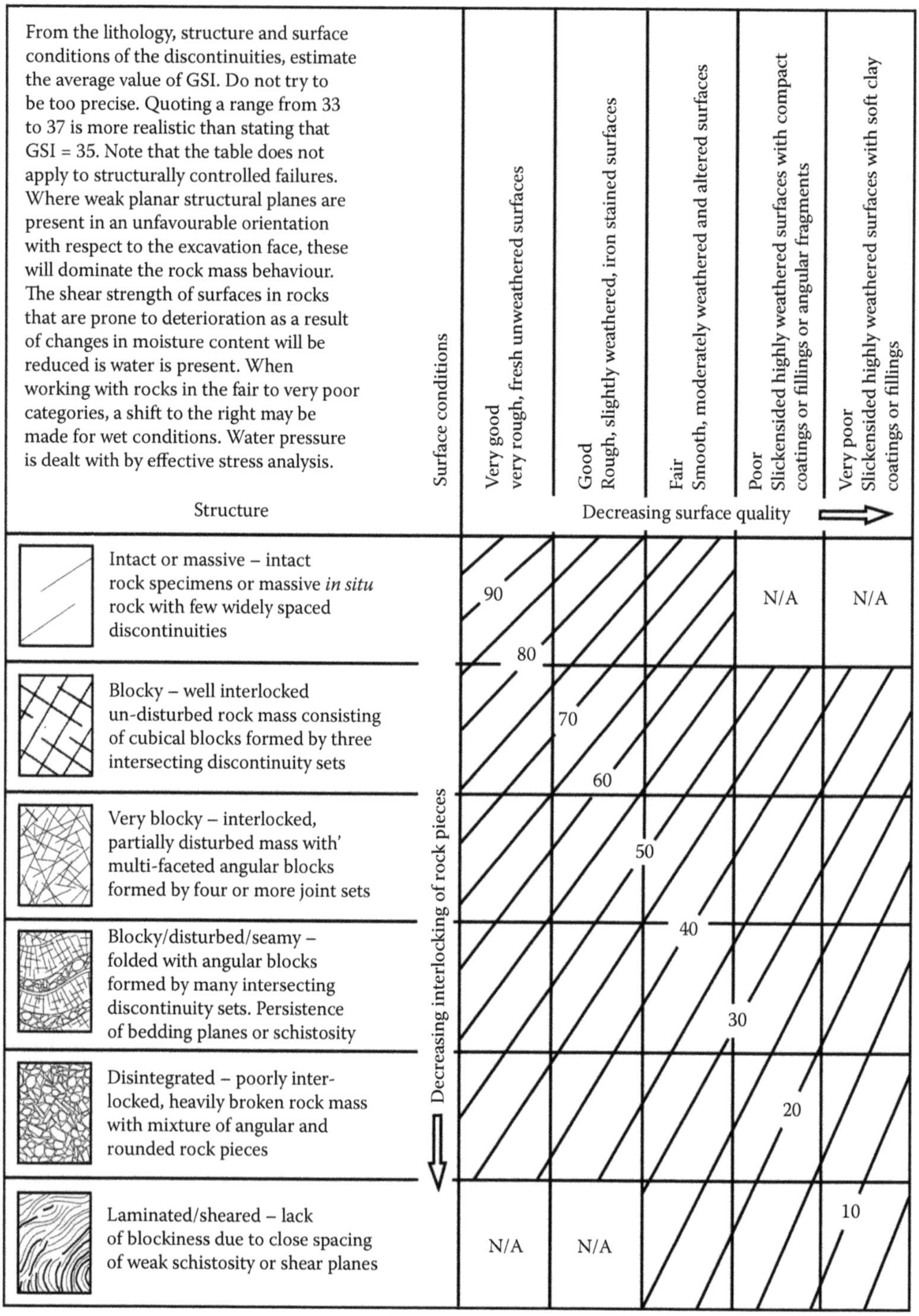

From the lithology, structure and surface conditions of the discontinuities, estimate the average value of GSI. Do not try to be too precise. Quoting a range from 33 to 37 is more realistic than stating that GSI = 35. Note that the table does not apply to structurally controlled failures. Where weak planar structural planes are present in an unfavourable orientation with respect to the excavation face, these will dominate the rock mass behaviour. The shear strength of surfaces in rocks that are prone to deterioration as a result of changes in moisture content will be reduced is water is present. When working with rocks in the fair to very poor categories, a shift to the right may be made for wet conditions. Water pressure is dealt with by effective stress analysis.

Surface conditions → Decreasing surface quality ⟹

- **Very good** — very rough, fresh unweathered surfaces
- **Good** — Rough, slightly weathered, iron stained surfaces
- **Fair** — Smooth, moderately weathered and altered surfaces
- **Poor** — Slickensided highly weathered surfaces with compact coatings or fillings or angular fragments
- **Very poor** — Slickensided highly weathered surfaces with soft clay coatings or fillings

Structure ⇩ Decreasing interlocking of rock pieces

Structure	Very good	Good	Fair	Poor	Very poor
Intact or massive – intact rock specimens or massive *in situ* rock with few widely spaced discontinuities	90 / 80			N/A	N/A
Blocky – well interlocked un-disturbed rock mass consisting of cubical blocks formed by three intersecting discontinuity sets		70 / 60			
Very blocky – interlocked, partially disturbed mass with' multi-faceted angular blocks formed by four or more joint sets			50		
Blocky/disturbed/seamy – folded with angular blocks formed by many intersecting discontinuity sets. Persistence of bedding planes or schistosity			40	30	
Disintegrated – poorly inter-locked, heavily broken rock mass with mixture of angular and rounded rock pieces				20	
Laminated/sheared – lack of blockiness due to close spacing of weak schistosity or shear planes	N/A	N/A			10

Figure 5.52 The geological strength Index. (After Hoek, E. and Marinos, P. 2000. Predicting Tunnel Squeezing. *Tunnels and Tunnelling International*, Part 1, **32/11**, 45–51, November 2000; Part 2, **32/12**, 33–36, December 2000.)

Figure 5.53 Blocky granite, Mount Butler quarry, Hong Kong; car for scale. Clearly there are preferential joint directions, which might render GSI inapplicable, but GSI might be applied for some zones. Blocky, good – perhaps GSI 55 to 65.

5.13.3.6 Application of GSI

Examples are given of the use of GSI in characterising rock masses in Figures 5.53 to 5.56.

The first thing to note from these examples is that it is difficult to apply GSI with any precision. That said, it would be even more difficult to assess Q values despite 100% exposure (not just boreholes) for which the main parameters (other than environmental variables relating to groundwater and stress conditions) are: RQD (rock quality designation), Jn (joint set number), Jr (joint roughness number) and Ja (joint alteration number).

From experience, it is found increasingly that engineers and engineering geologists use the GSI classification (or Q, RMR or RMi) directly for description and characterisation of rock masses, which is a very unhealthy trend despite ease of application.

That said, the Hoek–Brown criterion is an extremely useful tool for estimating strength of rock masses where there is no viable alternative as discussed in more detail in Chapter 6.

5.14 INTERPRETING GROUND CONDITIONS AND REPORTING

5.14.1 Design interpretation of ground conditions

It is up to the design engineer and his team to investigate the project site, carry out adequate testing and make sure that the works can be constructed safely and to meet the long-term requirements of the Client/Owner. As part of this he needs to prepare geotechnical models that can be used to represent the site and so that structural elements such as piles, diaphragm walls and cuttings can be analysed to ensure they have adequate FoS or meet the requirements re a limit state approach.

Figure 5.54 Faulted and folded Carboniferous schist, near St. Catarina, Algarve, Portugal. Height of face about 8 m. UCS of arenaceous (sandstone) horizons typically 70–100 MPa; shaley horizons weaker. It would be a challenge to assign Q or RMR values to this profile. GSI = 30 might be appropriate but note the major thrust faults that would control stability in particular directions. This exposure warrants careful description and characterisation for stability assessment.

Figure 5.55 Borrowdale volcanics in old mine, north of Grasmere, Lake District, Cumbria, England. GSI somewhere between massive and blocky with high strength incipient discontinuities so perhaps 60–80.

Figure 5.56 Faulted sedimentary rocks with faint bedding dipping obliquely into face; partially broken wall at foot of slope about 2 m high. North side Izmit Bay, Turkey. Probable GSI about 20–40 but at depth, away from the exposed slope with deteriorated face the value might be higher.

Figure 5.1 was used to illustrate the difficulties associated with characterising some complex sites. As stated earlier, to understand this particular site involved more than 5 km of drilling, installation of many piezometers, inclinometers and other instruments, many of which are now continuing to be monitored to check that the road is safe.

The production of geological models is sometimes quite straight forward, drawing straight lines between strata or 'rock quality' units and often such an approach is adequate because the site is 'forgiving'. In the case example this would not have worked and to get the models right required input from many specialist geotechnical engineers, numerical modellers, geologists and hydrogeologists together with instrument specialists.

This process was conducted on a 3D basis, supplemented by laboratory and *in situ* tests allowing conceptual models to be prepared as in Figure 5.57 and numerical models to be tested as in Figure 5.58 (after Wentzinger et al., 2013).

5.14.2 Fracture network modelling

For many applications there is a need to attempt to model the rock mass – that is represent the rock as a series of rock blocks with fractures. This is inevitably going to be imprecise and an approximation because of a lack of data. We cannot even determine the extent or nature of a single discontinuity away from an exposure or borehole in reality. All we can do is make observations and then try to interpolate and extrapolate these based partly on statistical approaches, tempered with a good geological understanding and interpretation of structure and lithological variability. For some situations, particularly when assessing mass

Figure 5.57 3D model prepared using Vulcan. (After Wentzinger, B. et al. 2013. Stability analyses for a large landslide with complex geology and failure mechanism using numerical modelling. *Proceedings International Symposium on Slope Stability in Open Pit Mining and Civil Engineering*, Brisbane, Australia, 733–746.)

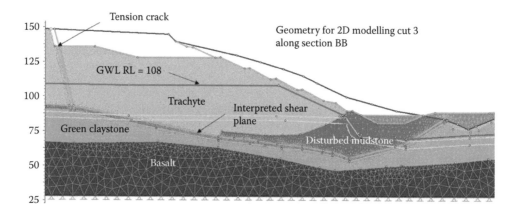

Figure 5.58 Numerical model used for analysis of stability using Phase2 (2D FEM model). (After Wentzinger, B. et al. 2013. Stability analyses for a large landslide with complex geology and failure mechanism using numerical modelling. *Proceedings International Symposium on Slope Stability in Open Pit Mining and Civil Engineering*, Brisbane, Australia, 733–746.)

strength and deformability and where there is no particular structural anisotropy relevant to the project then it might be assumed that the rock mass is essentially a continuum. This is implicit in the Hoek–Brown models as expressed in the geological strength index (GSI) discussed earlier. However, in other situations there is clear anisotropy with various joint sets and other fractures defined and these might be modelled explicitly. Discrete fracture network models (DFN) are at the heart of many different modelling approaches including UDEC, Fracman, Napsac and Elfen. Zhange et al. (2002) and Pine et al. (2006) describe attempts to model fractures in rock as realistically as possible given the current state of knowledge.

5.15 CONTRACTS FOR CONSTRUCTION

5.15.1 Introduction

It is realised that this section somewhat deviates from what one would usually expect to see in a 'Rock Mechanics' textbook but one of the original reviewers for the proposed book suggested that, for completeness, Baseline Reports should be dealt with and certainly what follows ought to be of interest to those 'practising' rock mechanics in rock engineering projects.

The GI Contractor usually presents ground investigation data as factual volumes of borehole logs and results from other field investigations together with the results of any laboratory data that have been commissioned by the design engineer. These factual data may be made available to bidding construction contractors. Other information collected by the design engineer including the results of exposure mapping and desk study including aerial photographic interpretation are generally used by the design engineer in preparing a ground model for the design of the works but are not usually given to the construction contractor.

Mostly contracts follow standard templates such as those prepared by the Institution of New Civil Engineers New Engineering Contract, NEC (ICE, 2005) or by the Fédération Internationale des Ingénieurs-Conseils (FIDIC), (Tottergill, 2006). Some contracts are suitable to engineer-designed contracts and others to design-build situations. Contractual relationships are discussed in some detail in Hencher (2012a).

The way in which ground risks are dealt with vary from contract to contract and in major and potentially difficult projects where there is a high chance that ground conditions might vary considerably from those anticipated, there is a tendency towards some kind of risk-sharing between the Owner/Employer and the Contractor. Deciding whether or not conditions were outside those to be expected can be difficult and lead to legal disputes.

5.15.2 Unexpected ground conditions

Many engineering projects are delayed by unforeseen ground conditions but there is no clear and routine method for resolving such claims and there is often dispute as to whether conditions were really unforeseeable. The relevance of the initial ground investigation report (and its scope and quality) will vary from contract to contract. The Red FIDIC (Employer design) and Yellow (Contractor design) books adopt a foreseeability approach for assessment of claims when adverse conditions are encountered as per the well-established ICE Conditions of Contract Clause 12. The Employer is supposed to make all relevant data in his possession on sub-surface conditions available, at least 28 days before submission of tender and the Contractor is deemed to have based his contract amount on such data. The Employer warrants the accuracy of the information provided and the Contractor is responsible for taking note of the data provided to him.

Elsewhere in the contract it is required that the Contractor has obtained all necessary information as to the risks that may influence or affect his tender for the works and it is his responsibility to inspect the site and examine other available information. The extent of his responsibility will however be limited to what is practicable taking into account cost and time.

Regarding the allocation of risks under these contracts, the Employer usually, and quite rightly, bears the risk of physical conditions that could not have 'reasonably been foreseen' by an 'experienced contractor' at the date of the tender. It is the Employer's site and his project. However what conditions were reasonable to expect and anticipate will often be disputed. One common line of argument is that if the Employer and his design engineer (responsible for the scope and nature of ground investigation) did not envisage particular physical conditions, then why should the Contractor, however experienced he is? (Dering, 2003). There may also be different views over the extent to which some certain conditions were 'predictable' or truly unforeseeable.

It is often the case that ground conditions were 'unforeseen' because the ground investigations were inadequate so that only a poor assessment of the ground conditions was obtained. Conversely, it is often argued that conditions were foreseeable even given a poorly conducted GI, following the rather purist view that, given enough thought and expertise, all geological conditions should be anticipatable (e.g., Fookes et al., 2000). This argument sometimes has merit depending on the complexity of the geological conditions, the degree of exposure and availability of other data. However, at many sites, extrapolating geological conditions from one point of observation to another is actually quite difficult even for experienced geologists with a good appreciation of the geological history. This is particularly true in weathered terrain and for relating discontinuity data (frequency, style and openness) from one structural domain to another.

For the FIDIC Silver (turnkey) contract, the employer is also obliged to make available to tendering contractors all information in his possession regarding the site conditions. For turnkey contracts however the Contractor has the responsibility to verify and interpret the available data so accepts the risk for adverse and perhaps unexpected ground conditions.

Nowadays for many contracts a 'Geotechnical Baseline Report' is set out for a contract to provide a reference document against which the validity of a claim can be judged.

5.15.3 Geotechnical baseline reports

5.15.3.1 Introduction

Geotechnical Baseline Reports (GBR) (or 'Reference Ground Conditions') are now commonly used as parts of the contract for geotechnical projects such as tunnels and deep excavations. This topic is addressed further in Chapter 9.

The purpose of the GBR is, or should be, to set out the design engineer's best estimate of the ground conditions that the Contractor will meet during construction. The philosophy is that, even where there are few boreholes and actual data, the setting out of the most likely, predictable, conditions will provide a baseline against which the supervising engineer can assess the validity of any claims for Unexpected Ground Conditions (or 'Differing Ground Conditions'). In the contract it may be set out that the GBR takes precedent against any alternative interpretation of the pre-existing ground investigation data that might be produced by any party. It is strictly a device for the Contractor to price the works and for assessing the validity of claims; whether the GBR was itself a good or poor interpretation of the ground conditions should be essentially irrelevant.

A good GBR should not be over-general though this is sometimes the stance taken by design engineers on the perverse logic that if the GBR allows for a broad set of geotechnical conditions then no conditions can be unexpected. This conflicts with the usually-intended partnering approach to contracts (at least initially); where specifics are avoided then the tendering contractor(s) will be unable accurately to price their bids for the works to be undertaken. In addition, in some cases only factual parts of the report (e.g., a range and scatter of UCS tests) are put forward as contractual; the rest of the report (e.g., general interpretation) is there for information only.

5.15.3.2 Contents of a baseline report

5.15.3.2.1 Introduction

Baseline reports should ideally give good guidance on the best interpretation of ground conditions that might be expected if they are to be of use to the Contractor. They should also be written clearly so that they can be interpreted unambiguously by those making claims and those assessing a claim's validity. They should comprise a number of elements as a minimum.

5.15.3.2.2 Introduction: General geological model

The general structure and distribution of geological/geotechnical units along the route should be briefly described, usually aided by a cross section (or several cross sections) for the site or route. It is often helpful to summarise the geological history of the various strata and their interrelationships because this should help the contractors geotechnical team to anticipate what conditions might be expected.

Rock head level will often be important for construction – excavation in rock is a very different matter to soil or to mixed ground. Rock head though is sometimes difficult to define – it is important to state the criteria (e.g. geological, rock strength, degree of weathering, percentage of rock to soil, degree of fracturing). Apart from individual soil and rock mass units, major structures such as faults should be addressed individually, perhaps by a number or name. Factors to be described include orientation, thickness and anticipated infill (gouge, breccia etc.) in the knowledge that faults can be essentially benign or cause major difficulties because of collapse and fluid inflow depending on the ground conditions and the way the construction is carried out.

It is the normal case that the extent of geological/geotechnical units and position and nature of faults is uncertain. The GBR should present the best interpretation of the ground conditions by the designers and state any limitations and reservations. In doing so the rationale should not be, somehow to outwit the Contractor contractually, but to allow the Contractor to select the right methods for construction, and to price and to programme his works adequately. Contractually, the reference conditions should be just that – something to refer to when considering whether some adverse ground was anticipated or anticipatable by an experienced contractor given the available information. The Contractor will have been expected to consider the site in a professional manner that would include examining any relevant rock exposures, say in quarries, adjacent to the route. Many contracts require the Contractor to satisfy himself of the ground conditions along the route but it would rarely be practical for him to carry out his own ground investigation at tender stage (with no guarantee of winning the work) and often that view is accepted by an arbitrator to a subsequent dispute.

5.15.3.3 Other considerations

It is impossible to suggest a blueprint GBR that would suit all projects in all geological conditions. It is recommended that the design engineer should follow a checklist approach to assessing hazards to the works following the three verbal equations of Knill as discussed in detail in Hencher (2012a) and where significant, these should be addressed in the GBR, setting out the likelihood of occurrence. It needs to be emphasised that the predictions might turn out to be incorrect – they often will due to lack of ground investigation data, testing or exposure. The purpose of the GBR is to set out the design engineer's best opinion (he was usually responsible for designing and supervising the site investigation) that can then be used to judge the validity of claims.

Some key subjects that would need to be considered for many sites include:

- Each geotechnical unit. Each geotechnical (usually geological) unit should be described in terms of its location, extent and physical and chemical properties. Care must be taken that terms are well defined with reference to local standards of description and classification.
- Environment and anthropogenic hazards. Any potential difficulties arising from environmental or anthropogenic considerations should be highlighted. Examples include contaminated land, pre-existing landslide hazards, mining, groundwater and flooding and active faults. The design engineer should follow established good practice in carrying out proper desk studies to identify hazards and these should be set out for the Contractor. It is not reasonable to omit information (however, obvious it might later be deemed that some hazard clearly existed).
- Construction considerations. Any features that might adversely constrain the method of working should be highlighted. These might include stability issues for temporary works such as piping and heaving and abrasivity that would cause damage to equipment and possible delays.

5.16 INSTRUMENTATION AND MONITORING

Instrumentation plays an important part in the characterisation of rock masses during ground investigation and then in monitoring behaviour with time, during construction and for long-term assurance. Often measurements are linked to predictions made at the design or analysis stage so that one can judge whether movements or levels are tolerable. For some projects Alert, Alarm and Action levels are set linked to appropriate investigation or mitigation measures.

Instruments can be installed and read manually and this is still commonly the case for many sites but on major projects instruments are installed to issue electronic data that can be sent to a receiver remotely and continuously. In particular, vibrating wire devices are often used for which frequency changes with displacement induced by changes in pressure or ground movement. Web-based monitoring and reporting systems are commonly employed for major projects and show instrument locations on maps and sections. The system will often provide real-time graphs of movements, loads or whatever else the instrument is designed to measure. Instruments may be damaged during construction or fail in the long term so where measurements are a vital part of the control process, then the engineer must ensure sufficient back up. This needs to be thought through carefully as there is often only a short time interval during investigation or construction that instruments can be installed, calibrated and tested so the instrumentation programme needs to be integrated with other aspects of the project. Many aspects of the selection, installation and use of instrumentation are discussed by Dunnicliff (1993).

5.16.1 Water pressure

Water pressure is an important parameter for many projects for example for slope stability or for monitoring drawdown caused by underground excavation. Piezometers of various types are installed routinely for many sites to monitor piezometric pressures below the water table or suction levels in the vadose zone and these can be monitored remotely so that data can be used as part of a hazard assessment monitoring system. There are several systems that can be used to monitor water pressures at different levels. A simple and cheap method to monitor transient positive water pressure is to install 'Halcrow buckets' which comprise thin plastic containers, installed sequentially, at perhaps 0.5 m lengths on fishing line, weighted with lead shot and suspended down a standpipe in a borehole. After a storm, the buckets can be retrieved and the highest one containing water gives an indication of the transient water level.

5.16.2 Displacement measurement

Devices to measure displacements are commonly used in slopes or in the walls of tunnels and other underground excavations. Vibrating wire strain gauges and tiltmeters are commonly used and data interrogated remotely.

In slopes, inclinometers are installed in boreholes and monitored regularly by lowering a 'torpedo' down grooves in two directions, allowing displacements to be resolved in three dimensions. Inclinometers work well generally and indicate levels and magnitude of displacements. These can be automated but often inclinometers are read manually, perhaps weekly or monthly, using the same torpedo on various boreholes each of which has the grooved tubes installed. A relatively recently developed device is a flexible inclinometer (ShapeAccelArray) comprising a series of sections that are instrumented separately. The tool can be installed down vertical holes, in the horizontal or even around a tunnel. The instrument is illustrated in Figure 5.59.

Other devices used in slopes include simple slip-indicators where a narrow tube is installed through a potential slip plane and two metal rods attached to string raised or lowered down the tube periodically. The level where the rods can neither be raised nor lowered gives an indication of the depth of movement. Tubes filled with gravel will give out acoustic emissions where the gravel rubs and these can be interpreted with respect to the level of movement; these were originally used in Japan and have been further developed recently in the United Kingdom (Dixon and Spriggs, 2007).

In tunnels and mining, much use is made of wire extensometers that are fixed in holes at some distance from the walls of the excavation. This allows movements towards the excavation to be measured and monitored.

Optical 'total stations' that measure distances and angles have been used for many years to provide accurate and reliable measurements of the displacements of prisms set on a slope

(a) (b)

Figure 5.59 Flexible strain measuring device. (a) Flexible inclinometer wrapped around drum to transport to site. (b) Inclinometer being lowered down vertical borehole.

or other structure. They provide accurate measurements of changes in x, y and z directions. Total stations can be programmed to seek each target prism at regular intervals (say hourly) thereby giving detailed records of on-going or accelerating movement.

As discussed earlier, laser scanners are being used more commonly to measure topography and to characterise rock mass remotely. Data collection is fast (100 k points per second with scan times of a couple of minutes typically). The instrument requires only limited operator training and they are now relatively portable. Surface accuracy is usually sub cm, especially for close range lidar scanners (i.e., <100 m range). Radar can also give measurements to less than 1 mm. These devices are being used more and more for routine monitoring of slopes, especially in open pit mining as an alternative to total stations.

5.16.3 Load cells

Load cells are sometimes installed both for temporary measurements during construction and for long-term monitoring perhaps behind a structural wall or the lining for a tunnel. They are especially important for checking loads on temporary works during construction to give some warning that structural members are being overstressed so that some remedy can be adopted. Load cells are also used in anchor systems, again to check on continued performance and the need for any action.

Chapter 6

Properties and parameters for design

Class 3: The rock can be broken by a kick (with boots on)...

Ollier (1975)

6.1 INTRODUCTION

Most rock testing is at the small scale. This is true of testing in field exposures, when logging rock core, down boreholes and in the laboratory. Rock-engineering design however needs to deal with large-scale geological and geotechnical units, with parameters assigned appropriately; such units are not usually testable themselves. Some properties can be measured at a larger scale indirectly via geophysics, by fluid-pumping tests or by using very expensive testing setups such as those employed in bored pile testing. Parameters are also derived by back analysis of the response of rock to excavation or loading beneath major structures or in response to fluid extraction (usually water, oil or gas). The measurement of hydrogeological parameters is discussed in more detail in Chapter 4.

6.2 SAMPLING

When selecting samples to test, one needs to bear in mind how the tests will be interpreted at the larger scale. Rock masses are often discontinuous, comprising blocks of intact rock separated by discontinuities. When determining the rock mass strength or modulus, one needs to measure the intact rock properties and also to characterise the discontinuities – their frequency, aperture and persistence.

A common dilemma with sample selection is that only the better and stronger sections of the core are recovered in a condition suitable for testing – testing these alone may give a wrong impression of the rock mass qualities *in situ*. The consequences can be severe, for example, in the incorrect selection of excavation equipment and method – such as in judging the need for blasting for excavation or the selection of the appropriate dredging equipment. For many laboratory or field tests, the sample needs to be as uniform as possible; in strength testing, samples with defects such as adverse incipient joints or large voids will probably fail at low values and the test data may be spurious and misleading with respect to engineering performance. A common limitation of laboratory tests is that they might not be applied directly to the rock mass without considerable interpretation and allowance for other factors. Nevertheless, the tests themselves and sample selection need to be as rigorous as possible if they are to be useful. The tests should provide reliable and repeatable data that can then be interpreted within the wider context of field characterisation and ground modelling.

Where the rock-engineering problem involves sliding on rock discontinuities as in the assessment of the stability of rock slopes or underground wedges in a tunnel or cavern, then laboratory tests can be used to define the basic frictional characteristics and some indication of the contributions from minor roughness and interlocking. However, strength at the larger scale will depend on characteristics of discontinuities at the field scale as discussed later.

6.3 ROLE OF INDEX TESTING

International standard agencies such as ASTM and the ISRM (Brown, 1981; Ulusay and Hudson, 2006) provide guidance on many difficult and complex tests where factors such as sample dimensions, rate of loading and stiffness of the machine need to be specified. Much information can however be gained from relatively simple laboratory or tactile tests in the field. For example, weathered rock profiles in exposures or suites of rock core are often best firstly differentiated into units by the eye – say based on colour, rock content and geological structure. The validity of these units can then be checked using simple tests of strength by trying to break pieces by hand or hammer or scratching with a knife. Hack and Huisman (2002) present a useful review and conclude that in many cases, an estimation of intact rock strength by 'simple means' is more useful for establishing the rock strength of a rock mass than by more elaborate testing. It is also a lot cheaper.

Other slightly more sophisticated index tests such as the Schmidt hammer and point load test that are specified by ISRM and discussed later in this chapter can be carried out routinely by the logging engineer/engineering geologist as part of the exercise of description and classification either for characterising exposures (differentiating between soft and hard areas) or in logging core in the case of the point load test. These are tests that should be carried out by the design team rather than items delegated to a contractor (at a considerable expense).

Once the preliminary geotechnical units have been identified either in the field or from borehole data, then, more sophisticated laboratory tests can be carried out as appropriate.

6.4 BASIC CHARACTERISATION

6.4.1 Introduction

There are numerous tests and procedures that can be useful for characterising rock material, including mineralogy, texture, density, abrasivity and absolute age, some of which are listed in Table 6.1. The need and application for each type of test depends on the nature of the project.

Many routine tests such as unit weight and density determination are dealt with in appropriate ASTM and other standards, including the recommended methods of ISRM (Brown, 1981; Ulusay and Hudson, 2006).

Mineralogy and chemical determination is a matter for geologists and specialist geochemical labs.

6.4.2 Suitability of aggregates

Mineral reaction and breakdown are important considerations for the selection of aggregate for concrete. Alkali–silica reactivity is always a possibility when using rocks that contain minerals such as volcanic glass and cryptocrystalline silica but has been reported for many

Table 6.1 Rock parameters: some fundamental properties and their significance

	Property	Approach	Details and comments	Examples
General description and characterisation; rock type	Mineralogy, texture and pervasive fabric	Visual examination and description, petrological microscope, x-ray and chemical analysis	Linked to the basic chemical and physical performance	Establishing the reactive nature of minerals in the aggregate Establishing the basic geological relationships (e.g., distinguishing between arkose and granitic rock)
	Absolute age	Radioactive dating	Carbon dating for younger sediments; potassium, argon and other methods for older rocks	Dating of landslide movements; establishing sediment rates Relative age of intrusions can be important in understanding complex geology
	Porosity, dry density and unit weight	Standard lab testing	Parameters are used for direct input to analysis Linked to design parameters	Dry density is a useful measure of the degree of material weathering Porosity links to strength, deformability and specific yield
	Abrasivity and durability	Measured quartz content Direct testing: Cerchar test The LCPC abrasivity-testing device measures potential wear from disaggregated material Slake durability test	Cemented sandstone and quartzite are especially problematic Wear on the pin mimics wear on the machine part Broken down slurry of silica-rich rock can cause unacceptable wear to components of TBMs Durability measured by wetting and drying and abrasion	Abrasivity of rock is important for wear of drilling bits and cutters Limestone can cause adverse wear to the dredging equipment – surprisingly considering low hardness of calcite

rocks including schist, gneiss and granite. Smith and Collis (2001) provide guidance on aggregate selection. The first stage is awareness of a potential problem, geological description and classification. There are quick tests to check the potential for a reaction; if there is a possibility identified, then trial concrete mixes are made up and mortar bars are checked for volume change over a period of time.

6.4.3 Age determination

Absolute age determination may be helpful in interpreting the geology at a complex site. This might seem rather remote from traditional 'rock mechanics' but the author has had cause to instruct carbon dating on samples to establish dates of sediment deposition as discussed in Chapter 3 and also to investigate the hydrogeological conditions in a major landslide. In the latter case, using historical natural background radioactivity as the measure, it was possible to date a trapped leaf in one particular clogged natural water channel, at a depth of about 20 m, to 40 years prior to the failure. This information proved useful in understanding the site history and the way that the hydrogeological conditions had changed over the years. At the Australian slope referred to in Chapter 5, radioactive dating was commissioned from the University of Queensland to help understand the sequence of igneous intrusions as part of the geological and geotechnical modelling (see Figures 5.57 and 5.58). The use of palaeontology as an aid to interpreting the geology for construction purposes can be traced back to William Smith in the late eighteenth century and it still has extensive application especially in the oil industry. Relative dating using macro and micro fauna was helpful for establishing units for the design and construction of the Channel Tunnel between England and France (Harris, 1996).

6.4.4 Abrasivity

Some rocks are far more abrasive than others and this will affect both the amount of time spent in drilling or excavation and the costs due to wear. The most common test used is the Cerchar Abrasivity Index (CAI). The test is carried out usually on freshly broken rock without weathering. Samples are held in a vice and then a conical, 90°, hardened steel pin, loaded by 7 kg, drawn 10 mm across the surface in 1 s. The test is repeated for five different pins. The minimum and maximum wear diameters are recorded for each pin as well as the shape of the worn area. The pins are examined using a microscope and two perpendicular diameters of the worn area are recorded for each pin. CAI is calculated as

$$\text{CAI} = 10 \ d/c \tag{6.1}$$

where d is the diameter of the area-worn flat and c is a 'unit correction factor' equal to 1 mm. The lower the CAI, the less abrasive is the rock for drilling and for cutters. A value of 1 is low whereas 6 is extremely abrasive. Methodologies and criteria are set out in ASTM D7625 and in the ISRM- recommended methods (see ISRM website).

Abrasivity of crushed material can be quantified using the *Laboratoire Central des Ponts et Chausées* (LCPC) abrasivity-testing device developed by the French LCPC for testing rock and aggregates. This test is useful for advance warning of the common problem of wear due to abrasive slurry in earth balance TBMs as described by Shirlaw et al. (2000). A steel impeller is rotated for 5 min at a speed of 4500 rpm in a cylindrical vessel containing a 500 g sample with size range between 4 and 6.3 mm. The LCPC–abrasivity coefficient (LAC) is calculated as the mass loss of the impeller divided by the sample mass (500 g) expressed as grams per tonne. A proposed classification based on CAI and the LAC is given by Käsling and Thuro (2010) together with examples of typical rock types and their abrasivity and reproduced here in Table 6.2. The abrasivity of silica-rich rock is highlighted but in practice, limestone often also causes adverse wear on cutters in dredging operations possibly due to adhesion rather than abrasion (Vervoort and De Wit, 1997).

6.4.5 Durability

Durability of rock in the long term might be a concern where rock is used as armourstone, as rock fill or as building stone. This property is generally quantified using a 'slake durability test' in which blocks of rock are dried and wetted and rotated in a perforated drum following defined procedures (see ISRM – recommended methods). The weight loss on each cycle is noted and provides an indication of durability. Additional procedures are set out in ASTM D5313/D5313M – 12 (ASTM, 2013). Tests specific to the deterioration of building stones with particular emphasis on ancient monuments are addressed in Siegesmund et al. (2002).

Table 6.2 Abrasivity of rock materials based on LAC and CAI

LAC (g/t)	CAI	Abrasivity	Examples
0–50	0–0.3	Not	Organic material
50–100	0.3–0.5	Not very	Mudstone, marl
100–250	0.5–1.0	Slightly	Slate, limestone
250–500	1.0–2.0	Medium	Schist, sandstone
500–1250	2.0–4.0	Very	Basalt, quartzitic sandstone
1250–2000	4.0–6.0	Extremely	Amphibolite, quartzite

Source: After Käsling, H. and Thuro, K. 2010. *Proceedings of the 11th IAEG Congress, Auckland, New Zealand, Geologically Active*, London: Taylor & Francis Group, 1973–1980.

6.5 ROCK STRENGTH AND ITS MEASUREMENT

6.5.1 General

A generalised strength envelope for intact rock is presented in Figure 6.1. It is convenient to consider three modes of loading:

1. Tensile
2. Uniaxial compression
3. Shear

6.5.2 Tensile strength

Tensile strength is important as a prime mode of failure (mode 1 of Atkinson (1987) – see Chapters 1 and 2). Most rock fractures develop as tensile joints and this is the way asperities

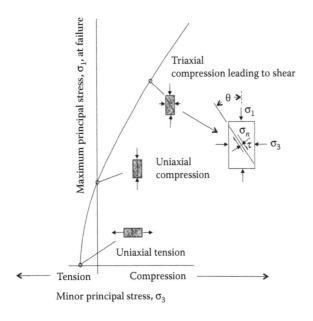

Figure 6.1 Generalised strength envelope for rock. (Based on Hoek, E. and Brown, E.T. 1980. *Underground Excavations in Rock*, London: Institution of Mining and Metallurgy, 527pp.)

Figure 6.2 Sandstone sample set up for direct tensile testing. Sample platens are glued to milled rock ends using strong glue and then pulled apart.

fail in direct shear and that rock is broken by disc cutters in tunnel-boring machines. An example of a sample prepared for direct pull-apart tensile testing is shown in Figure 6.2.

The Brazilian test is an indirect method of tensile testing and involves loading discs of rock across their diameter under compression. This test is used essentially as an index test particularly in assessing excavation conditions for mechanical equipment. The test is carried out on discs of rock cut to a 0.5 length-to-diameter ratio and a typical diameter of 54 mm and follows the procedures of ASTM D3967 (2008) or the ISRM-recommended methods (Brown, 1981; Ulusay and Hudson, 2006). The specimen is loaded until failure, as illustrated by a series of photographs taken using a high-speed camera in Figure 6.3.

Tensile strength, σ_t, can be calculated from a Brazilian test using the following formula (Mellor and Hawkes, 1971):

$$\sigma_t = 0.636 \; P/D \; t \; (\text{MPa})$$

where P is the load at failure (N), D is the diameter of the test specimen (mm) and t is the thickness of the test specimen.

6.5.3 Compressive strength

6.5.3.1 Uniaxial test

UCS is one of the basic parameters used in rock engineering and one of the most common strength determinations performed on rock core. It forms the basis for most rock mass parameter assessments, including rock mass classifications and is used in most description schemes for defining classes of rock (weak, strong, etc. as discussed in Chapter 5). This is despite the fact that in most circumstances, including laboratory testing, rock fails in tension or by shear or as a combination of both (see Chapter 2). The test is generally conducted following procedures as recommended in ASTM D2938 or by the ISRM (Brown, 1981; Ulusay and Hudson, 2006). Samples are typically NX in diameter (54 mm). Samples need to be of

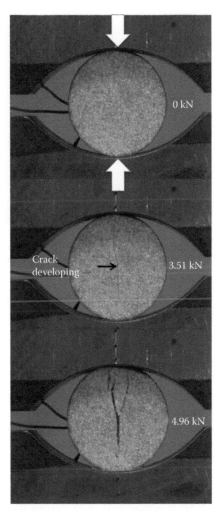

Figure 6.3 Three stages of the Brazilian test on a sample of Carboniferous sandstone from Black Hill Quarry, Leeds, UK, photographed using a high-speed camera. The progressive development of the final fracture can be seen in the central photograph at about 70% of the ultimate failure load.

particular dimensions usually with a diameter ratio of 2–3 partly to avoid unwanted stress concentrations at the loading platens. There are also strict requirements regarding the accurate end preparation of samples to prevent stress concentrations that might lead to failure at low stress. An example of a UCS test is presented in Figure 6.4.

Unconfined compressive strength is dependent on any contained flaws and weaknesses and where failure occurs on an included pre-existing discontinuity, the test result should be disregarded. Systematic testing of specimens with ubiquitous fabric such as cleavage in slate shows, not surprisingly, that where the weakness direction corresponds to the potential shear (or vertical tensile) fracture, the measured strength will be low as illustrated in Table 6.3. Similar directional variation can be found in tests on rocks containing preferential microcracking such as 'rift' and 'grain' found in some igneous rocks (e.g., Douglas and Voight, 1969).

Scale effects are sometimes reported for uniaxial compression tests with larger samples failing at lower loads, proportionally. Many data were compiled by Hoek and Brown (1980)

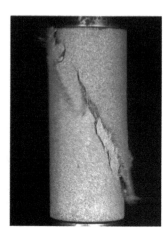

Figure 6.4 Three stages of UCS test photographed using a high-speed camera. Visible fractures (shearing mode) are first seen at about 70%–80% peak compressive strength.

and it seems likely that this reflects the higher possibility of significant contained flaws in larger specimens. This is a different phenomenon to that of scale effects in field-scale shear strength of rock discontinuities as discussed later.

6.5.3.2 Point load test

The point load test was introduced in the 1970s as a quick and cheap alternative to the UCS for estimating intact rock strength. Core pieces and even irregular fragments are broken between conical points advanced by a jack as illustrated in Figure 6.5. Quite often, tests are unsatisfactory in that failure occurs locally at the point of contact or along some incipient discontinuity and such test data should be disregarded. Depending on the size of the sample, corrections are made to derive an equivalent value for a 50-mm diameter sample. According to the ISRM-recommended method, UCS can be predicted by multiplying by a factor of about 25 although in practice, this multiplying factor can vary considerably and can be below 10 for some sedimentary rocks. The strength index is widely used for strength classification (generally assuming the correlation factor of 25) and indeed has been adopted in preference of UCS in strength classification in the Australian standard for rock classification. This is despite the need to convert PLI back into UCS (based on some correlation factor) in order to be input into the numerous rock mass classifications and rock mass parameter expressions that use UCS as their basis. In the author's view, for many situations, an estimate of strength, based on index tests such as breakage with a hammer is adequate as discussed earlier and in Chapter 5; if more precision is required, then UCS testing is probably the best way forward. Point load testing is useful as in index because many tests can be done very quickly and cheaply by a logging engineer to gain an impression of variability through the rock mass but the lack of consistent correlation factors with UCS seems to preclude using PLI as a primary measure and classification of rock strength.

6.5.3.3 Schmidt hammer

The Schmidt hammer was originally developed for estimating the compressive strength of concrete but is also used to estimate the strength of rock. The main types of hammer used are the L and N types, the N type giving much higher impact energy. The test is described

Table 6.3 Rock parameters: Compressive strength (small scale)

Details	Comments
Intact:	
Tactile tests (hammer/knife/break by hand)	Useful estimates when logging and field mapping
Schmidt hammer	Good as an index test in the field
Point load	Cheap, can do many. Best done and interpreted by the design team, not commissioned at great expense to a laboratory
Uniaxial compressive strength	Standard reference for classification
Triaxial test	Used to establish shear strength of weak rock at the project scale. For strong rock rarely done other than to establish input parameters to the Hoek–Brown model

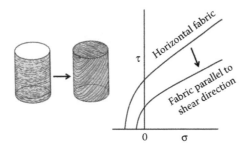

When intact rock has incipient fabric such as cleavage, schistosity, bedding or gneissic banding then the measured strength will vary according to the orientation of the applied load and degree of the weakening effect of the fabric (see Hoek and Brown, 1980 for a summary).

It is sometimes reported that UCS decreases with the increasing size of the sample in the laboratory. When this is the case it is probably related to the occurrence of significant flaws in larger samples that can provide the focus for microcrack development.

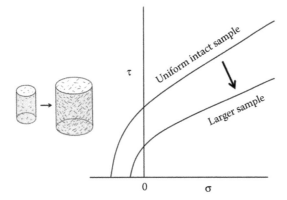

This scale effect is in addition to the general weakening of the rock mass caused by included mechanical discontinuities. The potential is rather difficult to quantify given the variability of rock masses.

in ASTM D5873 and by the ISRM. There are many reasons why measured rebound values can be too low relative to the unconfined compressive strength of the rock. In particular where the surface is rough, initial strikes by the hammer will give low rebound values until the roughness is flattened. The tool does come with a carborundum block supposedly to be used to grind surfaces flat prior to testing but in practice, it is easier just to start hammering and ignore data until the point where the rebound value remains constant. The author has

Figure 6.5 Rock core sample set up in point load apparatus.

demonstrated this to many groups of MSc students by testing the concrete columns in the university laboratories. Initial rebounds are always low (30-ish) because of the thick paint, increasing rapidly to a consistent value of about 55, which is some indication of the strength of the underlying concrete (with paint flattened). For rock testing, low values will similarly be measured where the rock mass contains joints which produce a 'drumminess' as loose blocks are shaken. The same is true for small rock samples and pieces of core which tend to bounce, even if clamped, and this severely limits the usefulness of the test as in the indicator of UCS other than on massive rock in the field.

In the author's experience, the most useful application of the Schmidt hammer is as an index of relative strength, especially for differentiating between different grades of weathered rock in the field. Figure 6.6 shows a Schmidt hammer being used to characterise weathered granite. The hammer is repeatedly used at one spot until consistent rebound values are achieved. One key parameter is zero rebound (not a flicker) despite repeated firing and evident densification. This behaviour tends to define weathered rock material that will slake

(a)   (b)

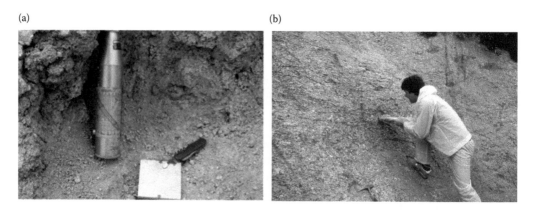

Figure 6.6 Schmidt hammer being used to characterise weathered granite, Cornwall, UK.

in water (Grade V or completely decomposed as defined in Chapter 5). The top of Grade IV where rock can just be broken by hand correlates to a rebound value of about 25.

There are published relationships between the rebound value and compressive strength but from experience, the hammer is insensitive for rock with UCS above about 100 MPa and unreliable for weak rocks but is still useful as an index measure in that the tool has a constant impact energy, provided it is regularly maintained.

There are various recommended procedures in the literature which either involve moving the location of impact between blows or averaging repeated rebound values in some way. In the author's opinion, as expressed above, the best way to use the hammer is to repeat strikes at single locations until the value becomes essentially constant and then to record that value. The test can be repeated at different locations in an exposure to demonstrate variability.

Barton and Choubey (1977) recommend using the test as one stage in a process for estimating the shear strength of rock discontinuities empirically. Vertical measurements are taken at 10 different locations (unprepared). The mean of the upper five values is then taken as a measure of the effect of weathering.

6.5.3.4 Shore scleroscope

The Shore scleroscope test is specified in ISRM (Ulusay and Hudson, 2006) and is used to test surface hardness in metals and is sometimes used to provide some indication of compressive strength for rock core. The test involves measuring the rebound height of a small diamond or tungsten carbide-tipped hammer dropped onto a horizontal smooth surface. An advantage of the test is that it causes no damage. A similar non-destructive device that has been used by various authors to characterise the surface hardness of rock samples is the Equotip.

6.5.4 Rock strength at the mass scale

At the very large scale under high stresses, it is quite astonishing that strength can be defined as essentially frictional with the effective friction angle being somewhere between 35° and 40°. Anderson (1951) adopted the value of 35° in his models for the development of faulting in the Earth's crust (Chapter 3) and this is essentially confirmed by the orientations of faults in nature (Price, 1966). Forty degrees is the friction angle measured by Byerlee (1978) for direct shear strength of rough rock joints at such high stresses that the roughness was sheared though. What makes this astonishing is that friction angles of 35–40° are typical of moderately dense sand and gravel and so, it is a surprise (to the author at least) that this can also be the effective friction angle through the Earth's crust (with no cohesion).

For engineering and mining situations, in brittle rock, effective strength is generally much higher than this and defined using Mohr–Coulomb (M–C) definitions (straight line with cohesion and frictional components) or a non-linear relationship as proposed by Hoek and Brown (1980). The strength will however be much lower than that of the intact rock as illustrated in Table 6.4. The Hoek–Brown criterion is discussed in more detail later when discussing shear strength estimation.

6.6 ROCK DEFORMABILITY

6.6.1 Small scale

Elastic moduli – Young's modulus, shear modulus and their equivalents for individual rock joints and Poisson's ratio are measured routinely as part of triaxial tests on intact rock samples or during shear tests (Table 6.5). These data have value as an input to numerical

Table 6.4 Rock parameters: Rock mass strength

Details	Comments
	For weak rocks (e.g., highly decomposed with UCS of hand sample <12.5 MPa), strength data derived directly from laboratory testing are often used for analysis without any further allowance for larger-scale rock fabric.
The strength of a fractured rock mass is usually much lower proportionally than intact laboratory samples, even in the absence of specific adverse discontinuities.	For stronger rock, strength estimates are usually based on the Hoek–Brown (H–B) criterion as a function of intact compressive strength and degree of jointing although the rock type is also part of H–B criterion. In older literature, there are attempts to use other rock mass classifications such as RMR to predict shear strength parameters – phi and cohesion. These are not recommended.

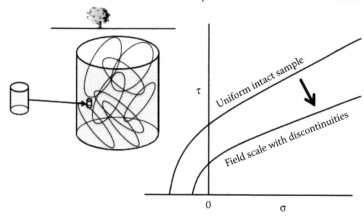

models such as UDEC or for basic research and can be broadly used as part of the estimation of rock mass modulus (Hoek and Diederichs, 2006). Moduli are measured down boreholes using, for example, the dilatometer and pressuremeter for weak rock. The Goodman Jack is often used in stronger rock and comprises steel plates that push against the sides of the borehole wall, but can only be carried out for sections that are essentially intact (Figure 6.7). Not surprisingly, results on intact lengths of rock, away from joints, tend to err on the high side compared to the moduli of the jointed rock masses. Moduli can also be measured using larger-scale plate loading and jacking tests in adits but the influence of mass fabric and especially any open discontinuities needs to be taken into account in their interpretation.

6.6.2 Mass scale

At the larger scale, moduli can be interpreted from compressional and shear wave velocities (Simons et al., 2001).

Moduli representative of the mass scale are also derived from large-scale tests, for example, on bored piles, either using jacks in-built to the piles or by jacking against large reaction loads of concrete blocks or steel as illustrated in Figure 6.8.

For design, however, resort is often made to an oft-published database of back-analysed case examples that are linked most readily to RMR (Figure 6.9).

Table 6.5 Rock parameters: Rock deformability

Laboratory test	Measurements are made of vertical shortening and lateral expansion due to Poisson's effect during compressive strength tests.	Generally, initially, the stress–strain curve is shallow due to closure of pores and other flaws; later, the sample stiffens until it approaches peak strength when it softens again. Different moduli can be defined for each stage of the test if considered necessary.

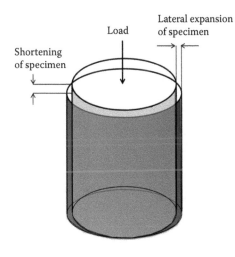
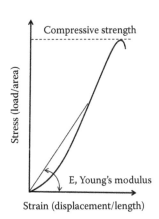

Rock mass deformability	There are numerous methods for measuring modulus in the field as discussed in the text. All suffer the same problem that they only measure conditions locally (as for *in situ* stress measurement) and so are strongly influenced by the local degree and openness of fracturing.	Because of scale effects and the difficulty in extrapolating from a local test to a large rock unit, reliance is generally made on empirical relationships. Commonly, engineers predict mass moduli using rock mass classifications as schematically illustrated below and as discussed in the text. There are some methods that use E (lab) as a starting point. All are prone to considerable potential error.

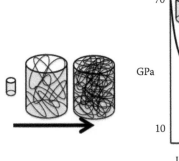
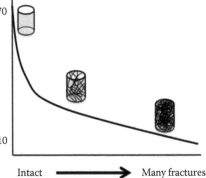

Section of unfractured
rock targeted for test

Plates to be expanded against side of borehole

Test under way

Figure 6.7 Goodman Jack test, Hong Kong.

Figure 6.8 Kentledge of concrete blocks set up as a reaction for a test on a large-diameter-bored pile, Singapore.

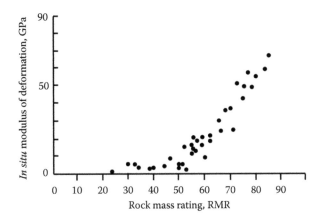

Figure 6.9 In situ modulus of deformation versus RMR from case studies reported by Serafim and Pereira (1983) and Bieniawski (1978).

The data in Figure 6.9 have been interpreted correlated to various rock mass classifications including RMR and Q, but even as stand-alone information, they are extremely useful for estimating likely deformation. Basically, any mass with RMR less than about 50 can be anticipated to have an *in situ* modulus between 0 and 10 GPa. Above RMR 50, there is almost linear increased stiffening up to about 70 GPa for RMR = 90.

Hill and Wallace (2001) however report that data from 15 pile load tests for various rock types, with RMR values ranging from about 30 to 70, indicate that rock mass moduli are an order of magnitude softer than those indicated by the data presented in Figure 6.9. The authors comment that, for a modified RMR value >40 (with zero weighting for water or joint orientation correction, which are generally considered essentially irrelevant at the base of a pile at depth), a rock mass modulus in excess of 1 GPa can be anticipated which would generally be insignificant for the design of building structures. Conversely, for poorer quality, rock-specific testing might be required to verify design assumptions. Examples of the comparative assessment of moduli for the foundation design are discussed in Chapter 7.

6.6.3 Prediction from GSI

The GSI was introduced in Chapter 5 as a way of characterising rock masses, largely on the basis of their broad geological characteristics. The GSI is linked to the Hoek–Brown rock mass strength criterion for rock masses that can be taken to be essentially isotropic, despite the contained discontinuities but where there is no preferential weakness direction as discussed later with reference to shear strength.

Hoek and Diederichs (2006) re-examined the numerous empirical equations that have been proposed to predict rock mass deformation modulus (E_{rm}) and concluded that, in particular, there were large errors for very poor rock masses and for very strong masses. Using high-quality data from China and Taiwan, they plotted these additional data against GSI. There was considerable scatter but the conclusion was that the following equation best fits the field data for undisturbed rock masses:

$$E_{rm}(\text{GPa}) = 100 \left(1/(1 + e^{(75-\text{GSI}/11)}) \right) \tag{6.2}$$

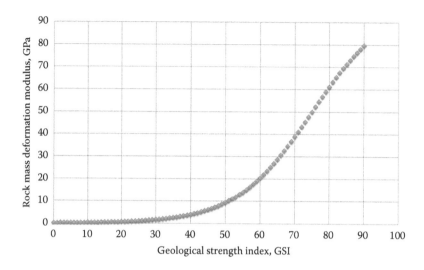

Figure 6.10 Rock mass modulus related to GSI. Note that, by eye, this curve fits quite well with the data presented in Figure 6.9. If the rock mass was disturbed, the predicted modulus would be lower than that shown here.

The equation is presented graphically in Figure 6.10. It can also be applied, with adjustments, to deal with rock that has been disturbed through blasting (e.g., to model conditions adjacent to a drill-and-blast tunnel) (see the discussion by Hoek, 2014d). Hoek and Diederichs also present ways to predict E_{rm} from intact rock modulus (E_i).

This will serve as a first and perhaps best approximation for prediction but engineers need to be aware that there can be considerable errors especially for weaker rock masses (as found by Hill and Wallace for pile tests and discussed earlier). Read and Richards (2007) applied the Hoek–Diederichs equation for an assumed GSI = 20 and intact E_i = 70 GPa which gave a predicted mass modulus of 0.7 GPa for the Waitaki Dam site in New Zealand. The back-analysed modulus was only 10 MPa. Similarly, poor agreement between predictions and measured values was found for soft, massive rock from Mohaka, New Zealand. Clearly, there is considerable work to be done to allow realistic prediction and for critical sites, site-specific testing and analysis may be justified. It is a salutary lesson that, even for simple engineering geological situations, as for predicting foundation settlement on relatively uniform sand and gravel and where there are site-specific strength tests (SPT and cone penetrometers), predictions can be in error by almost ±200% (Burland and Burbidge, 1985); so, perhaps, one should not expect better for fractured rock masses.

6.7 ROCK SHEAR STRENGTH AT MASS SCALE

6.7.1 Classes of problem

Several different classes of problem can be defined with respect to rock mass strength and potential shear failure. This is illustrated for rock slope stability analysis in Chapter 8 and summarised as follows:

Class 1. Discontinuities are rare and not adversely orientated. Rock is weak compared to the stress levels so that intact rock failure might occur.

Class 2. There are adverse discontinuities so that sliding or toppling can occur constrained by the geometry of the fracture network. In this case, shear strength of the discontinuities is the parameter that needs to be determined.

Class 3. Discontinuity network considerably weakens and softens the rock mass but does not define the geometry of failure. The degree to which the fractures will reduce the strength and stiffness below that of the intact rock is not amenable to direct measurement although attempts can be made to back-analyse deformation caused by excavation using numerical simulation.

A fourth, difficult category, is a rock mass comprising a mix of materials of different strength such as boulder colluvium as illustrated in Figure 6.11 and some heterogeneously weathered rock masses. Clearly, the interlocking of the strong blocks greatly influences strength but other factors include the cementing strength of the matrix. The options for assessing shear strength for such rock units in slope stability assessment are discussed in Chapter 8.

6.7.2 Class 1: Isotropic masses

For weak rock masses that can be taken to be isotropic, testing of intact material by direct shear and triaxial testing may be appropriate for defining the strength envelope. This would be the case typically for weathered rock of intact grades IV and V where the rock is weak enough to be broken by hand. It might also be appropriate for weak rocks such as poorly cemented sandstone, possibly chalk and weak mudstones such as London clay. It must be emphasised however that such rocks may still contain adverse discontinuities (e.g., 'greasy-backs' in London clay), which can dominate behaviour and cause failure unexpectedly.

6.7.2.1 Direct shear testing of intact material

Direct shear testing is extensively used in determining the shear strength of soil. Its application to intact rock is generally limited to relatively weak material because of limitations from the available equipment. It is appropriate for testing materials such as highly weathered rock that

Figure 6.11 Steep road cut in boulder colluvium partially cemented with iron oxides, Cape Province, South Africa.

can be broken by hand. Ebuk (1991), for example, measured cohesion in Grade IV granite up to 300 kPa, and friction angles of about 38° using direct shear machines designed for soil.

6.7.2.2 Triaxial testing

Triaxial testing is used especially for weak and weathered rock to help define the strength envelope for civil-engineering purposes. It is also used for stronger rocks in situations where confining stresses are important, as in deep mining or for basic research with respect to geological rock mechanics (e.g., Scholz, 1968; Farmer, 1983). An advantage of triaxial testing over direct shear testing is that water pressures can be measured and strain can be more carefully controlled. Whether laboratory parameters will apply at the field scale is a matter for engineering judgement. Tests can be designed to follow specific loading paths, for example, by keeping confining load constant and inducing failure by increasing water pressure to simulate the development of a landslide in intact, weak and weathered rock (Brand, 1981).

6.7.3 Class 2: Shear strength of rock discontinuities

6.7.3.1 Options for assessing shear strength of rock discontinuities

There are two ways of estimating the shear strength of rock discontinuities. The preferred approach is through field measurement of discontinuity characteristics, laboratory testing of samples of the natural discontinuity and analysis and that will be discussed in detail below. An alternative is to use empirical models such as that of Barton-Bandis (1990), which is incorporated as an option into the numerical software UDEC and widely used. The model strictly requires project-specific measurement and characterisation (e.g., Kveldsvik et al., 2008) but the formulaic approach means it is easy for engineers to simply judge the various parameters to derive an estimate or range of estimates. The model will probably give reasonable estimates of shear strength for rough interlocking joints and will have built-in conservatism where the discontinuities at the project scale are discontinuous. Such empirical methods are not however applicable to joints that are coated, infilled or polished, and can be in error for many other reasons as discussed by Hencher and Richards (2015).

6.7.3.2 The testing and analytical approach

6.7.3.2.1 Introduction

Shear strength of discontinuities is an important parameter for many rock-engineering projects and requires the determination of fundamental frictional parameters together with characterisation and quantification of geological factors for the discontinuity *in situ* as illustrated in Figure 6.12. The factors contributing to strength include

1. True cohesion contributed by the strength of local 'rock bridges' consisting of intact rock or incipient defects.
2. Roughness at a relatively large, field scale, causing interlocking and dilation. Large-scale undulations and waviness of the order of metres are termed first-order; smaller-scale roughness is termed second-order (Patton and Deere, 1970; ISRM, 1978; Wyllie and Norrish, 1996).
3. Smaller asperity interaction and textural friction (basic friction) at the scale of rock core and laboratory test samples.

The contributions to shear strength made by persistence and first-order (typically >0.5 m wavelength) and second-order (typically 50–100 mm) roughness need to be characterised by field measurement as discussed in Chapters 3 and 5. The contributions from relatively

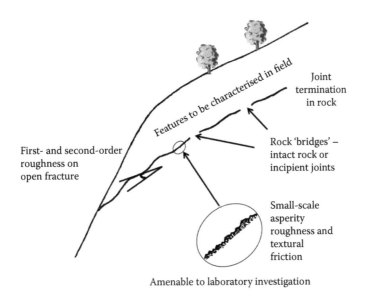

First- and second-order roughness on open fracture

Features to be characterised in field

Joint termination in rock

Rock 'bridges' – intact rock or incipient joints

Small-scale asperity roughness and textural friction

Amenable to laboratory investigation

Figure 6.12 Factors contributing to shear strength of rock discontinuity.

minor asperities over short profile lengths of, say, 100 mm and from basic friction can be investigated using direct shear testing in the laboratory.

Most direct shear tests at the laboratory scale are carried out on open or infilled joints with zero or very low tensile strength, mostly because open defects are more critical for stability assessments.

6.7.3.3 Basic friction

Open, planar (non-dilating) discontinuities exhibit purely frictional shear resistance that is directly proportional to normal stress and shear resistance is derived from adhesion at the true points of contact and textural interlocking (Hencher and Richards, 2015).

The textural (tactile) component of basic friction can be increased (by roughening) to the point at which the upper block is forced to dilate under some particular applied normal stress. Conversely, it can be reduced by polishing so that the basic friction is derived from adhesion alone, which is probably about 10° for most silicate minerals and rocks. Figure 6.13 shows the effect of repeated shearing on a saw-cut surface with consequent reduction in frictional resistance from 31° to 16°.

The basic friction angle is a function of the surface texture, weathering and the mineral coating of the surface and can be very variable even for apparently planar surfaces and is not a unique value measureable from a test on a saw-cut surface (which is the 'basic friction' reference point for empirical strength methods).

6.7.3.4 Direct shear testing of rock discontinuities

6.7.3.4.1 Introduction

Laboratory testing is used for

1. Determination of the basic friction angle for naturally textured and coated disconti- nuities.

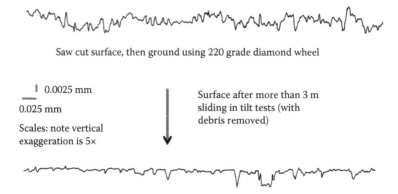

Saw cut surface, then ground using 220 grade diamond wheel

0.0025 mm

0.025 mm

Scales: note vertical
exaggeration is 5×

Surface after more than 3 m
sliding in tilt tests (with
debris removed)

Figure 6.13 Polishing of the saw-cut surface leading to a reduction in frictional resistance for Delabole slate. (After Hencher, S.R. 1977. *The Effect of Vibration on the Friction between Planar Rock Surfaces.* Unpublished PhD thesis, Imperial College Science and Technology, London University.)

2. Identification of surface features such as mineralogy and degree of polishing that affect shear strength over the applicable normal stress range.
3. Observation of damage to minor asperities during shear at particular stress levels, thereby allowing judgement of which roughness features to allow for in the design.

The International Society for Rock Mechanics (Brown, 1981; Muralha et al., 2013) and the American Society for Testing Materials (ASTM, 2008) give some information on laboratory-testing procedures but little guidance on the need for recording data at small increments of displacement or on how to analyse the results.

Firstly, it needs to be stated that testing rough rock joints and interpreting the results is not straightforward. The author has often been asked to review sets of tests carried out by accredited commercial laboratories. The main use of most such reports is to recycle into a nice cardboard box. An example of a fairly recent set of results, from commercial tests for a major project in Hong Kong, is presented in Figure 6.14.

It should be readily apparent that these data are useless for design though no doubt were obtained at a considerable expense. Many authors would suggest that this type of scatter plot is the inevitable result of testing samples of variable roughness (see the discussion by Hencher and Richards, 2015). On the contrary, providing care is taken to measure horizontal and vertical displacements and shear loads throughout the tests, simultaneously at very small increments of displacement (<0.05 mm), then such tests can provide reliable design parameters and allow insight into the factors controlling shear strength at a larger scale.

6.7.3.4.2 Test methodology

It has to be accepted that it is not possible to test in the laboratory samples that will be truly representative of an *in situ* discontinuity.

Discontinuities typically vary in dip on a scale of millimetres, centimetres and metres and include variable roughness features that are specific to each sample.

Samples should be selected so that they are representative of the *in situ* discontinuity in terms of surface texture, coating and mineralogy. Well-matched samples should be used if that is the nature of the discontinuity *in situ*. The recommended analysis corrects for the effects of roughness and therefore, individual sample roughness is of only secondary

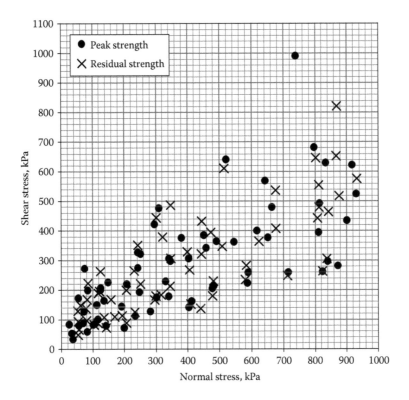

Figure 6.14 Commercial test results from direct shear tests on granite discontinuities, Hong Kong.

importance. Samples may be selected to contain particular roughness features of interest, for example, step features from cross jointing might be targeted to see if they survive during shear at design stress levels. Observations of overriding or shearing of asperities during shear at different normal stress levels in the laboratory will help in judging the roughness angles to allow for at the field scale.

6.7.3.4.3 Documentation

Samples should be described and photographed before and after testing together with all pertinent details of the test programme itself. The descriptions should record details of surface mineralogy, morphology and tactile texture. The general roughness of the surface can be recorded as illustrated in Figure 6.15. These measurements are taken for documentation purposes and for grouping similar samples. After testing, the surfaces and the nature of damage should be described and sketched.

6.7.3.4.4 Test procedures

There are two types of direct shear test – (1) constant normal load and (2) constant normal stiffness. The latter applies where the discontinuity is confined within a rock mass, or in a rough pile socket, so that dilation leads to increasing normal stress and therefore strength. Some workers have carried out research for these conditions (e.g., Saeb and Amadei, 1992; Seidel and Haberfield, 2002), and this can be modelled using software such as UDEC, but such conditions will not be discussed further here.

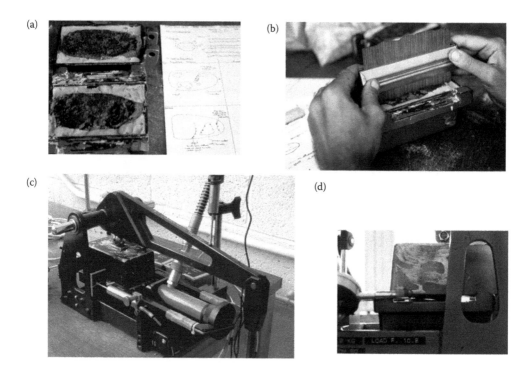

Figure 6.15 (a) Samples set in dental plaster and sketched. (b) Characterisation of roughness using a needle profiler. (c) Golder Shear Box set up. Normal load is applied through the lever arm and vertical displacement is measured on the same lever arm using an LVDT (magnified). Horizontal displacement is measured using LVDT. Shear load is measured through a load cell between the shearing yoke and jack. The yoke is also strain gauged. (d) Dilating joint during a test.

6.7.3.4.5 Testing under constant normal load

The following section applies to tests carried out under constant normal load, which is appropriate to rock discontinuities in rock slopes. Following the initial sample setup and descriptions, the shear box should be assembled and the first normal load should be applied. The normal stress range should be selected to match the field conditions. Tests are generally carried out at natural water content, particularly if the discontinuity is clay filled, although tests can also be carried out with the joint under water. Such a procedure might be appropriate, for example, when testing a discontinuity through weak and weathered rock.

Testing can be carried out following either single-stage or multistage procedures. Multistage testing involves testing the same discontinuity sample at a series of different normal loads. Such a testing strategy allows one to obtain maximum information from each sample. The normal load is generally increased after the shear resistance has remained constant or falling over a few mm or so of displacement. Another strategy is to reset samples at their original position prior to changing the normal load. This allows surfaces to be examined and photographed and for loose debris to be removed. Multistage tests can also be carried out with decreasing normal load at each stage, repeated stages at the same normal load or perhaps by repeating test runs at the same low load as earlier stages in between runs at higher loads to investigate how damage is occurring and affecting dilation. In all such tests, it is inevitable that damage caused at earlier stages will affect results at later stages but provided this is recognised and allowed for in analysis and interpretation this is fine.

When planning a series of multistage tests on different samples, it is a good practice to commence each individual test at different normal stress levels so that first stages (undamaged) can be used in isolation to interpret peak strength parameters.

Rates of shearing below a few millimetres per minute do not affect test results generally.

In order to be able to analyse results properly, horizontal and vertical displacements must be recorded throughout the test frequently and simultaneously.

6.7.3.4.6 Test data

Shear and normal loads are converted into average engineering stresses by dividing by the gross contact area of the samples, which usually varies throughout each stage of the test. For samples taken from rock core, the area of contact can be calculated from the following equation (Hencher and Richards, 1989):

$$A = \pi ab - \frac{ub\sqrt{4a^2 - u^2}}{2a} - 2ab\sin^{-1}\left(\frac{u}{2a}\right) \tag{6.3}$$

where
A = gross area of contact
$2a$ = length of the ellipse
$2b$ = width of the ellipse
u = relative horizontal displacement

Other samples can generally be represented as a rectangle or similar and changes in the gross area can be calculated to an adequate accuracy.

The shear stress versus horizontal displacement for a test on a tensile fracture through limestone is presented in Figure 6.16.

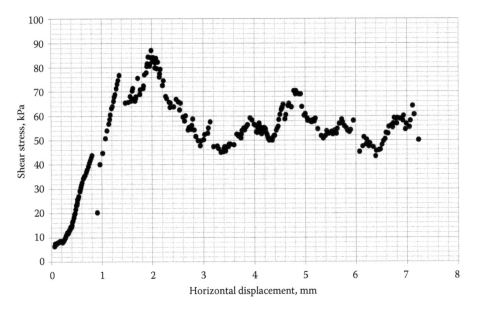

Figure 6.16 Shear stress versus horizontal displacement for a single-stage test on a rough limestone discontinuity.

6.7.3.4.7 Measuring dilation

Rough, matching joints dilate during shear at relatively low loads and work is done in lifting the upper sample. This increases the force required to continue shearing. The corresponding graph of vertical displacement versus horizontal displacement for the test stage illustrated in Figure 6.16 is presented in Figure 6.17.

The roughness angle can be calculated throughout the test by considering the incremental vertical (dv) and horizontal (dh) displacements as follows:

$$dv/dh = \tan i° \tag{6.4}$$

The vertical versus horizontal displacement curve in Figure 6.17 appears quite smooth so that an overall dilation angle, $i°$, might be drawn at peak strength as per ISRM recommendations. In detail, however, the dilation continuously varies throughout the test as minor asperities come into contact, are damaged and overridden and this is reflected in the measured shear strength and should be accounted for. It can be seen by comparing Figures 6.17 and 6.16 that the dilation peaks and troughs correspond quite well with variations in shear strength.

Each shear stress data point needs to be taken in turn and the dilation angle, leading to that strength measurement calculated over a selected horizontal displacement increment. If the increment selected is small, then scatter may be great because of experimental reading errors; if it is too large, then important detail is lost. Consider Figure 6.18, which presents data from an actual test run. These data show individual measurements of shear stress and vertical displacement plotted against horizontal displacement taken at intervals of about 1 s. For the data within the circle, it can be seen that the dilation angle varies markedly between consecutive data points. The sudden increase in horizontal displacement post-peak is partly attributable to release in strain energy. The potential errors are compounded by reading errors in the linear variable differential transformers (LVDTs) when taking data at very small increments of change.

If we consider the data at peak strength, the vertical and horizontal displacement data for the final increment just before peak strength define a dilation angle of 27°. If data are

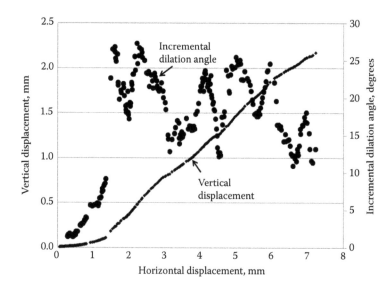

Figure 6.17 Vertical displacement and incremental dilation angle versus horizontal displacement.

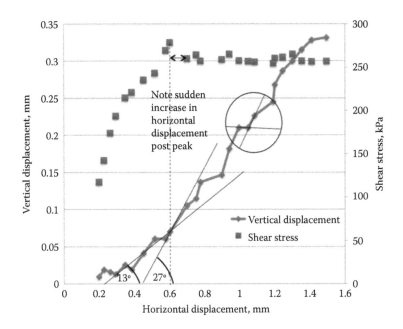

Figure 6.18 Details of measurements during a direct shear test (shear stress and displacements). Individual displacement data are affected by reading errors (sensitivity of LVDT) and sudden release of strain energy during shear.

measured over a larger interval of horizontal displacement (about 0.2 mm in this case), the calculated angle is 13°. In practice, optimal smoothing is best achieved by trial and error using an Excel spread sheet to vary the horizontal increment for calculation. The optimal horizontal increment employed for calculating the dilation angles in Figure 6.17 was 0.2 mm but this may vary from test to test.

6.7.3.4.8 Correcting data for dilation

In a direct shear test, shear load is measured in the horizontal plane and normal load is measured vertically. These loads are converted into engineering stresses, τ and σ by dividing by the gross area of contact of the samples as discussed earlier.

Where the sample is dilating or compressing, the horizontal and vertical stresses can be resolved tangentially and normally to the plane along which shearing is actually taking place using the equations below. The equations are for positive dilation (uphill movement); the signs should be reversed for downhill sliding.

$$\tau_i = (\tau \cdot \cos i - \sigma \cdot \sin i)\cos i \qquad (6.5)$$

$$\sigma_i = (\sigma \cdot \cos i + \tau \cdot \sin i)\cos i \qquad (6.6)$$

where τ_i is the dilation-corrected shear stress, σ_i is the dilation-corrected normal stress and $i°$ is the incremental dilation angle.

The effect of such corrections can be seen in Figure 6.19, which is a plot of the ratios between shear and normal stress, both uncorrected (as measured) and corrected, plotted against horizontal displacement together with the dilation angle throughout a test stage.

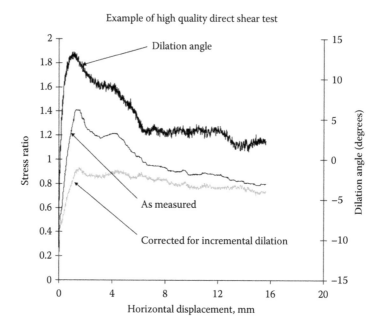

Figure 6.19 Shear/normal stress ratio (uncorrected and corrected) and dilation angle, both plotted against horizontal displacement.

For the uncorrected data, the peak stress ratio, τ/σ, is >1.4, which equates to 'friction' plus dilation angle in excess of 54°. Where data are corrected for incremental dilation, the vast majority of data throughout the test provide corrected stress ratios of 0.9–0.8 (42°–38°).

Figure 6.20 shows uncorrected shear stress against normal stress for a six-stage test on a single limestone discontinuity sample. The apparent change in normal stress throughout each test run reflects the calculated change in gross apparent contact area between the samples (under constant normal load). Presenting the data in this way helps visualise the test history.

The dilation-corrected data for the same test are presented in Figure 6.21 and these data provide a reliable indication of the available basic friction for the naturally textured joint with dilational roughness effects removed. In this case all stages indicate a non-dilational basic friction angle of >45°.

6.7.3.4.9 Typical basic friction angles for natural joints

Basic friction values from dilation-corrected tests on natural joints are presented in Table 6.6. Tests on rough, natural joints through silicate rocks, once corrected, typically give a textural frictional resistance of approximately 38–40°. This value is the same as the value for friction of rough joints measured for a variety of rock types at high stress levels by Byerlee (1978) ($\tau = 0.85\sigma$) where dilatancy was suppressed. Where discontinuities have a smooth textural roughness and major asperities do not come into contact during shear, the basic friction angles can be much lower. Mineral coatings and infill will also influence strength particularly at low stress levels typical of many engineering works, which makes it imperative that, when dealing with such discontinuities, a suitable programme of testing is carried out.

It should be noted that the basic friction angles given in Table 6.6 derived from tests on natural joints are often higher but sometimes much lower than published 'basic' or

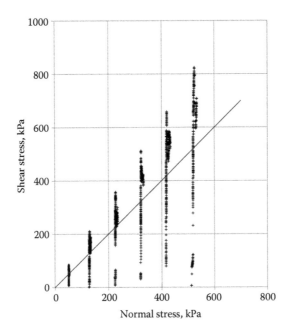

Figure 6.20 Shear stress versus normal stress, data as measured.

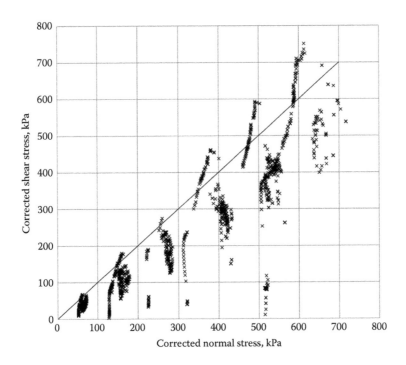

Figure 6.21 Dilation-corrected data from multistage test ($\phi_b > 38°$).

Table 6.6 Example of basic friction angles for natural discontinuities (dilation-corrected or non-dilating)

Rock type	Details	ϕ_b	Reference
Granite	Natural joints, weathering Grades II to IV, including iron and manganese oxides coating. Wet and dry	38–40°	Hencher and Richards (1982)
Quartz monzonite	Chlorite coated (hard, thin) Iron oxide coated	17° at low stresses 38°	Hencher (1981, 2012)
Andesite	Matched, interlocking	38–40°	Unpublished
Volcanic tuff	Iron oxide coated The lower values were for planar natural joints (non-dilating during shear)	32–38°	Hencher (1984)
Schist	Malaysia	26–40°	Unpublished
Limestone	SW England bedding plane discontinuities, FeO stained	40–45°	Unpublished
Sandstone	Darleydale sandstone, UK	42°	Unpublished
Sandstone	Mostly coal measures, South Wales	40°	Swales (1995) for much
Siltstone bedding	Note that the naturally planar,	26–40°	of the coal measures
Shale bedding planes Korea	non-dilating joints with a low tactile texture have lower ϕ_b	26° wet to 33° dry	data, Korean data from Lee and Hencher
Claystone bedding		10–34°	(2013) and
Coal (joint)		9–25°	unpublished data
Polished mudstone		10°	
Clay infill	Various (South Wales and SW England) Mainly illite/kaolinite: low-to-intermediate plasticity	14–30°	Unpublished
Clay infill	Montmorillonite shear zone through mudstone (Australia)	8–9°	Wentzinger et al. (2013)

'residual' values mostly derived from tests on saw-cut surfaces that are typically 25–35°. Despite advice in the literature to the contrary (e.g., Hoek, 2014d; Simons et al., 2001), the dilation-corrected values measured from tests on natural joints listed in Table 6.6 are *not* interchangeable with the basic friction angle used in empirical strength criteria such as the Barton–Bandis empirical strength criterion.

6.7.4 Assessing shear strength at the field scale

6.7.4.1 Persistence and rock bridges

Zhang and Einstein (2010) provide a recent review of what we know about the planar shape and extent of rock joints and the answer is, not very much. Some certainly terminate against other joint sets and this defines their possible extent; others terminate in rock as illustrated in Figure 6.22. Currently, there is no way of estimating persistence away from a point of observation. One can infer potential persistence by examining exposed joint surfaces from the same set but what is visible in exposure is the final result of fracture propagation – it is highly unlikely that joints from the same set, hidden within the rock mass will have the same persistence. It is a common practice to assume that discontinuities are generally persistent although that might be an ultra-cautious approach. It might be possible to justify less-onerous assumptions from field observation combined with inference and judgement as illustrated in Chapter 8 for a large rock slope in the Himalayas.

Figure 6.22 Impersistent joints terminating in the volcanic tuff, Hong Kong. Note the seepage from joints (darker).

The concept of rock bridges – representing discontinuous sections of otherwise persistent joints – makes intuitive sense but there are very few examples in the literature of what these actually look like; so, again, the common approach is to assume that joints are fully persistent unless one can establish otherwise through field studies. An example of a rock bridge exposed after shearing a weathered piece of core containing an incipient discontinuity is shown in Figure 6.23.

6.7.4.1.1 Allowance for roughness

The Fecker and Rengers plate method discussed in Chapter 5 intrinsically encompasses the variation of roughness with scale across a discontinuity surface and can therefore characterise the variation in roughness characteristics from one area to another. If the variation of

Figure 6.23 Rock bridge (grey area) in an otherwise open, weathered joint, coated with FeO. It was exposed following a direct shear test. (After Hencher, S.R. 1984. *Three Direct Shear Tests on Volcanic Rock Joints*. Geotechnical Control Office Information Note, IN 5/84, Hong Kong Government, 33pp. [unpublished].) Centimetre scale.

dip can be established to have similar characteristics over a wide surface area at different scales, that is, there are many roughness features that will jointly act to cause dilation, then this effectively deals with any scale effect. Observations of dilation angles during shear tests can assist in the decision over what roughness angle to allow as discussed in Hencher and Richards (2015) with respect to a particular case example. If the roughness characteristics vary across the plane, so be it – this is a similar dilemma as for a rock engineer dealing with different structural regimes within a single project.

Allowance for larger-scale roughness (first-order) can be made where waviness can be established (Richards and Cowland, 1982, 1986) although care must be taken because, just as waves might resist shear, locally, they can increase the effective dip of a plane. A design example is discussed in Chapter 8.

6.7.4.1.2 In situ testing of rock-joint shear strength

Some workers suggest that *in situ* tests will provide more realistic shear strength parameters than laboratory tests because of the slightly larger sample size but generally, most such tests are expensive and poorly constrained and the level of instrumentation and control is far less than is possible in the laboratory.

6.7.5 Class 3: Generalised failure surface through fractured rock

6.7.5.1 Hoek–Brown criterion

The Hoek–Brown criterion can be used for estimating the shear strength of rock masses where there is no controlling adverse discontinuity or set of discontinuities. The criterion is linked to the geologically friendly GSI as discussed in Chapter 5 albeit that GSI might be difficult to predict or establish within perhaps 10%–20%. The generalised Hoek–Brown failure criterion for jointed rock masses is defined by

$$\sigma_1' = \sigma_3' + \sigma_c(m_b\sigma_3'/\sigma_c + s)^a \qquad (6.7)$$

where σ_1' and σ_3' are the major and minor effective principal stresses at failure, σ_c is the UCS of the intact rock between the joints (averaged or judged in some way) and m_b (which varies with rock type), s (varies with fracture spacing) and a are dimensionless empirical constants (Hoek, 2014d). Eberhardt (2012) in a review of the development of the criterion comments that large values of m_b lead to steeply inclined strength envelopes which are analogous to friction in the M–C strength criterion; s decreases in value with increasing degree of fracturing and relates to the cohesional intercept in M–C.

6.7.5.1.1 The empirical constants

The parameter m_b is for the rock mass and is a function of m_i for the intact rock and GSI (rock quality). The original values for m_i were derived from the shapes of stress–strain curves from triaxial tests for intact rock and indicative values are graphically presented in Figure 6.24. Generally, higher values are quoted for more massive, brittle rocks and lower values are quoted for more ductile rocks. That said, if there is a scientific reason for the 'trend', it seems rather surprising that tough rocks such as hornfels and metaquartzite have values much lower than gneiss and granite. Some of these values are described as 'estimates' and the ranges given are typically ±30%. Hoek (2014b) recommends that values for m_i should be determined for a project using at least five well-spaced data points from triaxial tests. Variation of m_b with GSI is presented in Figure 6.25.

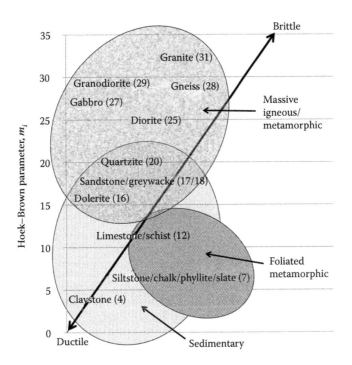

Figure 6.24 Typical values of the parameter m_i for use within the Hoek–Brown failure criterion. High values are typical of brittle, massive, igneous and metamorphic rocks; they are lower for more ductile sedimentary rock. (Values derived from Hoek, E. and Brown, E.T. 1988. The Hoek–Brown failure criterion – A 1988 update. *Proceedings of 15th Canadian Rock Mechanics Symposium, Toronto, 31–38.*)

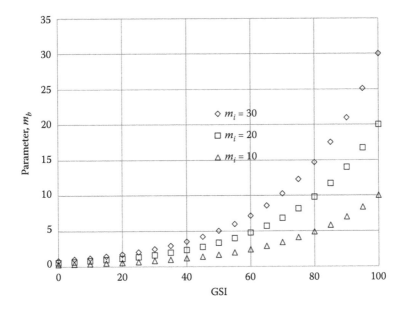

Figure 6.25 Variation of m_b with GSI for three intact m_i values. $m_i = 30$ is appropriate for rocks such as granite and gneiss, 20 for breccia, conglomerate, norite and porphyry; 10 for marble, and some limestone and schist.

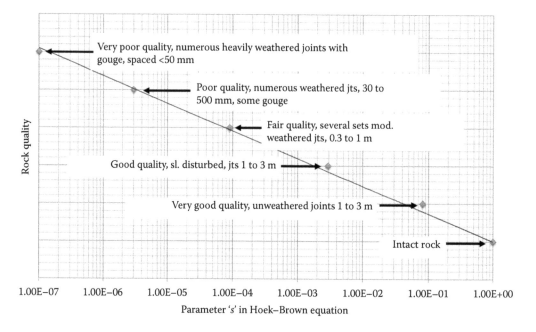

Figure 6.26 Variation of parameter, s, with GSI.

The parameter, *s*, relates to fracture spacing primarily and is defined relative to GSI as in Figure 6.26.

The final empirical variable is *a* which varies with GSI as in Figure 6.27.

6.7.5.1.2 Hoek–Brown strength envelopes

An example of strength envelopes calculated from the Hoek–Brown criterion are presented in Figure 6.28. At Roclab (part of Rocscience web page), similar curves can be generated

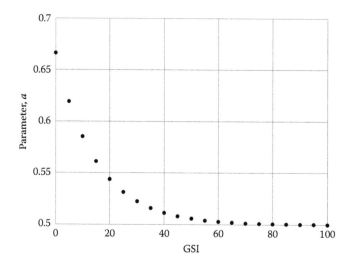

Figure 6.27 Parameter *a*, varying with GSI.

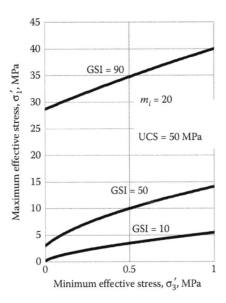

Figure 6.28 Hoek–Brown strength for UCS = 50 MPa, m_i = 20 and various GSI values.

and 'instantaneous' equivalent M–C parameters c' and phi', can be calculated at particular stress levels.

Inevitably, given the difficulty of the problem of dealing with strength of hugely variable, fractured rock masses, where there is no real knowledge of the extent of fractures and stress conditions, the Hoek–Brown criterion has to be seen as a general estimate, but the only real option for such difficult situations. The only viable alternative is numerical simulation where all the components in the geological model can be 'best guessed' and then performance correlated with the numerical outcomes.

6.7.6 Conclusions over applicability of GSI and other classifications

The example given in the previous section highlights the difficulty of the problem, which perhaps should not come as a surprise to geotechnical engineers. As noted elsewhere, while there is always a need in engineering to put numbers to geology (Hoek, 1999), conversely, there are great dangers in trying to reduce complex geology to single numbers be it using Q, RMR or GSI or some weathering classification. Remember again that 85% of tunnel collapses are because of geological reasons rather than some error in assigning parameters (Pells, 2011).

6.8 HYDRAULIC CONDUCTIVITY AND RELATED PARAMETERS

For most rock masses, fracture conductivity dominates. Measurements are made from rising, falling or constant head tests in individual boreholes, by flow tests between boreholes, by measuring flow into underground openings or by using large-scale-pumping well tests as addressed in Chapter 4. Matrix permeability is rarely important for rock engineering, and many rocks have extremely low permeability as intact materials. An exception is in the oil and gas and water industries and potentially for nuclear waste disposal investigations. In these cases, tests can be carried out at the lab scale using gas permeameters and the results can be interpreted linked to a sophisticated porosity measurement. The detailed fabric and directional properties of rock can be investigated using electron microscopy, ultrasound scanning and x-ray analysis but such techniques are rarely applicable to rock-engineering projects either in civil engineering or mining.

Chapter 7

Foundations on rock

It isn't that they can't see the solution. It is that they can't see the problem.

G.K. Chesterton
Scandal of Father Brown

7.1 INTRODUCTION

Foundations on rock can be classified into two categories:

1. Shallow foundations
2. Deep foundation

Shallow foundations are obviously preferable to deep foundations where the situation allows because of ease of investigation and construction costs. As illustrated in Figure 7.1, in the case of foundations for bridges, foundations will be constructed on land where possible, perhaps using a temporary cofferdam to work in the dry. For some bridges large diameter caissons are constructed on land or in dry docks and then floated and towed to position where they are sunk onto prepared rock surfaces. Caissons are also used sometimes on soil surfaces, either sunk into the soil or placed onto the soil that has been prepared and perhaps reinforced by piling (as at the Rion-Antirion bridge).

Where the ground is incapable of carrying the load at shallow depths or where there is a need to design for high lateral loads then deeper foundations may be adopted, generally using piles or barrettes (Figure 7.2). Piles may be driven to bear on rock (as in H-shaped steel piles) or constructed by rotary excavation down to and into rock. A critical part of the design and construction of such bored piles is the degree to which reliance is placed on 'skin friction' from the walls of the bored socket and/or on 'end bearing' of the toe of the pile, resting on the rock.

7.2 DESIGN OF SHALLOW FOUNDATIONS

7.2.1 Building regulations/empirical approaches

Foundation design for rock as for soil is governed by two conditions, the ultimate limit state (ULS), that is, the condition at failure, and the serviceability limit state (SLS) that is an acceptable deformation criteria, related to the use and tolerance of the structure.

The design of rock foundations is often governed by national standards that provide guidance on presumed or allowable bearing pressures for different classes of rocks based on long experience of the behaviour of foundations locally. Sometimes values permitted by authorities

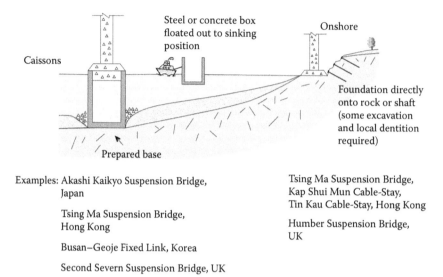

Figure 7.1 Shallow foundation options for major bridge structures and examples of their use.

seem rather low, relative to the compressive strength of intact rock quoted in the regulations, but this may be due to the local experience with some allowance for high degree of fracturing, weathering and workmanship.

Prescriptive* tables are provided for presumed bearing pressures for different classes of rocks generally based on intact rock strength and discontinuity spacing as illustrated in Figure 7.3. Examples of recommended values are presented in Table 7.1. Guidance tables tend to be specified in terms of rock strength using terms like 'weak' and 'strong', so care must be taken that the terminology used at a site in ground investigation is the same as that

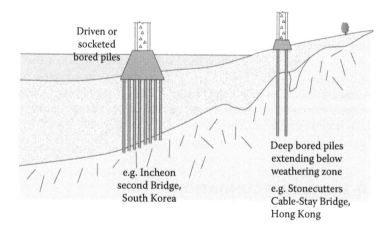

Figure 7.2 Deep pile options for heavy structures.

---

* The terms 'presumed' and 'allowable' are often used interchangeably even in the same document (e.g. Appendix G of Eurocode 7). 'Presumed' values are presented in guidance documents for preliminary design purposes, as generally safe whilst limiting settlement to some nominal value such as 25 mm. Allowable values should strictly be structure-specific, the maximum that can be adopted to satisfy both the ultimate and serviceability state requirements (no failure and acceptable deformation that will vary from structure to structure).

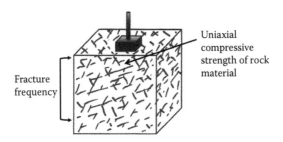

Uniaxial compressive strength of rock material

Fracture frequency

Figure 7.3 Allowable bearing pressures for shallow foundations on rock are generally prescribed according to uniaxial compressive strength of the rock and the closeness of fracturing.

Table 7.1 Examples of presumed bearing pressures for rock masses[a]

Uniaxial compressive strength of intact rock, MPa	Discontinuity spacing, m	Presumed bearing value, MPa
>50	>0.2	10–12.5
	0.06–0.2	5–10
12.5–50	>0.6	8–10
	0.2–0.6	4–8
	0.06–0.2	1.5–4
5–12.5	>0.6	3–5
	0.2–0.6	1.5–3
	0.06–0.2	0.5–1.5
1.25–5	>0.6	0.75–1

Source: After Tomlinson, M.J. 2001. *Foundation Design and Construction*, 7th edition. Prentice Hall, 569pp.

Note: These values are for horizontal foundations not exceeding 3 m width and not more than 10 m long, under vertical static loading; settlement of individual foundation not to exceed 50 mm. Higher values may be appropriate for square foundations up to 3 m wide.

[a] Figures are for igneous rocks, well-cemented sandstone, indurated carbonate mudstone, oolitic and marly limestone, and metamorphic rock including schist/slate with flat-lying cleavage/foliation. For other rock types, see Tomlinson (2001). In particular note that presumed bearing values are lower for 'uncemented mudstone' and shales.

If discontinuities are open, bearing pressure must not exceed half the unconfined compressive strength of the intact rock. For rock masses with tight or incipient discontinuities higher values might be justified up to the UCS of the intact rock.

in the guidance document because strength classifications vary across the world and have changed with time as discussed in Chapter 5.

An alternative recommendation is the chart produced by Peck et al. (1974) and presented in Figure 7.4. This chart is used widely, for example, forming part of the American Association of State Highway and Transportation Officials (AASHTO) guidelines on bridge design (2007) that were specified for the construction of the second Incheon Bridge, in South Korea. The guideline is very simple, relating allowable bearing pressure to RQD% alone. This guidance is specified for 'competent' rock, which elsewhere in the AASHTO guidelines is defined as a rock mass with joint openings <3.2 mm. There is no reference to the strength of intact rock other than a stipulation that the allowable bearing pressure should not exceed the compressive strength of the intact rock.

Eurocode 7 governs practice in Europe and provides guidance for presumed/allowable bearing resistance for rock via a series of charts in Annex G. These are derived from BS 8004 (1986) the data from one of which is reproduced in Figure 7.5.

The charts in the Eurocode/BS are very different from the guidance by Peck et al. For example, for 12.5 MPa rock, if the joint spacing was 200 mm (could still be RQD 100%), the allowable bearing pressure in Europe is only about 3.5 MPa whereas the Peck et al. chart

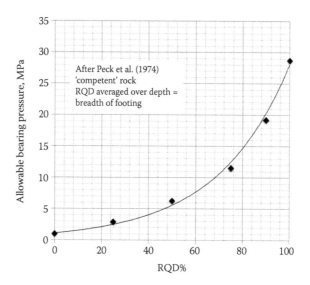

Figure 7.4 Allowable bearing pressure for 'competent' rock. (After Peck, R.B., Hanson, W.E., and Thornburn, T.H. 1974. *Foundation Engineering*, 2nd edition. New York: John Wiley and Sons, 514pp.)

would allow much higher bearing pressures, probably 10 MPa or so assuming an RQD of 70%. This simply demonstrates the empirical nature (and generally inherent conservatism) of these design guides. Basically, provided the site geology is well understood and not too complex then following the national guidelines will probably result in safe foundations that do not settle too much.

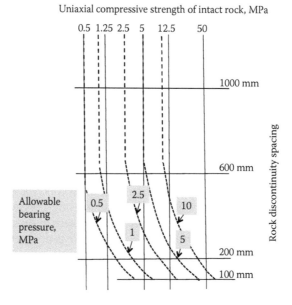

Figure 7.5 Allowable bearing pressure for square pad footings on rock (igneous rocks, oolitic limestone, well-cemented sandstone). There are similar charts for other rock types. For uncemented mudstone and shale, even where joint spacing is very wide (RQD 100%) the allowable stress is only about 50% of UCS. (Data from British Standards Institution. 1986. BS 8004:1986 *Code of Practice for Foundations*, 186pp.)

In Hong Kong, values of presumed bearing pressure for foundations are provided in a Code of Practice (HKBD, 2004) and values for rock are given in Table 7.2. The table is provided as an alternative to 'rational design' but in practice is generally adopted to justify higher loading would probably be quite difficult and involve site as specific testing.

The categories of rock are defined using a range of criteria including weathering grade (defined in Geoguide 3 as set out in Chapter 5), unconfined compressive strength (or point load strength) and core recovery (but not degree of fracturing or RQD). Note that the rock quality has to be proved to a depth of at least 5 m below the founding level.

The values in Table 7.2 are generally lower than those allowable according to the AASHTO guidance or European practice. One of the reasons for this apparent conservatism is to allow intrinsically for some variability in ground conditions when dealing with weathered rocks as is the normal situation in Hong Kong. A problem is that Table 7.2 has a number of ambiguities allowing different interpretations. For example the term '85% of the grade' might be interpreted as meaning that provided 85% of the core is sufficiently strong, then the nature of the other 15% (or non-recovery) does not matter and this is probably the way the guidelines are intended. In practice supervising engineers interpret this as meaning that all recovered core needs to be of the strength specified and force founding levels deeper than really necessary. Confusingly, for 1(c) rock, the grade is specified as Grade III or better (UCS >12.5 MPa) whereas there is an additional specification for UCS >25 MPa for which there is no simple recognised index test that can be used in the field (see Chapter 5). These mixed messages lead to practical problems especially for inexperienced site supervisory staff.

Table 7.2 Presumed allowable vertical bearing pressure under foundations on horizontal ground

Category	Description of rock or soil	Presumed allowable bearing pressure (MPa)
I(a)	Rock (granite and volcanic): Fresh strong to very strong rock of material weathering Grade I, with 100% total core recovery and no weathered joints, and minimum uniaxial compressive strength of rock material (UCS) not less than 75 MPa (equivalent point load index strength PLI_{50} not less than 3 MPa)	10.0
I(b)	Fresh to slightly decomposed strong rock of material weathering Grade II or better, with a total core recovery of more than 95% of the grade and minimum uniaxial compressive strength of rock material (UCS) not less than 50 MPa (equivalent point load index strength PLI_{50} not less than 2 MPa)	7.5
I(c)	Slightly to moderately decomposed moderately strong rock of material weathering Grade III or better, with a total core recovery of more than 85% of the grade and minimum uniaxial compressive strength of rock material (UCS) not less than 25 MPa (equivalent point load index strength PLI_{50} not less than 1 MPa)	5.0
I(d)	Moderately decomposed, moderately strong to moderately weak rock of material weathering grade better than IV, with a total core recovery of more than 50% of the grade	3.0

Source: After Hong Kong Buildings Department (HKBD). 2004. *Code of Practice for Foundations*. The Government of the Hong Kong Special Administrative Region, 67pp.

Note:

1. Minimum socket depth along the pile perimeter is 0.5 m for categories I(a) and I(b), and 0.3 m for categories I(c) and I(d).
2. Total core recovery is the percentage ratio of rock recovered (whether solid intact with no full diameter, or non-intact) to the length of 1.5 m core run and should be proved to a depth of at least 5 m into the specified category of rock.
3. Point load strength index of rock quoted in the table is the equivalent value for 50 mm diameter cores.

This approach might seem to lead to robust design but the danger is that the designer and the owner become complacent, and if the ground model is wrong, taking foundations deeper might not compensate. Furthermore, the time and effort that it takes to excavate good quality rock can lead to difficulties for the contractor and subsequent claims. In Hong Kong several high-rise buildings had to be demolished before they were occupied because the foundations were constructed inadequately partly because contractors acted fraudulently to avoid constructing foundations to the levels required by the design engineer (Hencher et al., 2005).

7.2.2 Settlement of surface foundations on rock

Generally for good quality rock, settlement will be small but may be significant on weaker and weathered rock. This may occur where there are particularly weak and potentially compressible zones or layers within the rock mass carrying the load (typically to a depth of about 1.5 times the width of the foundation).

Methods of predicting elastic moduli, based on rock mass rating and calculated with reference to GSI are presented in Chapter 6.

Deformation modulus for the rock mass can also be estimated with reference to uniaxial compression of the intact rock, and degree of fracturing from the following equation:

$$E_m = jM_r\sigma_c \tag{7.1}$$

where E_m is the deformation modulus for the rock mass (MPa); M_r is the ratio between the deformation modulus (E) and compressive strength of the intact rock σ_c, and j is a mass factor related to the degree of fracturing. Hobbs (1974) presented M_r values for different rock types as reproduced in Table 7.3.

The 'mass factor', j values are given in Table 7.4. These values were used in preparing the presumed bearing pressure charts presented in Tomlinson (2001) and in the British Standard (BSI, 1986) and now in Eurocode 7.

Example: For a foundation on Grade III granite with $\sigma_c = 25$ MPa and RQD 55%,

$$E_m = 0.3 \times 300 \times 25 = 2.5 \text{ GPa}$$

This value might then be used in rational design using appropriate elastic theory.

Table 7.3 Modulus ratios

Group	Rock types	Modulus ratio, M_r
Group 1	'Pure' limestone and dolomite Carbonate sandstone of low porosity	600
Group 2	Igneous Oolite and marly limestone Well-cemented sandstone Indurated carbonate mudstone Metamorphic rock including slate and schist (shallow dipping cleavage)	300
Group 3	Very marly limestone Poorly cemented sandstone Cemented mudstone and shale Slate and schist (steeply dipping cleavage)	150
Group 4	Uncemented mudstone and shale	75

Source: After Hobbs, N.B. 1974. General report and state-of-the-art-review. In *Proceedings of a Conference on Settlement of Structures*, Pentech Press, Cambridge, 579–609.

Table 7.4 Rock mass factors

Rock quality	RQD%	Fracture frequency per metre	Mass factor
Very poor	0–25	15	0.2
Poor	25–50	8–15	0.2
Fair	50–75	5–8	0.2–0.5
Good	75–90	1–5	0.5–0.8
Excellent	90–100	1	0.8–1.0

Source: After Hobbs, N.B. 1974. General report and state-of-the-art-review. In *Proceedings of a Conference on Settlement of Structures*, Pentech Press, Cambridge, 579–609.

7.2.3 Rational design

7.2.3.1 Calculation of allowable bearing pressure

The settlement of a rigid foundation or average settlement of a flexible foundation at the surface of an isotropic elastic half-space (not necessarily a particularly realistic representation of geology) is given by Equation 7.2:

$$s = (q_a B(1 - v^2)I)/E_m \qquad (7.2)$$

where

s is settlement in mm
q_a is the average bearing pressure on the rock foundation
B is the width or diameter of the foundation
v is Poissons' ratio
E_m is modulus of the rock mass
I is an influence factor depending on the shape of the foundation as in Table 7.5.

Then by combining Equations 7.1 and 7.2 the allowable bearing pressure can be calculated for a given allowable settlement, by the equation

$$q_a = \sigma_c(s/B)jM_r/I \qquad (7.3)$$

So, for example, if the allowable settlement for a structure was 10 mm, and it was proposed to construct a rigid square pad footing of 3 m diameter, for the 25 MPa Grade III granite used earlier for calculating modulus, the allowable bearing pressure (SLS) would be

$$q_a = (25 \times 10/3000 \times 0.3 \times 300)/0.82 \text{ MPa} = 10 \text{ MPa}$$

Table 7.5 Influence factor *I* for foundation analysis

	Circular	Square	Rectangular (length/breadth) 2	5	10
Rigid	0.79	0.82	1.1	1.6	2.0
Flexible	0.85	0.95	1.3	1.8	2.2

Source: After Lysmer, J. and Duncan, J.M. 1969. *Stresses and Deflections in Pavements and Foundations*, 4th edition. Berkeley: Department of Civil Engineering, University of California.

7.3 DIFFICULT SITES

Most sites where foundations are constructed on rock have few difficulties. Problems might be encountered however especially when dealing with complex geology, faulted ground or weathered rock masses as illustrated in Figure 7.6.

1. Where the rock is very variable, for example, where the rock is weathered, geology is complex or good rock at the surface is underlain by weaker and more compressible horizons. Problems may occur where the ground investigation does not provide specific information on the foundation conditions at particular footings.
2. In areas affected by natural subsidence or by underground mining and quarrying.
3. Where there is a potential mechanism for failure or deformation, for example, where a structure is built adjacent to a slope.

Each of these situations is discussed in more detail below.

7.3.1 Foundations on variable and complex rocks

Difficulties may arise in constructing foundations where the rock is of variable quality particularly where the design is based on widely spaced boreholes that may have preferentially sampled good quality rock. As ever there is need to check design assumptions during construction, which would include visual inspections and testing for surface footings.

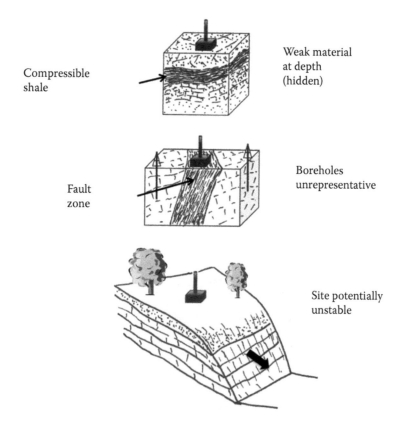

Compressible
shale

Weak material
at depth
(hidden)

Fault
zone

Boreholes
unrepresentative

Site potentially
unstable

Figure 7.6 Potentially problematic sites for shallow foundations on rock.

Figure 7.7 Construction of buildings at Kornhill, Hong Kong. A major fault associated with deep weathering runs along the valley and necessitated caisson foundations to be taken much deeper than anticipated.

Material that is weaker than required for the design would normally be excavated further and replaced with concrete. In weathered terrains the engineer needs to be aware that rock of good quality might be underlain by poorer quality material and, if critical to design, further investigation will be required. Sowers (1988) reports several case examples of difficulties associated with founding on weathered rock. These include founding diaphragm walls too shallowly where hard bands were underlain by weaker material, which resulted in failure, by collapse of weathered materials as a result of groundwater lowering and heaving of weathered rock rich in smectite minerals and sliding on weak horizons through saprolite.

The author is familiar with one case where large pad footings for a power plant were constructed on very poor quality, faulted rock whereas the preliminary design was based on the assumption of better rock quality as interpreted from widely spaced boreholes, trial pits and field observations. There was no investigation at the sites of the pads that failed and no adequate check during construction that the ground conditions at the pad footings matched the design assumption. Severe settlement occurred as the steelwork was erected and this necessitated dismantling the works, redesign and reconstruction adopting deep bored piles instead of surface footings.

In a similar case in Hong Kong, a fault with deep weathering was encountered unexpectedly and required founding level to be dropped by 20 m using caissons rather than the originally planned pad footings (Figure 7.7).

7.3.2 Dissolution, piping and underground openings

Whereas the main considerations for foundations are usually settlement and possibly bearing capacity for weak rocks, in weathered rock there may be other factors to consider. At the site illustrated in Figure 7.8, three complex raft-like foundation structures, 60 m long were founded predominantly on highly and completely weathered granite. Following construction of the foundations, these were undermined due to groundwater piping. This was brought about by high rainfall, flooding and poor drainage provisions so that the weathered granite was weakened and eroded. The foundations settled and tilted necessitating costly remedial works.

Figure 7.8 Raft foundations for switchgear at a nuclear power station site, South Korea.

There are regular dramatic news reports worldwide where sinkholes suddenly open up and damage buildings or swallow vehicles. Such problems are usually associated with soluble rocks such as limestone and dolomite including chalk, and salts that may be interbedded with otherwise inert rocks like sandstone and mudstone. Dissolution by groundwater leads to voids that can cause problems during construction with collapse of borings, potential for environmental pollution and difficulties in concreting. Voids may migrate to the ground surface causing dangerous holes and damaged infrastructure. The main way to avoid this is through desk study to establish the likelihood of a hazard and then to investigate using drilling and possibly geophysics. Natural voids are usually associated with particular geological situations and a history of occurrence so that they are predictable and hazardous areas can be avoided (e.g., Culshaw and Waltham, 1987; Cooper and Waltham, 1999; Edmonds, 2008).

More rarely collapse can occur in rock types that one would not usually associate with this type of failure. For example a failure occurred in a hillside in Hong Kong, underlain by granite (Hencher et al., 2008). Surface drainage had migrated underground to exploit disturbed ground associated with an ancient landslide (probably thousands of years ago). The subsurface streams caused subsurface erosion that eventually 'chimneyed' up to the ground surface (Figure 7.9). The cave system through the hillside was large enough to walk through. On researching this case, similar examples were found in weathered rocks, mainly in the United States, where the term 'pseudokarst' has been proposed. Such conditions are extremely difficult to predict unless similar incidents have occurred previously nearby.

Foundation failure is also caused by the collapse of underground workings such as quarries and mines. These can usually be avoided through good desk study and geomorphological interpretation of a landscape. There are usually histories of mining in an area and, if not, then careful interpretation of topographic features such as old bell pits, and an awareness of the local economic geology (e.g., presence of coal or other valuable minerals at depth), should provide a warning. Where voids, either natural or man-made, are suspected at a site then this can be investigated using probe holes (not necessarily cored), down hole cameras, and sonar and radar devices as well as micro-gravity geophysics. Old workings often have a magnetic signature that can be picked up using geomagnetic surveys and voids can

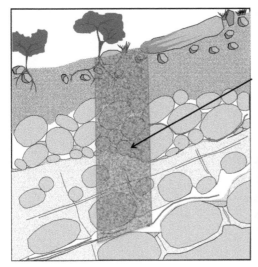
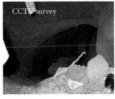

Cavity due to undermining by sub-surface streams in weathered granite profile

CCTV survey

Figure 7.9 Schematic representation of cavity collapse in weathered granite terrain caused by sub-surface stream erosion at a depth of about 10 m, Hong Kong.

be identified using resistivity surveys. Where economically viable, voids can be backfilled or otherwise engineered although with natural sinkholes the problem might be difficult because of the potential for progressive deterioration.

The need for proper desk study is really entrenched in good engineering geological practice but such practice is not always followed. In one case, familiar to the author, one long section of high-speed rail track in South Korea had to be relocated several kilometres from its planned course after tunnellers blasted into old mine workings unexpectedly. Embarrassingly, the old buildings and lifting gear associated with the mine workings were still there to be seen and visited on the hillside above, so no excuses.

7.3.3 Incorrect ground model

Foundation failures can often be attributed to inadequate ground investigation that leads to incorrect interpretation of the ground conditions and then adoption of 'over-aggressive' parameters in analysis. The author is familiar with major cases in Malaysia and Singapore where ground conditions were assumed to be far better than they turned out to be. In each case the foundations for the buildings needed to be rectified through the installation of additional piles and jacking up. In the Malaysian case the depth and severity of weathering was far worse than assumed for the design. In the Singapore case a unit of clay with boulders (the Fort Canning Boulder Bed) had been wrongly interpreted as a weathered sedimentary rock sequence of the Jurong Formation. Hencher (2012a) describes other examples where incorrect interpretations of site investigations had major consequence for the construction and performance of structures.

7.3.4 Pre-existing geological mechanism

When constructing a structure on sloping ground care must be taken to identify any pre-existing mechanism on which failure might occur, for example where the foundations are built above a rock slope and where there are adverse discontinuities as in Figures 7.10 and 7.11.

Figure 7.10 Building constructed on columns founded on rock mass containing extensive sheeting joints. Near Stanley, Hong Kong.

A famous case of foundation failure in rock caused by sliding on discontinuities is Malpasset arch dam, France. The foundations in one abutment failed shortly after the reservoir level began to be lowered to prevent overtopping (Maurenbrecher, 2008). The failure occurred on a pre-existing, daylighting wedge of rock, bounded by three major discontinuities (two of which were probably faults, possibly brecciated and infilled with clay-rich gouge), which slid out from the abutment. According to Post and Bonazzi (1987) severe weathering in the abutment and a low rock mass modulus contributed to the failure due to

Figure 7.11 Structural columns founded on rock containing adversely oriented discontinuities. Kwun Tong, Hong Kong.

preferential loading in the abutment that failed. Londe (1987) noted that the foundation had a high degree of fracturing at all scales, with high deformability. He also suggested that the permeability of the fractured rock was highly sensitive to stress.

In the Malpasset dam case, the wedge was formed by rare, discrete discontinuities. There is always a possibility that such important features might be overshadowed by more numerous but relatively unimportant discontinuity data following routine stereographic interpretation and analysis (Hencher, 1985). Clearly for such important structures imposing high stress onto the rock foundations it is imperative to develop a correct structural model prior to analysis and this needs a geological approach rather than statistical. When constructing on rock where there is any potential for sliding failure, any adverse discontinuities need to be investigated and characterised individually as discussed in Chapter 5, and shear strength parameters determined as explained in Chapters 6 and 8. Then the situation requires proper design as for a rock slope, allowing for the forces imposed by the structure. In the case of a dam, particular care needs to be taken of potential water pressure in the discontinuities. If the FoS is too low, or there are residual uncertainties with the ground model, then foundations might be taken deeper, below the level of potential sliding planes. Alternatively the rock mass may need to be strengthened or supported with dowels or anchors and probably drainage measures installed if the water table is potentially high.

Another hazard for foundations on sloping ground is that the site might be broadly unstable due to previous landsliding. This possibility needs to be assessed by field mapping and probably by air photograph interpretation of geomorphological features.

7.4 DEEP FOUNDATIONS

7.4.1 Driven piles to rock

Piles may be driven to found on rockhead essentially as end-bearing structures. The common pile types are steel H piles, steel tubes and reinforced concrete. Piles are usually driven 'to set' whereby the penetration per blow by a hammer of known energy is specified, allowing for any temporary elastic rebound of the pile. Care must be taken not to overdrive piles as this can cause a pile to break or be damaged, especially when striking a sharp rockhead interface and where that interface is sloping.

7.4.2 Bored piles to rock

Rational design involves proper ground investigation, testing and selection of parameters, and analysis of the potential design solution based on proper engineering principles. Usually bored piles are cased through overlying soil horizons and then a socket cut into the underlying rock prior to concreting using a 'tremie' pipe to send the concrete to the toe of the file; the concrete then rises to displace any liquids (typically bentonite or water).

The key issues are how much 'skin friction' to allow for in the rock socket and how much 'end bearing' to allow for from the horizontal interface between the concrete and the underlying rock. Both of these factors are difficult to estimate realistically and are dependent on the way that the pile is formed. There are additional factors in that skin resistance in the shaft will be mobilised long before any load is transmitted to the toe. For major projects it makes good sense to carry out some preliminary pile tests that are instrumented along the pile so that realistic design parameters can be ascertained.

For piles in weak rock, potential settlement may be an issue and pile tests can give confidence regarding a design. Alternatively there are numerous pile design methods that might be adopted but these in turn depend on the selection of parameters including mass modulus. Gannon et al. (1999) present a number of design examples for various weak rock types that demonstrate the uncertainties and need for engineering judgement. For example for piles in mudstone they show how using different approaches the rock mass stiffness might be variously estimated as (a) 525–5,541 MPa (based on RMR), (b) 430 MPa, (c) 200 MPa, (d) 2,500 MPa (based on pressuremeter tests) or (e) 7,560–17,992 MPa (cross-hole geophysics). The authors chose in this example to use a design value of 1,000 MPa with a check using 500 MPa.

7.4.2.1 Skin friction

It is evident that socket resistance will depend on the roughness of a socket as well as the strength of the rock itself, but establishing the roughness of a socket is not easy. There are numerous published ways for estimating adhesion factors and many of these are reviewed and discussed for piles socketed in weak rock by Gannon et al. (1999).

GEO (2006) suggests that mobilised shaft resistance might be calculated from the following equation:

$$\tau_s = 0.2\sigma_c^{0.5} \qquad (7.4)$$

where τ_s is the mobilised shaft resistance and σ_c is the uniaxial compressive strength of the rock. GEO notes that it is Hong Kong practice to limit the contribution of shaft resistance in a rock socket to a length twice the pile diameter or 6 m, whichever is less. Alternative guidance is given by the Hong Kong Buildings Department for the allowable bond between rock and concrete as presented in Table 7.6. In some cases, designers will not allow for any skin friction and design solely on the basis of end-bearing.

7.4.2.2 End bearing

GEO (2006) suggests that the values given in Table 7.2 for foundations generally can also be used for calculating end bearing of piles. In practice, great care must be taken that the base of the bored pile excavation is cleaned carefully. There are many cases where, even after cleaning, sediment falls to the base of the pile before concreting, resulting in 'soft toes'. In Hong Kong it is routine to drill through bored piles (using a tube, incorporated into the reinforcement cage), after construction, to check that the toe conditions are as required for the design. If not, remedial measures will need to be taken such as cleaning and pressure grouting.

7.4.3 Examples

An example of the design of deep foundations is for the second Incheon Bridge (Figure 7.12) which used a rational process rather than presumed parameters. Figure 7.13 shows

Table 7.6 Presumed allowable bond or friction between rock and concrete for piles

Category of rock as defined in Table 7.2	Presumed allowable bond or friction between rock and concrete or grout for piles (kPa)	
	Under compression or transient tension	Under permanent tension
I(c) or better	700	350
I(d)	300	150

Source: After Hong Kong Buildings Department (HKBD). 2004. Code of Practice for Foundations. The Government of the Hong Kong Special Administrative Region, 67pp.

Figure 7.12 Incheon second Crossing under construction 2010 (longest span cable – stay bridges in South Korea).

the core boxes for one of the boreholes below one of the major footings. The plastic boxes are samples from SPT tests. There is then a metre or so of strongly discoloured and weak material (Grade IV/III). The design took account of skin friction within these weaker materials as well as end bearing at the toes of the piles. Piles were founded typically about 3 m into jointed Grade II rock. The design was carried out with confidence provided by a high-quality testing programme. This included the testing of 3-m diameter piles with loads in excess of 30,000 tonnes provided by Osterberg cells built into the test piles (Cho et al., 2009b).

By comparison, the foundations for Stonecutters Bridge in Hong Kong were designed to be end bearing with no allowance for shaft resistance. The design called for the piles to be founded on 'sound bedrock', which resulted in some of the large diameter bored piles being excavated to depths in excess of 100 m (Tapley et al., 2006).

7.5 CASE EXAMPLE: THE IZMIT BAY CROSSING: ROCK ENGINEERING FOR THE ANCHORAGE OF A MAJOR SUSPENSION BRIDGE

7.5.1 Introduction

A practical example of the impact of rock mechanics on a large and complex civil engineering project involving rock foundations and anchorages can be found in some aspects of the design and construction of the Izmit Bay Crossing. The Izmit Bay Crossing is a very long

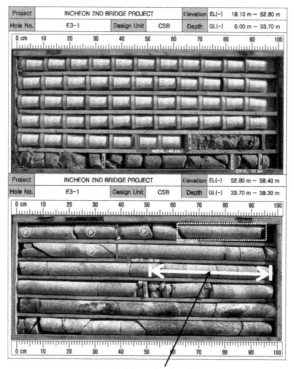

Typical founding level for piles

Figure 7.13 Samples and core from beneath one of the piled foundations for Incheon second Crossing. Plastic boxes are SPT test samples. P indicates samples tested by point load. Sample with white dotted box surround was tested for UCS. (From Hencher, S.R. 2009. *Proceedings of International Commemorative Symposium for the Incheon Bridge*, ROTREX 2009, Incheon, South Korea, 456–465.)

suspension bridge, which is being constructed, at the time of writing, in Turkey. The location is shown in Figure 7.14.

Izmit Bay is the easternmost part of the Sea of Marmara. The bridge will be part of the BOT (Build-Operate-Transfer) project to construct and operate a new motorway connecting Gebze with Bursa and eventually Izmir, in Western Anatolia and will cross Izmit Bay by connecting the Diliskelesi Peninsula on the north shore with the Hersek Peninsula on the south shore. A major technical challenge to the safe realisation of the bridge is the very high earthquake loading that the structure is likely to experience during its life. This is due to the close proximity of the North Anatolian Fault, which crosses the Hersek Peninsula just south of the bridge and which produced the 1999 earthquakes of nearby Izmit and Duzce (of magnitudes Mw = 7.4 and Mw = 7.2, respectively).

7.5.2 Design concept

The basic design of the bridge is a suspension bridge as illustrated in Figure 7.15.

Suspension bridges can span remarkable distances by exploiting the very high tensile resistance of metal cables. A key structural function is assigned to the pair of steel cables that are supported by tall pylons (or towers) and anchored at the two opposite ends of the bridge, in the regions often referred to as 'shoulders'. The deck is then connected to the cables using vertical suspenders. In terms of foundations the main loading components are represented by

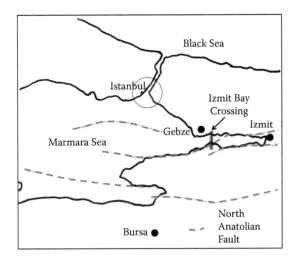

Figure 7.14 Location of Izmit Bay Crossing Suspension Bridge with active segments of the North Anatolian Fault.

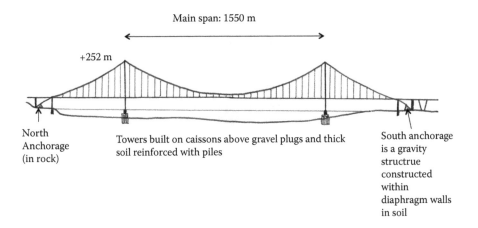

Figure 7.15 Izmit Bay Crossing. General design concept.

high compression of the ground under the pylons and high tension at the anchorages, which are large concrete bodies connected to the cables and embedded in the ground. Anchorages resist the large pull-out forces exerted by the cables thanks to a mix of own weight and stress exchange with the soil or rock in which they are embedded. The relative contribution of these two factors varies from case to case and is influenced by the local geology.

For the Izmit Bay Crossing, the suspended part of the deck, which will provide three lanes of traffic in each direction, has a total length in excess of 2.5 km; the central span between the two 290 m-tall pylons is slightly more than 1.5 km long. The depth of water is about 40 metres. The Crossing will be the fourth longest suspension bridge by central span upon completion.

7.5.3 Seismic issues

As it is impractical to construct the bridge to resist extreme earthquakes without experiencing any damage, a performance-based design was adopted. In performance-based design it is accepted that the structure can respond with a lower performance level to less likely, larger loads. For the Izmit Bay Bridge, three reference seismic events were fixed as (i) a functionality evaluation earthquake (FEE) which has a mean return period of 150 years, (ii) a safety evaluation earthquake (SEE) which has a mean return period of 1000 years and (iii) a no-collapse earthquake (NCE) which has a mean return period of 2500 years. The bridge has been designed to experience an FEE without suffering significant damage and an SEE with damage that is easily and quickly repairable. In the case of the NCE extensive damage is acceptable provided the structure does not collapse.

To resist earthquake vibrations the towers are designed to rest on foundations that allow controlled sliding on a bed of gravel (a 'gravel fuse') during major shaking so that the energy transmitted to the superstructure is reduced. Such an arrangement, which also includes strengthening of the natural soil with driven steel tubes below the gravel bed where necessary, was first adopted for the cable-stay, Rion-Antirion Bridge, in Greece (Combault et al., 2000; Pecker, 2004).

7.5.4 Rock engineering for the North Anchorage

A sketch of the North Anchorage of the Izmit Bay Crossing is given in Figure 7.16. The main anchorage block is approximately 33 m long, 22m deep and 50 m wide. Two legs depart from the main block and terminate with a front pad each. These hold the cable at a convenient angle while providing an efficient structural arrangement. At the main block, the cable is separated into its individual strands, each with its own connection to the mass of reinforced concrete. The transition from cable to individual strands takes place at the splay saddle, which is located near the vertex of the slanted elements that constitute the legs. The deviation imparted by the splay saddle to the strands, which form a larger angle with the horizontal direction than the incoming cable, is beneficial to the anchorage stability; the arrangement results in a large compressive force in the front element of each leg. Provided the bearing capacity of the ground is sufficient, the pads, which are prevented from sliding by a proportionally large frictional resistance, become the location around which the entire anchorage structure tends to rotate. The safety of the system is then achieved by making sure that the stabilising moment (weight of the main block multiplied by the distance between its centre of gravity and the axis of rotation at the front pads) is larger than the overturning moment (cables pull multiplied by the distance between the splay saddle and the axis of rotation at the pads).

However, checking the capacity of the rock mass in front of the main block is also essential. In particular, for a jointed rock mass, the designer must verify that the cable forces, directed upward and seawards, do not have the possibility to generate translational failures along weak discontinuities.

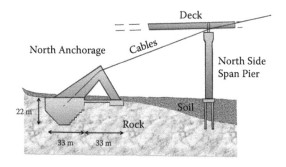

Figure 7.16 Sketch of arrangement for North Anchorage for Izmit Bay Crossing.

7.5.4.1 Preliminary ground model

At the northern end of the bridge, there was a large flat area with a cut rock cliff and higher area to the rear. There were therefore two preliminary options for where the anchorage might be sited – one in the flat-lying area and the other in the hillside. A first set of 10 boreholes was put down to explore these options. A section showing interpreted geological conditions based on boreholes and site mapping is presented in Figure 7.17.

Following this work it was decided to select the site closest to the sea below the flat area covered with fill from previous quarrying activities. There were two main issues to be resolved. Some of the boreholes had poor recovery and gave indications that there was at least one fault across the site. Although the faults in this peninsular are regarded as inactive, the geometry of fault(s) was unknown as was the quality of rock, which would affect the design parameters. Furthermore, from the tender investigation it was known that three sets of joints were present in the rock mass in which the anchorage was to be embedded. One of these was bedding parallel and dipping shallowly to the north. The Palaeozoic limestone beneath the anchorage footprint was described in the borehole logs as inter-bedded with marl and there was therefore concern over the shear strength (fundamental frictional properties and persistence) of these bedding parallel discontinuities. From the analysis it was considered that the other sets of joints could act as release surfaces allowing development of a wedge failure with sliding predominantly on bedding-parallel discontinuities (see Figure 7.18). This issue was important because it might limit the size of concrete structure (whether or not the rock mass would contribute to the tensile resistance necessary to anchor the suspension bridge cables, especially during an earthquake).

Estimates of the geometrical and mechanical properties of the joint sets were available but with a large uncertainty.

7.5.4.2 Stage 2 investigations

Following initial review of the information from the preliminary investigation and its implications, a joint site visit was carried out by the design team (COWI), their specialist geotechnical advisors (GeoDesign) and the design checkers (Halcrow, now CH2M[HILL has gone]).

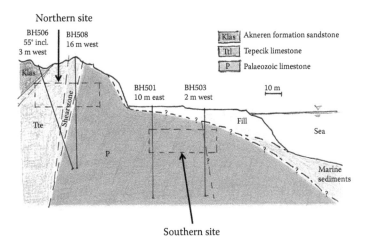

Figure 7.17 Preliminary ground model for Northern Anchorage based on field mapping of exposures and ten preliminary boreholes.

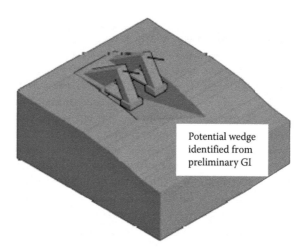

Potential wedge
identified from
preliminary GI

Figure 7.18 Isometric model of anchorage with possible wedge formed from discontinuities identified from first stage ground investigation.

The inspection of rock exposures confirmed the orientation of the joints and provided a better understanding of their persistence, large-scale undulation and small-scale roughness. The inspection of the rock cores clarified the nature of zones of drilling-induced breakage, which were reported in the borehole logs but had been difficult to interpret on the basis of core photographs. The direct access to the cores confirmed that such discontinuities were unlikely to be naturally open in the undisturbed rock mass. It also suggested that the fears over adversely oriented marl layers and potentially clay-infilled discontinuities were unlikely to be realised. Samples were selected for direct shear testing to allow an improved assessment of shear strength. There remained uncertainty of the extent and nature of faults, which might impact the design as well as issues such as hydraulic conductivity and the need for pumping during excavation below sea level.

7.5.4.3 Stage 3 investigations

A second and a third targeted ground investigations were therefore carried out to improve the understanding of the geology in the rock mass in front of the anchorage, which is key to its performance under loading. The second ground investigation (NA-100 series) consisted of four vertical boreholes (Figure 7.19a). Borehole NA-101 confirmed good rock quality in the central part of the main block and assured that fault F1, if present, must have modest thickness. NA-102 confirmed the presence of fault F2, as did NA-105 and NA-106, which were, however, short holes with the main purpose to define the elevation of bedrock and the thickness of the loose fill in the area. Despite allowing a refinement of the geological model this second campaign still left several open questions regarding the number and thickness of faults at the site.

To help resolve these issues a third ground investigation was commissioned, comprising four borings, three of which were inclined, was specified (Figure 7.19b).

The purpose of the inclined boreholes was to investigate the sub-vertical features such as the joints set which delimits the sides of the potential wedge failure and the faults. The holes were bored at an inclination of 50° (dip) and in the direction perpendicular to the presumed strike of the faults. Boreholes NA-203 and NA-205 substantially confirmed the interpretation proposed after the second ground investigation but added complexity to the picture. Zones of high fracture frequency were encountered where fault F1 was predicted but these were

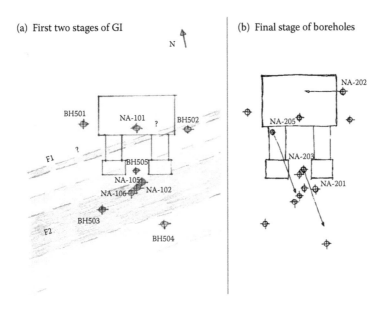

Figure 7.19 (a) First two stages of boreholes. (b) Inclined boreholes to explore the nature of faults and sub-vertical release joints.

associated with the opening up of incipient joints. There were no other indicative signs of faulting such as gouge, slickensiding or mineralisation. The main fault F2 now was interpreted as an anastamosing system of at least two features with better quality rock in between. This picture is consistent with a system of parallel faults associated with a common, large-scale shearing mechanism. A general cross section through the anchorage with orientations of main faults as determined following the third campaign of boreholes is shown in Figure 7.20.

During the final ground investigation (boreholes) major trial trenches were excavated through the fill in the footprint of the anchorage as illustrated in Figure 5.4. These trenches allowed the geological model to be reviewed and checked, particularly with respect to the persistence and nature of bedding-parallel discontinuities and faults in a way that is simply not possible from boreholes.

The final geological interpretation did not necessitate a change of design for the North Anchorage structure and indeed provided great confidence that the design was adequate. It also highlighted the potential constructability problems due to the high permeability of the faulted rock mass in front of the main block. Large water ingress was indeed experienced during excavation and was successfully managed thanks to the level of preparedness

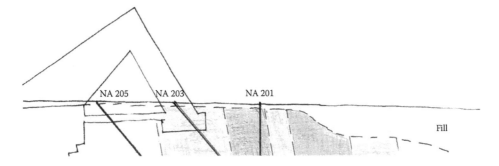

Figure 7.20 Final geological model for the North Anchorage.

Figure 7.21 North Anchorage (close to rock cliff) and North Side Span Pier foundation under construction.

informed by the geological understanding of the site. Figures 7.21 and 7.22 show the anchorage under construction. Note that for this major infrastructure project it was not considered necessary or appropriate to carry out large-scale *in situ* tests such as plate loading tests. It was considered adequate to rely on our knowledge of the ground conditions together with published data to adequately constrain the design parameters (moduli in particular). Shear strength became a non-issue following field characterisation of discontinuities at large scale as exposed in the large trial trenches.

Figure 7.22 North Anchorage under construction, February 2014. Workers show scale.

7.5.5 Conclusions

The design process of the North Anchorage of the Izmit Bay Crossing provides a good example of how the conceptual ground model should evolve and be refined, potentially through several phases of ground investigation and interpretation, until the risk posed by the geological uncertainty is satisfactorily low. This iterative nature of the geological and geotechnical investigation may pose some challenges, partly because the procurement of major infrastructures is complex and involves many players which may be responsible for the definition of the geological model (and its translation in design calculations and construction specifications) at different stages of the process. The iterative approach, moreover, intrinsically conflicts with the pre-defined sequence of activities, which is desirable in terms of project management. To achieve, before construction, the necessary level of geological understanding, which is instrumental to a safe and successful project, it is paramount that competent geological and rock mechanics specialists are involved, by each of the organisations involved, as early as possible.

Two additional ground investigations were necessary, after the tender ground investigation, to produce a satisfactory geological model for design. The use of inclined boreholes was vital to allow a reasonable understanding of the sub-vertical features present at the site. No realistic sampling of sub-vertical joints and determination of faults thickness would have been possible without this technique. The trial trenches allowed the key geological structures to be examined and characterised in a much better way than was possible from boreholes alone.

7.6 SITE FORMATION

Site formation often needs rock to be blasted or mechanically excavated prior to removal and factors to be considered include time, cost and safety restrictions, especially on blasting. Aspects of rock excavation are dealt with in Chapter 8 (rock slope formation) and these are generally applicable to site formation as well.

Chapter 8

Rock slopes

Believe we're gliding down the highway
When in fact we're slip slidin' away

Paul Simon

8.1 CIVIL ENGINEERING

8.1.1 Introduction

Engineering works such as the construction of highways or site development in hilly terrain necessitates the cutting of slopes (Figure 8.1). These need to be excavated safely and economically, designed not to fail and to avoid any ongoing risk from rockfall or erosion. In addition, rock-slope stability is important to quarrying and open-pit mining; some of the important related issues are addressed here.

Generally for rock-slope design, a Factor of Safety (FoS) is used to express the degree of stability. FoS is the ratio between available shear resistance and the forces driving failure for any particular failure mechanism. Rock slopes are often complex geotechnically, so a FoS approach, in which the designer assesses the many contributing factors with respect to how well each is characterised and understood, makes good sense. A partial-factors approach, as used for soils in EC7 is too simplistic, and implies a level of knowledge of forces, geological structure, hydrogeological conditions and, especially, origins of shear strength, that is generally unwarranted. There is a move towards adopting a statistical approach to rock engineering within EC7 but in this author's view this is unlikely to be generally applicable due to the complexity of potential failure mechanisms in rock slopes that necessitate a deterministic and judgemental approach. Generally for rock slope design, where ground conditions are well understood, and consequence of failure low, a FoS of 1.2 is adopted. Where the potential consequence is greater, then a FoS of 1.4 is adopted. However, it must be emphasised that adoption of a high FoS does not necessarily provide any protection, where one is designing on the basis of an incorrect ground model.

A FoS approach provides a measure of the ultimate stability, generally through limit equilibrium analysis. Some numerical software, notably FLAC and UDEC, allow rock masses to be modelled so that deformation in the slope can be expressed visually and numerically, whether or not the slope eventually fails. These also provide a FoS similar to that produced by limit equilibrium software.

An initial consideration is whether the potential failure mechanism will involve sliding on discontinuities. For the various geotechnical scenarios in Figure 8.2, there are really two main options for analysis:

1. For scenarios 1 to 3, analyse using isotropic or gradually varying shear strength for each geological unit. This is appropriate for weak rock, and also for masses like

Figure 8.1 Rock slopes under construction in Hong Kong.

heterogeneous, weakly-cemented colluvium, where there is no directional variation in strength, and for Hoek–Brown type highly fractured rock masses.

2. For scenario 4, the failure mechanism is controlled by individual discontinuities or parallel sets. When analysing the slope, it is the strength of the discontinuities and their orientations through the rock mass that are important, rather than the rock itself.

For weak and weathered rock containing discontinuities, it may be necessary to carry out separate analyse for 1) potential failure through intact material and 2) potential failure involving pre-existing structures.

8.1.2 Analysis of slopes in rock that can be treated as isotropic/homogeneous

The analysis of slopes in homogeneous material is achieved, generally, either by using one of the various 'methods of slices' in limit equilibrium software, or by using finite element, or discrete element analysis, and there are many software packages that are available that can be used for this purpose. This is illustrated in Figures 8.3 through 8.5 for a slope in fairly uniform, highly weathered granite, analysed using the method of slices (Morgenstern and Price method) in the software SLIDE, and using FLAC/Slope.

The slope is about 10 m high and standing at about 70°. The rock is weak (Schmidt Hammer rebound value <25). To analyse stability, using limit equilibrium software, the user needs to identify the potential failure surface. Generally, this is done by setting out a grid for the centres of trial slip circles, as illustrated in Figure 8.4. In this case, a wide grid has been specified, and FoS calculated for each trial surface, for adopted shear strength parameters $c' = 10$ kPa and $\phi' = 38°$ (dry slope). It can be seen that the slope would be marginally unstable, with a failure involving the full height of the slope.

In Figure 8.5, analyses are shown for the same slope, but using a slightly higher cohesion value of 25 kPa. The limit equilibrium analysis in Figure 8.5a now indicates a FoS of over 1.3. The analysis for the same slope, using FLAC is shown in Figure 8.5b. The calculated FoS (minimum) is 1.27. FLAC calculates the displacement in the slope that would have occurred during cutting of the slope, and provides 'velocity vectors', even though the slope

Geotechnical scenario	Schematic diagram	Parameters and analysis
1. Weak, uniform rock without adverse discontinuities	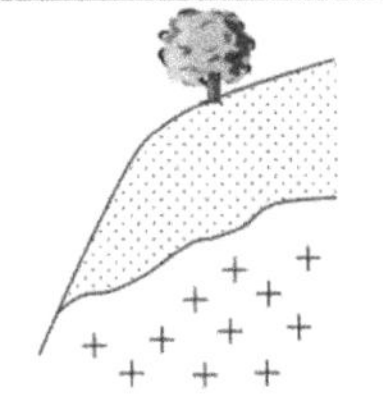	Parameters from laboratory or *in situ* tests taken to be representative of unit. Assume isotropic.
2. Weakened by discontinuities but essentially isotropic	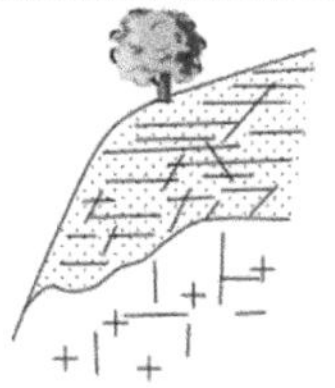	Use Hoek–Brown; assume isotropic or varying strength with stress level.
3. Heterogeneous rock mass (e.g. cemented colluvium or corestone bearing)	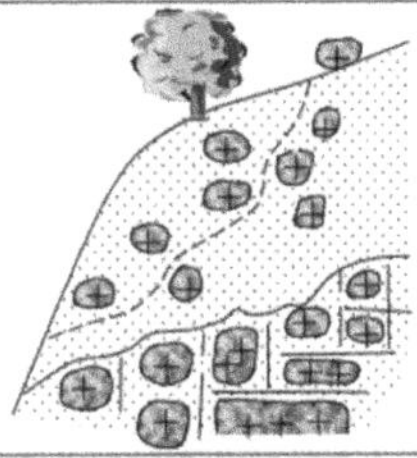	Consideration might be given to influence of strong inclusions. Beware layering or relict joints that might control failure.
4. Rock mass with adverse discontinuities	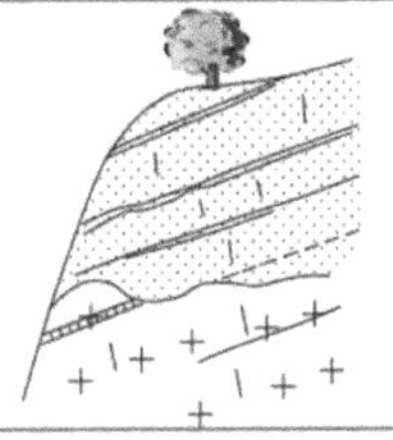	Discontinuity controlled: characterise network including strengths and identify mechanisms.

Figure 8.2 Scenarios for cut-slope analysis. (Modified from Hencher, S.R. and McNicholl, D.P. 1985. *Quarterly Journal of Engineering Geology*, **28**, 253–266.)

is stable. The analysis plot illustrates that there is potential for a deeper failure mechanism than readily apparent from the SLIDE analysis.

All of these software packages are highly sophisticated, allowing modelling of water pressures, dynamic loads and reinforcement, and in some, probability and sensitivity studies can also be carried out.

8.1.3 Analysis of slopes in stronger rock

8.1.3.1 Introduction

In stronger rock slopes, the mechanism of failure will usually be controlled by the presence and orientation of discontinuities relative to the slope geometry. Rock-slope failure mechanisms are split into several types, as illustrated in Figure 8.6:

1. Translational sliding on a single discontinuity or single set, day-lighting from the slope – otherwise known as planar failure

Figure 8.3 Cut slope in highly weathered granite, Mount Nicholson Road, Hong Kong.

2. Wedge failure, where two discontinuities intercept and allow sliding along the line of intersection
3. Toppling failure
4. Rockfall/ravelling erosion
5. Non-structural failure (analysed as isotropic as discussed in Section 8.1.2)

In many large slope failures, there may be various geological features and mechanisms operating together to define the critical geometry, sometimes including failure through intact rock at the toe and other zones, where the discontinuity framework is not adversely oriented. Similarly, even for the same overall geometry, different potential failure mechanisms may be possible, as illustrated in Figure 8.7.

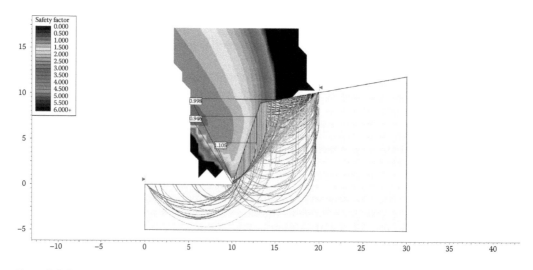

Figure 8.4 Analysis of uniform weathered rock slope using SLIDE (Rocscience) ($c' = 10$ kPa, $\phi' = 38°$).

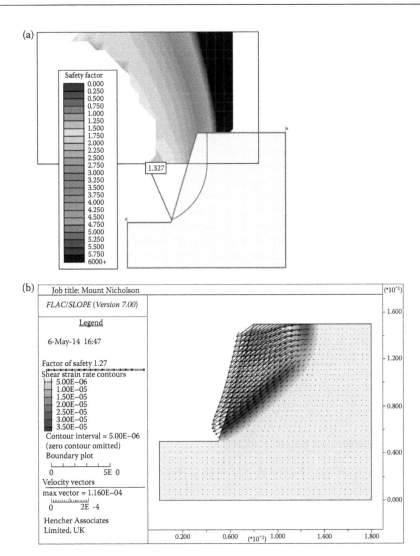

Figure 8.5 Comparison analyses of uniform weathered rock slope using SLIDE (a) and FLAC/Slope (b) ($c' = 25$ kPa, $\phi' = 38°$).

8.1.4 Planar and wedge failure

Planar and wedge failure geometries are illustrated in Figure 8.8. In my experience, planar failures are relatively common; wedges rare. I once was to give a short course on rock-slope stability in Hong Kong, and spent a day hunting unsuccessfully for any example of wedge failure in the many exposed rock slopes.

The methods of analysis for these types of failure are well documented in Hoek and Bray (1974) and Wyllie and Mah (2004). The first question is whether adverse discontinuities are present, and this follows from rock characterisation, determining dips of discontinuities, and then considering the fracture network geometry in conjunction with the existing or proposed slope geometry (Figure 8.9). Basically, if you can 'see' individual joints dipping out of a face, or there are combinations of discontinuities that allow a sliding wedge, then stability needs to be assessed. If there are no adverse structures, then consider analysing it as homogeneous material, as discussed in Section 8.1.2. Determining whether or not these

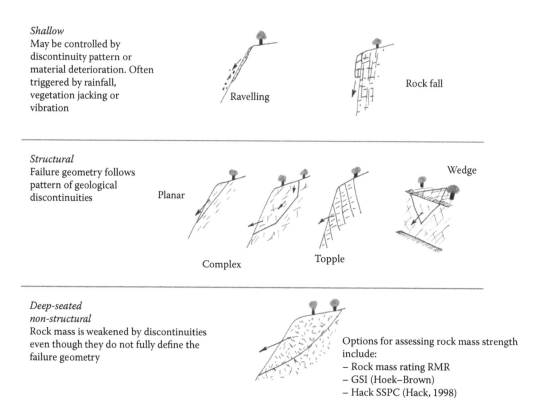

Shallow
May be controlled by discontinuity pattern or material deterioration. Often triggered by rainfall, vegetation jacking or vibration

Ravelling

Rock fall

Structural
Failure geometry follows pattern of geological discontinuities

Planar

Complex

Topple

Wedge

Deep-seated non-structural
Rock mass is weakened by discontinuities even though they do not fully define the failure geometry

Options for assessing rock mass strength include:
– Rock mass rating RMR
– GSI (Hoek–Brown)
– Hack SSPC (Hack, 1998)

Figure 8.6 Modes of failure in rock slopes.

mechanisms can be 'seen' may be simply through direct observation in the field for an existing slope, or by using stereographic projection techniques, either by hand (Chapter 5), or using software such as Dips (Rocscience).

8.1.5 Analysis using stereographic projections

The condition allowing sliding on a single discontinuity is illustrated in Figure 8.10. Great circles are plotted, representing the slope, and then, any discontinuities that might

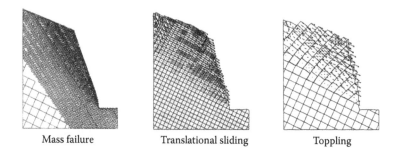

Mass failure   Translational sliding   Toppling

Figure 8.7 In these three UDEC models of a slope, the orientations and strengths of joints were kept constant, and only the fracture spacing varied, giving rise to different styles of failure. (After Hencher, S.R., Liao, Q.H. and Monighan, B. 1996. *Transactions of the Institution of Mining and Metallurgy*, **105**, A37–A47.)

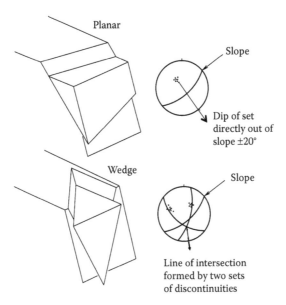

Figure 8.8 Planar and wedge failures.

daylight. Then the 'friction angle' that is considered representative of shear strength is plotted as a full circle, with angle measured in from the circumference (friction acts in any direction). If the central point of the discontinuity-plotting great circle falls within the zone steeper than friction, and shallower than the slope-plotting great circle, then there may be a problem for stability. Generally, 'planar' failure is restricted to where the strike-of-failure plane is fairly close (say ±20°) to that of the slope; otherwise, other joints will need to be involved, and will contribute to the shear resistance. This kind of plotting is rather tedious, and becomes confusing when many joints are shown on the stereogram. It is generally easier to plot all discontinuities as poles to their great circles. This can be done quickly, using a polar stereonet, as explained in Chapter 5. The 'danger zone' for day-lighting discontinuities, dipping at greater than the friction angle, and ±20° of the slope azimuth, is transferred into the polar plot, as illustrated in Figure 8.10. This is done manually, by drawing the slope-plotting great circle on tracing paper, and then rotating

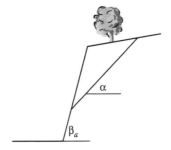

Figure 8.9 Primary test: failure can only occur where the dip of the single plane or line of intersection of a wedge (α) is shallower than the apparent dip of the slope in the direction of potential sliding (β_a), and steeper than the effective angle of frictional resistance (ϕ). Note that this simple test ignores factors such as water pressure, and additional loads such as earthquakes – all it is doing is alerting to the fact that further and more detailed analysis is required.

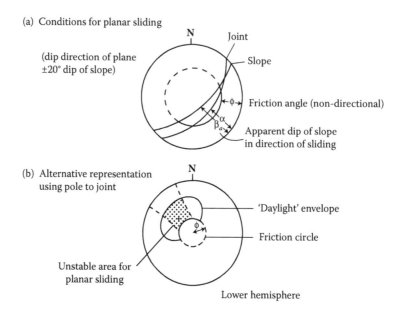

(a) Conditions for planar sliding

(dip direction of plane
±20° dip of slope)

Joint

Slope

φ — Friction angle (non-directional)

Apparent dip of slope
in direction of sliding

(b) Alternative representation
using pole to joint

'Daylight' envelope

Friction circle

Unstable area for
planar sliding

Lower hemisphere

Figure 8.10 Test for planar discontinuity sliding using (a) great-circle representation for plane and slope and (b) using pole plot.

the tracing paper and marking points 90° away from the great circle, along its complete length, to construct a 'daylight envelope'. The friction circle marks the lower boundary of the 'danger zone', this time drawn from the stereonet centre, rather than from the circumference. The ±20° criteria can be drawn as radii from the centre, or along convenient great circles, as shown here. Obviously, if a major set is shown dipping straight out of the slope, and much steeper than the assumed friction angle, then the problems are likely to be far more severe than discontinuities that are more oblique and shallow. It must be emphasised that this type of stereographic analysis is only the first stage – it is not slope-stability analysis per se.

For assessing the potential for wedge failure, discontinuities are usually plotted as poles, primarily to determine what sets are present (Chapter 5). Sets are identified as clusters with mean orientations, and then these poles are redrawn as great circles, as in Figure 8.11.

The 'danger zone' is the same as for planar failures as shown in Figure 8.10a, but without the ±20° restriction. Wedges can also be identified in polar plots and, indeed, every potential wedge intersection from any two planes can be plotted, and then these intersections themselves treated statistically. However, just because intersections are shown on the stereogram (individual data points having been recorded in the field somehow), does not mean that they will form a wedge in reality. Once potential mechanisms have been identified, then it is time to revisit the field data and the photographs, and maybe make another site visit. Toppling as a mechanism is discussed later. On a stereonet, the potential for toppling is identified by a cluster of discontinuity poles that plot between the great circle for the slope and the circumference of the stereogram, again within about ±20° of the slope azimuth. There are other conditions that can be analysed using a stereographic projection, including taking account of the interface angle of friction within the joint set, and this is illustrated in Wyllie and Mah (2004) and other textbooks; but in my opinion, this is not usually warranted because it is rather unrealistic.

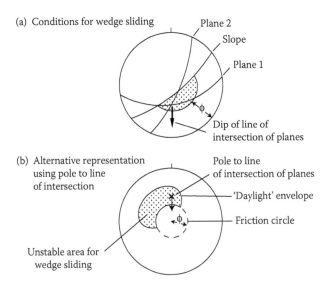

(a) Conditions for wedge sliding

Plane 2

Slope

Plane 1

ϕ

Dip of line of
intersection of planes

(b) Alternative representation
using pole to line
of intersection

Pole to line
of intersection of planes

'Daylight' envelope

ϕ

Friction circle

Unstable area for
wedge sliding

Figure 8.11 Assessing wedge intersection using two great circles to represent set concentrations (or individual planes).

8.1.6 Summary regarding stereographic methods

Table 8.1 provides a list of dos and don'ts regarding the use of stereographic analysis as part of slope-stability assessment.

8.1.7 Detailed analysis for planar failure

The analysis method for planar failure (essentially 2D) is amenable to hand calculation, and allows one to examine the sensitivity of the slope to various assumptions, say, with respect to shear strength and water pressure. Such sensitivity calculations are extremely useful for focussing attention on the nature and importance of the uncertain parameters (which generally includes all of them, apart from the geometry of slope). An example of such analysis is presented below.

8.1.7.1 Introduction

This example is based on an actual slope design, but the calculations are not those used for the real design. They are for illustration only.

The slope is above the spillway for Kishanganga Dam in Kashmir, India. The main slope view in Figure 8.12 is about 75 m in height and the intention is to cut a further 40 m below the toe of the exposed joint, down to the spillway.

8.1.7.2 Geological model

There are a few boreholes. These show that there are joints dipping in a similar manner to the plane exposed in the photograph (about 45°) at depth, but there is insufficient data to characterise the discontinuities, in particular their persistence. For design, it is assumed that there will be persistent joints at depth, with similar characteristics as the exposed plane. Measurements from 3D photogrammetric models indicate a wavelength, for the first-order waviness, of about 20 m, with a first-order dilation (*i*) value of 10°. This means that for some

Table 8.1 Use of stereographic projections for slope analysis

Uses for rock slope stability

1. Measuring angular relationships.
2. Identifying sets of joints.
3. Setting out relationships between discontinuities and slope geometry to establish potential for failure mechanism.
4. Comparing failure geometry with likely sliding angle (expressed as equivalent friction).
5. To show roughness variation on individual discontinuities.

Limitations

1. All discontinuities just shown as single poles, without distinction as to persistence and other properties.
2. Single poles do not represent roughness and waviness variability of individual discontinuities.
3. Spatial relations not shown (wedges might not exist in reality).
4. 'Cohesion' due to impersistence not readily taken into account using effective friction approach.
5. Water pressure and other forces not readily shown.

Common misuses

1. Used as 'stability analysis' rather than as tool to sorting discontinuity data and identifying what further analysis is required.
2. Contouring to define 'sets' without consideration of fundamental geological controls – whether sets 'make sense' with respect to geological history or not.
3. Contouring too few and unrepresentative discontinuity measurements (sampling bias).
4. Contouring hides important discontinuities (e.g., faults that are not part of discontinuity set).

Hints

1. Only contour data to get a better understanding of fracture network.
2. Always check nature of any adverse discontinuities removed by contouring exercise.
3. Critically assess representativeness of data and whether they can be extrapolated.
4. Deal with each geological structural domain separately (identify these first).
5. Once critical discontinuities have been identified, put away stereoplots and analyse properly.

Source: Modified from Hencher, S.R. 1987. *Slope Stability – Geotechnical Engineering and Geomorphology*, John Wiley & Sons Ltd, UK. 145–186.

10 m lengths, the effective dip will be reduced by 10° (from 45° to 35°). For others, it will be increased from 45° to 55°. This is illustrated in Figure 8.13. In this example, the design is considered for the first cut (7 m high) below the exposed plane.

8.1.7.3 Design conditions and parameters

The site is prone to earthquake-shaking and a design acceleration of 0.5 g is adopted, acting in the horizontal direction.* Regarding shear strength, a basic friction angle of 38° plus dilation angle of 7° (from small scale, second-order roughness) is adopted, as discussed in Chapter 6 and Hencher and Richards (2015). The joint is fully persistent, so there is zero cohesion. Unit weight of rock, γ_R, is 27 kN/m³ and for water, γ_W, is 10 kN/m³.

---

* Assuming a pseudo-static horizontal force to represent earthquake-shaking is a gross simplification but, given all the other uncertainties, a reasonable approach. Similarly, there is little point in searching for the most adverse direction to apply the force. In the analysis, the forces of weight and earthquake load could be expressed as a combined vector, reducing the number of calculations. However, separating the forces in the calculations allows the results to be judged rather better.

Figure 8.12 Distant view of slope after initial cleaning-off (and failures).

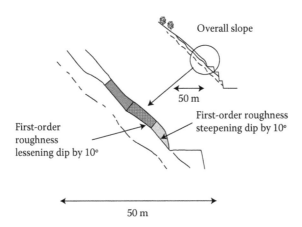

Figure 8.13 General cross section of slope with four cuttings down to spillway level and close-up of geological model assumed for design of topmost rock-cut. It is assumed that a thick slab, undulating over a 20 m wavelength, day-lights in the cut, and that there are cross-joints allowing each 10-m block to slide independently.

8.1.7.4 Factor of safety

A factor of safety of 1.4 is to be adopted, reflecting the importance of the structure and its remoteness, making maintenance and repair difficult, and probably delayed, in the event of a failure.

The potential failing slab is considered as a series of blocks (some steeply dipping, others shallowly, because of the first-order waviness). It is assumed (conservatively) that the lowest day-lighting block that will be exposed in the cut, is on a down warp, i.e., dipping at $45 + 10°$. The lowest two blocks (A and B) are shown in Figure 8.14.

Each of the blocks will be considered below.

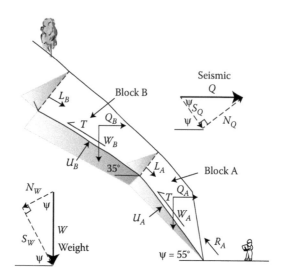

Figure 8.14 Force diagrams for analysis of lower two blocks.

8.1.7.5 Analysis of Block A

There are several forces potentially acting on Block A as follows:

W_A Weight of block
Q_A Horizontal, inertial force due to earthquake-loading
L_A Cleft water force in cross-joint to rear of Block A
U_A Uplift force due to water pressure on shear surface
T Shear strength (from friction and cohesion as appropriate) resisting shearing stresses
R_A The additional resistance that might be required to ensure the required FoS

The analysis is carried out for a 1 m width of the block.

Length of the sliding block, l, is 10 m.
Angle of dip, ψ, is 55°.
Length of cross-joint is about 3.5 m.
Cross-sectional area of Block A is about 26.25 m² so weight of block, $W_A = 26.25 \times$
    27 (γ_R) × 1 kN, i.e., 708.75 kN.
Friction (basic plus second-order roughness) ϕ_b is 45°. Cohesion is zero.

8.1.7.5.1 Dry case, no earthquake

$$\text{FoS} = \frac{\text{Available shear resistance}}{\text{Shearing forces}}$$

By the 'triangle of forces', the weight of the block (W_A) can be resolved into two components: one component acting normal to the plane (N_W) and one acting parallel to the sliding direction (S_W), as illustrated in the lower left diagram in Figure 8.14. From geometry:

$$N_W = W_A \cos \psi$$

$$S_W = W_A \sin \psi$$

i.e., $N_W = 708.75 \times 0.574 = 406.5$ kN, and $S_W = 708.75 \times 0.819 = 580.6$ kN; this is the shearing component of weight, causing the block to slide.

The available shearing resistance $= N_W \times \tan \phi_b$. As $\tan 45 = 1.0$; the available shear resistance here is also 406.5 kN. If cohesion (c, kN/m^2) was allowable (not here), this would be added by multiplying $c \times$ length of joint under consideration $\times 1.0$ to derive a resisting force in units of kN.

So, for Block A, in the dry case, without earthquake-loading,

$$\text{FoS} = \frac{406.5}{580.6}$$

i.e., FoS = 0.7.

This is not acceptable, of course, and failure would take place during excavation, so reinforcement would be required prior to cutting.

To calculate the requirement for reinforcement:

$$1.4 = \frac{406.5}{580.6 - R_A}$$

Here, R_A is the resisting force, acting directly up the shear plane.

Rearranging shows that to provide a FoS of 1.4, a force, R_A of 290.2 kN (29 tonnes) would need to be applied to ensure safety. This might be achieved by dowelling (a 36 mm diameter steel dowel has permissible shear strength of about 22 tonnes, so a couple of dowels in this block would suffice for the dry condition). Two dowels (repeated every metre along the width of slope) might be installed to ensure safety during the preliminary excavation works, although, if blasting was to be used, then vibrations and gases would need to be considered for temporary stability. Usually, one would install all reinforcement measures required for the permanent works, and for ensuring stability during temporary works, before proceeding with cutting below.

8.1.7.5.2 Allowing for groundwater pressure

The worst-case scenario is to assume water pressure up to the ground surface. As discussed in Chapter 4 (and Richards and Cowland, 1986), this might be very conservative, and, in practice, drainage measures would normally be adopted.

We have assumed that there will be a release cross-joint to the rear of Block A, and this is full of water. At the point where the cross-joint intersects the sliding plane, the vertical depth, Z_W, is

$Z_W = 3.5 \times \cos 55°$, that is, about 2 m; so, the water pressure will be about 20 kPa.

Then the water force acting on the cross-joint will be

$L_A = 0.5 \times 3.5 \times 2 \times 10$ (integral of the stress triangle) $= 35$ kN

There will also be an uplift force, reducing the effective normal stress across the shear surface. Assuming a triangle of stress, petering out to zero at the cut face, then

$$U_A = (2 \times 10 \times 10)/2 \text{ kN} = 100 \text{ kN}$$

So, for the wet case, combining forces:

$$\text{FoS} = \frac{(406.5 - 100)}{580.6 + 35} \tan \phi$$

i.e., FoS = 0.5.

8.1.7.5.3 Earthquake load plus water pressure

The horizontal earthquake load, Q_A is $0.5 \times W_A$ (354.5 kN), and can be resolved into normal and shear components as before; at the worst moment during the loading (which reverses during real shaking), the normal component will be away from the plane, reducing effective normal stress (and frictional resistance). The shear component acts with the other shear forces to encourage failure. (Blasting forces can analysed in the same way.) So,

$$N_Q = Q_A \sin 55 = 290.4 \text{ kN},$$

and

$$S_Q = Q_A \cos 55 = 203.3 \text{ kN}$$

So, to achieve the required FoS for the case with full water pressure and earthquake loading:

$$1.4 = \frac{(406.5 - 100 - 290.4)\tan \phi}{580.6 + 35 + 203.3 - R_A}$$

Resolving, R_A is calculated as about 800 kN (80 tonnes), i.e., four dowels of shear capacity of 22 t each. Even if the slope were fully drained, the required resistance would be 736 kN (i.e., 74 t), so the dowelling requirement would only be reduced slightly.

8.1.7.6 Analysis of Block B

The same process can be repeated for Block B (and others up the slope). The length of the cross-joint above Block B is about 4.7 m, and the cross section area of Block B is about 41 m², so $W_B = 1107$ kN. L_B is about 90.5 kN, and U_B about 292.5 kN (integrating the parallelogram of water pressure).

Considering the components of the various forces, one arrives at:

$$1.4 = (906.81 - 292.5 - 317.4)\tan \phi/(634.95 + 90.5 + 453.4 - R_B)$$

$$1.4 = 296.8/1178.9 - R_B$$

i.e., $R_B = 966.8$ kN, which is much the same requirement as for Block A. This is a slightly surprising result, given that the plane is only dipping at 35° (compared to the friction angle of 45°) and FoS for dry condition without earthquake is $\cos \psi / \sin \psi$, which is 1.43 (stable). The difference is the size and weight of Block B compared to Block A, which increases the required support.

8.1.8 Detailed analysis of wedge failure

The analysis of 3D wedges is more complicated, because the strength contributions from the two or more planes involved depend upon the stress levels, and upon the amount of frictional resistance available from each discontinuity. Charts to determine the 'wedging factor' are presented in Hoek and Bray (1981) and reproduced in Wyllie and Mah (2004), but often engineers will make use of software such as Swedge (Rocscience) that allows any geometrical wedge to be visualised and analysed, and to optimise any preventive measures (such as ground anchors) that might be necessary to achieve an adequate FoS. In practice, a shallow, flat day-lighting wedge with internal angle of more than 100°, looking up the line of intersection, will have a FoS similar to that of a single plane dipping in the same direction as the line of intersection. For tighter wedges, the FoS can increase markedly to twice or even three times that of an equivalent plane, but this also depends upon whether the wedge is tilted, and by how much. It also, of course, depends upon the shear strength characteristics of the discontinuities making up the wedge.

8.1.9 Toppling

Toppling failure can occur when discontinuities dip steeply into a rock face, as illustrated in Figure 8.15 (and in Figures 8.6 and 8.7). Again, in my experience, this type of failure is quite rare by itself, but parts of larger failures often show localised toppling, and it is also a feature of mass wasting in natural hillsides, rather than specific landsliding. Similarly, steep columns of rock often collapse, for example, in coastal cliffs, but that is often a case of undercutting or weakening at the toe. Methods of analysing toppling failure are summarised in Wyllie and Mah (2004).

If there is a potentially toppling set, then the main questions follow: (a) whether they form a set of true mechanical discontinuities (often not); (b) whether there are cross-joints, allowing freedom of rotation; and (c) for an existing slope, whether there is evidence of deformation. Quite often, there are steep orthogonal sets of joints in rock slopes that would plot out as a toppling risk, but are innocuous, other than as a source of potential shallow rockfalls that can be dealt with by netting or by other minor support measures.

8.1.10 Rockfall

8.1.10.1 Introduction

Rockfall is the commonest type of rock-slope failure, essentially an ongoing erosion process in rocky terrain (Figure 8.16).

Rockfall is a particularly important hazard adjacent to roads and railways. This is partly due to difficulties in prediction, but also because a rockfall can have consequences out of proportion to the small volume of original failure. This is largely because of knock-on risks associated with traffic accidents and derailment of trains. Hungr and Evans (1989) report 13 deaths due to rockfalls in Canada in the previous 87 years, mostly on highways.

Figure 8.15 Steep joints dipping into a slope face with the potential for toppling, Hong Kong.

Figure 8.16 House builders seem to have been oblivious to risk from rockfall, Western Cape, South Africa.

Figure 8.17 Boulder-fall in Hong Kong. Stubbs Road, Hong Kong 1985. Since 1926, there have been at least nine fatalities due to rock/boulder falls in Hong Kong. (From ERM-Hong Kong Ltd. 1998. *Quantitative Risk Assessment of Boulder Fall Hazards in Hong Kong: Phase 2* [GEO Report No. 80]. Report prepared for the Geotechnical Engineering Office, Hong Kong, 61pp.)

In Hong Kong, in the period 1978 to 1995, there were 74 rock/boulder fall incidents reported to the government (Figure 8.17).

Rockfalls are commonly triggered by heavy rainfall, where water causes increase in cleft water pressures, and are the commonest type of failure in many earthquakes (Keefer, 1984). They also occur unexpectedly, as part of gradual slope deterioration, often dislodged by growth of tree roots or, conversely, by the death of vegetation that had been retaining loose rocks.

Rockfalls may destabilise other blocks as they travel down a slope, triggering rock avalanches of much larger volume than the original block fall (entrainment). Minor rockfalls are also often the first sign of a larger failure, and should be considered as a warning – an example is given in Figure 8.18.

8.1.10.2 Rockfall hazard assessment

Usually, assessment of rockfall hazard at a specific site will involve field-mapping, together with remote examination using binoculars, photographs and, possibly, terrestrial radar. Radio-controlled helicopters with cameras can be used to inspect dangerous areas remotely. If discontinuity data are available, then key-block analysis (Goodman and Shi, 1985) can be used to identify key blocks, which, if they failed, could lead to progressively large rockfalls. Using computer algorithms, the removability of blocks can be tested, using detailed trace maps of actual rock faces. To assess the size of potential failures, *in situ* block-size distribution (ISBD) can be a useful parameter (Lu and Latham, 1999). Golder Associates (1998) used the computer software FRACMAN to estimate the volumes of potential rockfalls statistically, with respect to the widening of Tuen Mun Road in Hong Kong. The data were used to determine pre-excavation support and permanent support requirements.

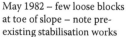
May 1982 – few loose blocks
at toe of slope – note pre-
existing stabilisation works

August 1982, 50 m³

March 1983, 500 m³

Figure 8.18 Minor rockfall in May 1982 as an indication of a progressively deteriorating rock slope, Tsing Yi Island, Hong Kong.

It can be very difficult to identify the potential for rockfall other than in broad terms; there may be an obvious hazard associated with a rock cliff, but it will be difficult to establish which block may become dislodged. Rockfalls can occur on natural or blasting-induced fractures not observed through systematic characterisation of discontinuities. In many circumstances, it will be necessary to take a broad-brush approach to either prevent falls (e.g., by netting or shotcreting faces), or to mitigate the risk, using fences, barriers and catch ditches.

8.1.10.3 Management of risk

Assessment of rockfall hazard usually involves rating slopes as critical or non-critical, assessing the potential for actual failures to occur, evaluating the associated risks and defining the requirements for rockfall mitigation. When a slope is assessed to be critical, detailed risk assessment, including formulation of mitigation strategies, should be carried out. Analysis of the rockfall trajectory is commonly conducted, using rockfall modelling software, such as Rocfall (Rocscience). Parameters are assigned or can be varied statistically. One of the most problematic input parameters for rockfall simulation is the 'coefficient of restitution', as this greatly affects the parameters of impact energy, bounce height and travel distance (Chau et al., 1998). Most programmes define the coefficient as the ratio of the velocity after to that before impact, and with different values applicable to the normal and tangential directions. However, one of the programmes (CADMA) uses a single coefficient applied to energy loss. Several of the programmes also apply a velocity scaling function to the restitution coefficients. Factors affecting the coefficient of restitution include impact angle, impact strength of rock block, surface condition of terrain, the shape of boulders, surface roughness of the rock blocks and the slope and the rotational energy of the rock blocks. In general, however, hard rock surfaces have high coefficients; thick gravel will absorb energy and has a low value. Azzoni et al. (1992) reported on the results of field-testing of rockfalls, and Azzoni et al. (1995) discussed practical approaches to modelling. They found that the main deficiency was the poor topographic definition of cross sections used for the analysis. They modelled cross sections that were surveyed, using 152, 79, 43, 25 and 16 topographic

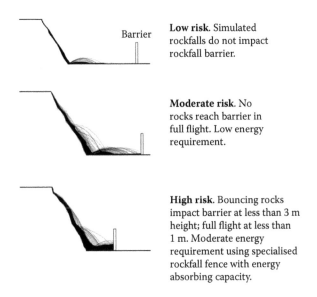

Low risk. Simulated rockfalls do not impact rockfall barrier.

Moderate risk. No rocks reach barrier in full flight. Low energy requirement.

High risk. Bouncing rocks impact barrier at less than 3 m height; full flight at less than 1 m. Moderate energy requirement using specialised rockfall fence with energy absorbing capacity.

Figure 8.19 Judging degree of hazard using rockfall simulation programmes. (Modified from Halcrow China Ltd. 2011. *Study on Methods and Supervision of Rock Breaking Operations and Provision of Temporary Protective Barriers and Associated Measures.* GEO Report No. 260, Geotechnical Engineering Office, Hong Kong Government, 250pp; Geotechnical Engineering Office. 2000. *Highway Slope Manual,* Hong Kong: Geotechnical Engineering Office, 114pp.)

points, and concluded that detailed topographical surveys are necessary to represent the field conditions as accurately as practically possible.

Rockfall simulations can be used to judge the need for rockfall barriers and nets, as per Figure 8.19, and discussed later in this chapter, although they are not foolproof. Sometimes, rockfalls occur progressively over an hour or two (and unpredictably), so that later rocks bounce on earlier debris, and travel higher and further than might have been indicated by simulations based on pre-fall geometry (Figure 8.20).

8.1.10.4 Hazard rating systems

Systems for assessing rockfall hazard are sometimes used to rate rock slopes for prioritisation of stabilisation and mitigation measures, and to allow general risk evaluation. The

Figure 8.20 Two views of rockfall in Dubai, 2006. Note that largest block has travelled furthest, sliding and bouncing on earlier debris.

systems are appropriate for broadly assessing the varying hazard along a new or existing highway. They also have potential uses at the planning stage of a project (and periodically throughout the project), to assess rockfall hazards and to estimate or check the adequacy of rockfall mitigation measures. They are not applicable for site-specific evaluation.

One of the most widely used systems is the Rockfall Hazard Rating System (RHRS) published by the United States Department of Transport. The RHRS is a formal hazard-assessment system used to support the management of rockfall sites adjacent to highways, and *'provides a legally defensible, standardised way to prioritise the use of the limited construction funds available by numerically differentiating the apparent risks at rockfall sites'* (NHI, 1993). The RHRS approach is to 'reduce' the risk of rockfall, but not to completely eliminate it. Note the terms 'legally defensible'. The system has been adopted in several states in the United States and Canada, and was considered for adoption in Hong Kong. A detailed and critical review of its potential use in Hong Kong is given in Halcrow China Limited (2011), and this review might be of use to other authorities that are considering adopting the system.

The RHRS approach involves point-scoring for categories such as slope-height, road geometry and usage, including decision-site distance, as well as geological assessments, potential block-size and rockfall history. The 'preliminary assessment' is based on professional judgement only. There is no rockfall simulation.

Some category parameters are deterministic, while others are subjective. Very limited guidance is provided on the identification of failure mechanisms in RHRS, and what does exist, is simple and subjective. Experienced engineering geologists would question whether it is possible to establish rock-block removability from a large rock mass 'by eye'. The main difference between RHRS and other systems is that the assessment of hazard is strongly linked to historical observations of 'actual' rockfalls. This makes sense in that, empirically, if there is a persistent problem at a location, then it needs to be resolved. The reliance upon historical records of rockfall is appropriate for the Oregon roadside slopes, where the system was developed because of the mountainous terrain, and because the slopes were formed using 'aggressive construction methods' in such a way that left them prone to numerous rockfalls. In areas where there are fewer regular occurrences of rockfall (as in Hong Kong), or poor maintenance records, there is a problem in defining the 'potential' for rockfall as fundamental to the RHRS system. The final stages of RHRS involve estimating costs for rockfall mitigation measures, to allow cost-benefit analysis for works on high-priority slopes. At some locations, as in Figure 8.21, the cost of making facilities safe would be impossibly high, and residual risks have to be accepted, perhaps justified by a quantified risk assessment (QRA) calculation based on road usage and number of previous recorded failures (or dents in the road surface since last repaired), nets to catch minor falls and warning signs, so that at least drivers are aware that they might encounter debris in the road and have time to stop.

For completeness, it should be mentioned that various classifications have been prepared for judging relative rock-slope stability, and these no doubt have a role in assessing the degree of hazard, especially for broad examination rather than on a site-specific basis (Selby, 1980, 1993; Hack, 1998; Liu and Chen, 2007).

8.2 DESIGN OF ENGINEERING WORKS

8.2.1 Assessing need for preventive engineering measures

As noted earlier, generally, the need for engineering preventive works is judged with reference to the existing FoS compared to that required for the risk level. A relatively recent approach to assessing the need for preventive measures is to use QRA as described by Pine

Figure 8.21 Examples of severe rockfall hazard (in terms of potential volume); cost of preventive works would be unjustifiable because of low road usage. Risk can be reduced by putting up warning signs and local netting, to minimise small block-falls. Spanish Pyrenees, NE of Leon.

and Roberds (2005). The project involved remediation and stabilisation of several sections of high cut and natural slopes, dominated by potential sheeting joint failures and the hazard of boulders bouncing down exposed sheeting joints, to impact the road below. The design of the slope cut-backs and stabilisation measures was based on a combination of reliability criteria and conventional FoS design targets, aimed at achieving an ALARP (as low as reasonably possible) risk target which, in actuarial terms, translated to less than 0.01 fatalities per year per 500 m section of the slopes under remediation. Reeves et al. (1999) recommend maximum allowable individual risk (i.e., frequency of harm to an individual per year) of 10^{-5} and 10^{-4} for new and existing developments, at the base of a slope, respectively.

8.2.2 General considerations

Remediation of stability hazards on high rock slopes is often very difficult, and implementation of the works can itself increase the risk levels, albeit temporarily. Factors that will influence the decision on which measures to implement include: the specific nature of the hazards; topographic and access constraints; locations of the facilities at risk; cost; and timing. The risks associated with carrying out works next to active roads, both to road users and to construction workers themselves, and how to mitigate these, are addressed in some detail in Halcrow China Ltd. (2011). Pre-contract stabilisation works will often be needed to allow initial access and preparation to the site. Preventive measures, such as rock-bolting, may be carried out at an early stage, to assist in the safe working of the site, and could be designed to form part of the permanent works, as discussed later. Options for the use of protective barriers and catch nets to minimise disruption to traffic during the works, also need to be addressed, as do contractual controls, and alternatives for supervision of the works. The use of a risk register, as piloted for tunnels (Brown, 1999 and BTS, 2003), with clear identification of particular risks and responsible parties, helps to ensure that all hazards and consequences are adequately dealt with during construction. Decision analysis is now widely applied at an early stage to assess whether to mitigate slope hazards (e.g., by rockfall catch nets) or to remediate/resolve the problem by excavation and/or support approaches. If construction of intrusive engineering measures to stabilise hazards might be unduly risky, then passive protection could be adopted instead. A hybrid solution is often the most pragmatic solution for extensive, difficult slopes, where some sections might be stabilised by anchors and buttresses, and other sections protected by nets and other measures (Carter et al., 2002; Pine and Roberds, 2005).

8.2.3 Engineering options

Some of the options for improving the stability of slopes are listed in Figure 8.22 and illustrated in Figure 8.23. These can be split into *passive* options that either deal with the potential failure by controlling surface deterioration at source, or installing preventative

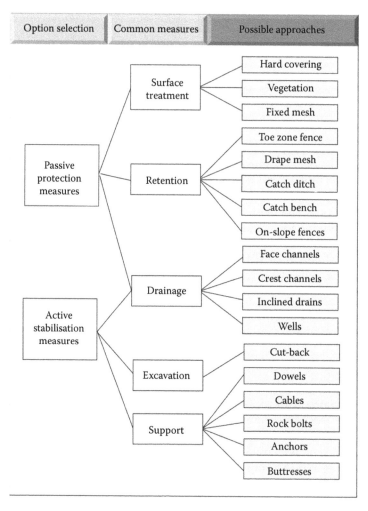

Figure 8.22 Options for engineering rock slopes. (Prepared by Trevor Carter of Golder Associates and modified from Hencher, S.R. et al. 2011. *Rock Mechanics and Rock Engineering*, **44**, 1–22.)

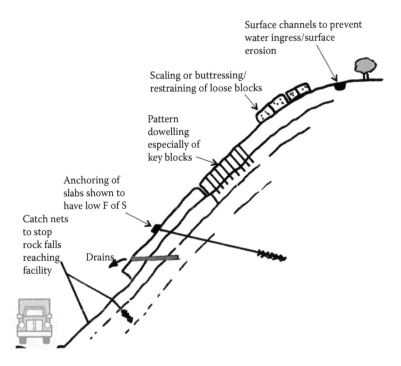

Surface channels to prevent
water ingress/surface
erosion

Scaling or buttressing/
restraining of loose blocks

Pattern
dowelling
especially of
key blocks

Anchoring of
slabs shown to
have low F of S

Catch nets
to stop
rock falls
reaching   Drains
facility

Figure 8.23 Some of the options for engineering works in rock slopes. (After Hencher, S.R. et al. 2011. *Rock Mechanics and Rock Engineering*, **44**, 1–22.)

reinforcement to increase local factors of safety, or adding walls or buttresses to restrain detached debris before it causes injury or damage, and *active* measures that enhance overall FoS of larger sections of slope by major engineering works, including cut-backs, or buttresses, or heavy tie-back cabling.

8.2.4 Surface treatment

Overall geometry of slope is important, not only for improving FoS, but also for reducing rockfall hazards, installing preventive measures and inspection and maintenance access. Berms are common on high permanent slopes, and may be an effective means of catching rockfall, providing they are of sufficient width (say greater than 3 m). Temporary berms are also formed during the excavation of major slopes. Optimum berm size is derived both from geotechnical considerations (overall slope angle), and from rock-trap consideration (often assessed by computer simulation). The geometry of standard flat berms can be modified to leave a wall of rock at the outside edge to act as a catch structure for rockfall, and increase the rockfall capacity of the berm, although it is often difficult to cut rock so precisely. This arrangement can be thought of as a rock-trap ditch located on the slope, rather than at the toe.

Many risks can be mitigated through surface treatment to stabilise or remove relatively small blocks of rock. There is a temptation to use hard slope treatments like shotcrete to constrain loose blocks at the slope surface, but such measures, if not properly designed, can restrict drainage from the slope, hide the geological situation from future investigators, and can themselves cause a hazard, as the shotcrete deteriorates, allowing large slabs of shotcrete to detach. Furthermore, shotcrete is increasingly an unacceptable solution for aesthetic reasons, and there is a push towards landscaping and bioengineering highly visible slopes, where such measures can be justified from an engineering point of view. In this context, it is to be noted that most bio-engineering solutions are unreliable for high-risk slopes because they may fail in the long-term if vegetation dies, for example, in fires. Where individual rockfall sources are identified, these can be scaled off, reinforced by dowels, bolts, cables or dentition, and/or netted, where the rock is in a closely jointed state. Removing large blocks can be difficult because of the inherent risks associated with breakage techniques, which can dislodge blocks unexpectedly. Care must be taken to protect the public and workers during such operations. The most difficult zones to deal with are those with poor access. Implementing passive or active protection approaches needs to start from safe ground and move progressively into the areas of more hazardous stability.

Surface drainage is a very important consideration for all slopes, but particularly for slopes comprising part rock, with very high runoff, and part soil sections, which might be eroded and undermined by surface flow concentrations.

8.2.5 Mesh drapes

Where slope heights are significant, and ramp or bench approach is difficult, mitigating hazards can be problematic, even using rope access techniques, because face stability may be too unstable to even allow rock-climbing personnel onto the face. Under such conditions, draping surface mesh may allow achievement of some effective protection, preventing progressive ski-jump-style bouncing of rock down the slope (Carter et al., 2002). Application of drape-mesh (varying from chain-link, triple twist, hex-mesh to ring-net, in increasing order of energy capacity) can be effected by a variety of techniques, ranging from climber-controlled unrolling of the mesh, to helicopter-access placement. Typically, crest restraint is provided by dowels or tie-back anchors, usually cabled back some distance from the crest zone to provide a safe anchorage. Netting and mesh can be anchored along the top of the slope, using galvanised rock dowels and cables, and is commonly employed for permanent slopes. An example is given in Figure 8.24.

Mesh is also used for temporary faces integrated with the sequencing of the excavation for the new rock slopes and upgrading works for existing slopes. The purpose of the mesh is not to prevent rockfall but to control it by trapping blocks between the mesh and the rock face. Bigger (1995) described in detail the practical experience of installing a large net system in the United States.

(a)   (b)

Figure 8.24 Slope in complexly folded and faulted schist (a) before and (b) after drape mesh. Near Sao Bras, Portugal.

8.2.6 Fences, catch-nets and barriers

Where there is the potential for repeated small-scale detachments impacting a highway, then catch-nets or diversion/stopping barriers can be the solution, as discussed with reference to reducing risk by Pine and Roberds (2005). Such catch-nets or fences and barriers can be positioned on-slope or in the toe-zone of the slope, depending on energy requirements and site restrictions as illustrated in Figure 8.25.

Where energies computed from rockfall analyses are too extreme for toe-zone protection alone to maintain risk levels below prescribed criteria for highway users, on-slope energy protection fences may be used to reduce total energy impact at road level. This was the approach adopted at Tuen Mun Road in Hong Kong for sections of the slopes, which were to remain in place, and where sheeting joint geometries were considered hazardous enough to allow potential release of blocks of sizes that could not be stopped by toe-zone fencing alone. On-slope 3000 kJ fences were designed and installed above the highway to catch rockfall blocks from the hundred metres of slope above the fence. Further details are given in Hencher et al. (2011). For the Tung Chung Road in Hong Kong, which cuts across mountainous terrain, nets were designed to stop debris from natural terrain landslides and boulders. The nets were constructed in lines, each line designed to cope with any failure in the hillside between it and the next line of defence up-slope, based on a 'design event' approach, using air photograph interpretation and field mapping, combined with dynamic analysis of the energy generated by each landslide event.

Rock-traps are also constructed as excavated ditches, deformable barriers such as banks of fill or gabion structures, and walls. A combination of several rock-trap types is often designed at the toe of slopes adjacent to highways. Un-compacted gravel is placed on the floor of traps to absorb energy from falling blocks. Empirical design charts were developed for the design of trap ditches in slopes (Ritchie, 1963; Mak and Blomfield, 1986).

Where the hazard is very great and the road cannot be rerouted, rock-avalanche shelters have been successfully used adjacent to steep slopes above narrow railways and roadways, as illustrated in Figure 8.26.

(a) (b)

Figure 8.25 High-capacity nets designed to catch debris from upslope. (a) Tung Chung Road, Hong Kong, (b) Tunnel portal, Spanish Pyrenees.

Figure 8.26 Shelters used to protect road users from severe rockfall hazards and periodic debris avalanche, both examples from Central Taiwan.

8.2.7 Drainage

8.2.7.1 Surface works

Surface finish is an important consideration to the long-term stability of slopes, and to reduce maintenance requirements. When designing layout of benches/berms (the terms are synonymous) and heights of individual faces, consideration should be given to long-term access and safety for inspections, as well as for maintenance works, including repair, cleaning of drains, tensioning, and perhaps, replacing support elements. It is often appropriate to construct permanent access ladders and protective barriers along edges of berms, to which workers can anchor themselves. As a basic stabilisation measure for rock slopes, drainage works are appropriate, both during and after excavation (Peckover and Kerr, 1976). Crest drains are generally needed, as well as channels along berms with upstands and step channels to carry water down the slope to a toe drain that channels water properly away from the slope. In some cases, berms are hard-surfaced without drains and with falls designed to carry water and debris away from the slope. A typical arrangement for surface drainage is illustrated in Figure 8.27. Details of drain dimensions are given in the Geotechnical Manual for Slopes (GCO, 1984a). Coverings to be adopted might be shotcrete with drainage relief holes or, more expensively, propriety drainage mats with netting, though which vegetation can grow (Figure 8.28). The numerous options for landscaping slopes are reviewed in GEO (2011).

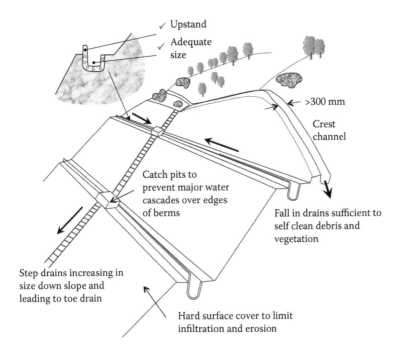

Figure 8.27 Surface drains along berms combined with step channels to take water off slope.

Exposed joints that may allow water pressure to be dissipated should be left exposed where possible; otherwise, free drainage may be impeded and water might dam up. If the exposed joint is weathered, the weak material may back sap and, possibly, a pipe may lead to destabilisation, partially caused by lack of free drainage. This can be avoided by installing closely spaced sub-horizontal drains with geotextile filter fabric sleeves so as to prevent

Figure 8.28 Nets with permeable bonded gravel mats used as surface protection (with some shotcrete). Bowen Path, Hong Kong.

blocking, together with protection of the weathered material. No-fines concrete, while appearing to be suitable to protect weathered zones, often ends up with lower permeability than designed, and should not be relied upon without some additional drainage measures.

8.2.7.2 Drainage of sub-surface water

Sub-surface drainage can be effective in reducing the potential development of adverse water pressures, but there is a need to target sub-surface flow channels, many of which will be shallow and ephemeral. Channels can be active during the wet season and then clog up and collapse in the dry season, so that new channels are exploited and developed when inflow and through-flow starts again. The paths may be tortuous and hard to identify, and drainage measures can therefore be rather undependable (Hencher, 2010). Regular patterns of long, sub-horizontal drain holes can be very effective, but it must never be expected that all drains will yield water flows, and the effectiveness of individual drains can change with time as sub-surface flow paths migrate.

In some cases, major measures are called for. Remediation of the Aberfan landslide in South Wales in 1966 included the construction of pressure relief wells soon after the failure. In 1980, a drainage gallery was constructed, by drill and blast below the rockhead. The gallery, with 28 drainage holes drilled from the tunnel up to 19 m in length, did reduce water pressures sufficiently to achieve the required FoS improvement, but did not reduce the piezometric surface to the elevation of the gallery, which was rather surprising. It was also found that flows were very localised, and predominantly along bedding parallel fractures. Siddle (1985) considers that a greater draw-down of the piezometric surface could have been achieved by drilling more drainage holes.

Following the Po Shan landslide that killed 67 people in Hong Kong in 1972, the slope was cut back, and 73 long, shallowly inclined drains installed, which successfully reduced water pressures. However, with time, decreasing outflows were noted, and there were also concerns that drains might be damaged by shallow landsliding. As a more robust solution, two tunnels of 3.5 m diameter were constructed, with a total length of about 470 m, and 172 drains drilled out of the tunnel up to 100 m in length. The proximity of the tunnels to properties ruled out the use of blasting, so the tunnels (one 190 m and the other 260 m) were constructed by TBM with the electric power fed to the machine by umbilical cables from an adjacent worksite.

Elsewhere in Hong Kong, lines of deep caisson drains have been used successfully, to intercept groundwater paths, and limit recharge into slopes below (McNicholl et al., 1985).

For a major and complex slope in Australia, described by Starr et al. (2010) and Wentzinger et al. (2013), pumping wells were and are being used to control water pressures, with the system set to extract water, as groundwater pressures rise to predetermined alarm levels. Similar systems are commonly used in deep open-pit mining to 'depressurise' slopes (Beale and Read, 2014).

8.2.8 Reinforcement

Slopes may be reinforced using passive and active anchorages.

8.2.8.1 Passive anchorages

Passive forms include 'soil nails' and dowels, which are installed in drilled holes, and then fully grouted without tensioning. 'Soil' nails are not only used for soil but for many weak rocks, as illustrated in Figure 8.29.

Figure 8.29 Soil nail-heads under construction in weathered granite with felsite and basalt dykes, Kwun Tong, Hong Kong.

Often, nails are just galvanised steel with reliance upon sacrificial diameter to cope with corrosion in the design life of the structure, but they can be protected with plastic coatings, as shown in Figure 8.30.

Pattern nails provide a general strengthening effect to the rock mass; they become active as the slope begins to deform, perhaps due to rise in groundwater pressure, or dynamic loading. The nails resist in shear, as well as in tension, due to their fully grouted nature. Optimal

(a)   (b)

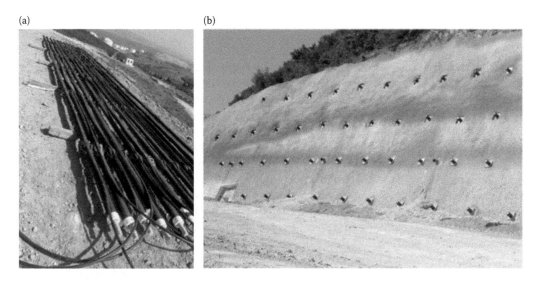

Figure 8.30 Plastic sheathed 'soil nails' with grout tubes attached (a) and installed (b).

support requirements can be investigated, using commercially available software, including SlopeW, Slide and Flac; prescriptive advice to design is also given in GEO (2009).

For slabs of rock underlain by rough joints, the interlocking nature of fractures provides considerable shear strength, and this needs to be accounted for in design, in order to avoid over-conservatism. If the joint can be prevented from sliding by reinforcing at strategic locations, then full advantage can be taken of the considerable natural frictional resistance. In such cases, the preferred method is usually to use passive dowel designs, rather than tensioned bolting, for necessary shear constraint. This is because it is considered that active reinforcement members are more subject to corrosion damage, and that passive dowels allow both mobilisation of a normal force (due to the restraint provided by the full column bond against asperity ride during shear), plus active shear restraint provided by the steel of the dowels resisting block-slide mobilisation (Spang and Egger, 1990).

The Geotechnical Engineering Office in Hong Kong has published some guidelines on prescriptive measures for rock slopes, and gives guidance on rock dowelling for rock blocks with volume less than 5 m^3 (Yu et al., 2005). In essence, it is advised to use pattern dowels with one dowel per m^3 of rock to be supported with minimum and maximum lengths of 3 m and 6 m, respectively, provided the potential sliding plane dips at less than 60°. The dowels are to be installed at right angles to the potential sliding plane, with the dowels acting in shear, while also enhancing the normal restraint, due to asperity dilation during sliding. In practice, dowels frequently need to be used in more variable orientations, and different dowel diameters and lengths may be adopted according to support requirements.

It is often difficult to identify the thickness and volume of a given block requiring support, and therefore, dowel patterns are frequently based on some assessment of cross-joint spacing. Along the Tuen Mun Highway, in Hong Kong, typical support layouts were adopted, based on field mapping of cross-joint spacing, and orientation with respect to the sheeting joint geometry, and inferred direction of sliding. The design used 40 mm dowels at 5-m spacing, based on analysis using the approach of Spang and Egger (1990) for definition of shear resistance. In areas of closer cross-joints, 25 mm dowels were used at 2-m spacing, split spaced between the wider pattern bolting layouts (Pine and Roberds, 2005). Field placement of reinforcement was, however, always double-checked against natural disposition of features, and decisions made by the engineer in the field for additional spot-bolting or dowelling, as required.

8.2.8.2 Active anchorages

Individual rock-bolts are more efficient than dowels, because they apply compression, thereby increasing shear resistance actively. Rock-bolts are typically used to support individual key blocks critical to potential larger scale failures, and are especially used in underground excavations. They are not generally used over a large scale in rock slopes. They are anchored at the base, using grout, resin, or mechanical devices, and then tensioned. Rock-bolting techniques are fully described in many references (e.g., Littlejohn and Bruce, 1975; BS 8081, 1989). Owing to the uncertainties involved with the rapid assessment of block geometries, and the need for quick temporary stabilisation, the use of pre-prepared bolt-load charts may be applicable, examples of which are given in Fookes and Sweeney (1976).

Sub-horizontal cable anchors can be installed to 10s of metres depth, and so are especially appropriate to large-scale failure mechanisms, rather than bolts and dowels that really deal only with relatively shallow failures. Like rock-bolts, they are fixed at the base (the fixed length), and the potential bond strength can be gained readily from the literature, or by field-testing on sacrificial test anchors. They carry high loads, and need careful design, especially to resist corrosion during their lifetime (Figure 8.31). They are often installed

Figure 8.31 Triple corroded, three tendon cable anchor pulled out under test; fixed length in foreground. A55 Coast Road, North Wales, United Kingdom.

with load-cells in line to monitor performance, and it is a requirement that they be tested periodically, using lift-off jacks.

8.2.9 Buttressing and larger retaining structures

Concrete or masonry structures are used to support local blocks of rock and overhangs, often strengthened with dowels into the rock mass; these are known as buttresses, and the general process of filling gaps is known as 'dentition work'. Drainage needs to be considered, where the buttress might block a drainage path, as illustrated in Figure 8.32.

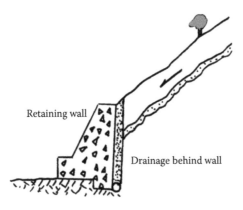

Figure 8.32 Buttress/retaining wall with drainage at rear.

(a) (b)

Figure 8.33 Rock failure, Magazine Gap Road, Hong Kong. Preliminary idea was to stabilise rock blocks with rock bolts but, a mass concrete wall was built, tied back into the rock face.

At a larger scale, reinforced concrete walls are used where the problem is too difficult to deal with by other measures, and especially where the rock is broken and complex, so that a clear failure mechanism is not obvious. For example, after a rock failure in Hong Kong (Figure 8.33), attempts were made initially to scale-off loose blocks, so that a final face could be rock-bolted and netted. In the event, scaling simply exposed more and more loose blocks, and there was a significant danger of making the failure even larger. In the event, a mass concrete wall was constructed, filling the gap.

A similar example is shown in Figure 8.34. Again, the rock mass was too variable to identify specific key blocks to be stabilised, or mechanism to be dealt with. The best option was lateral restraint, using a buttressing retaining wall.

8.3 SLOPE FORMATION

8.3.1 Safety and contractual issues

A detailed report on excavation of rock adjacent to highways was prepared by for the Hong Kong Government, and this is freely available and downloadable (Halcrow China Ltd., 2011). The report goes into great detail on the risks associated with excavation and construction on rock slopes adjacent to highways, and how these may be mitigated. The work has wider application than just for highway work, and is also applicable outside Hong Kong. Apart from technical guidance on rockfall assessment, it includes case studies, and has sections and appendices that deal with issues such as traffic management, contracts and supervision requirements. Some of the following observations are summarised from that report.

The safety of site personnel and road users is of paramount importance when forming, upgrading, maintaining, or repairing rock slopes. Several factors need to be considered carefully, including geotechnical conditions, traffic restrictions, safety, including statutory regulations, access constraints, land issues and the environment.

(a)

(b)

(c)

Figure 8.34 Failure in variably jointed and hydrothermally altered rock. Solution was to clear off and construct a mass concrete retaining wall.

Mixed rock and soil profiles can cause particular difficulties where large corestones are surrounded by soil. The rock will need to be broken up for removal, while the rest may be rippable (Chapter 7). Where rock dominates, the task is essentially one of breaking up the mass into blocks of a size suitable for safe removal.

Whatever method of excavation is used, there is always the potential for spillage or dislodging of loose material, and this must be recognised. Furthermore, there is a danger that excavation can cause the opening up of hidden weaknesses, leading to instability.

Before re-grading slopes or removing loose rock, it may be necessary to shield and/or strengthen adjacent structures and services against blast damage and resultant rockfall. Both potential blast damage and rockfall may be assessed beforehand and reviewed on-site during construction in the light of the encountered geological and geotechnical findings. It may be necessary to erect either temporary fences (which may also serve as permanent fences in some cases) or barriers, in combination with catchment ditches, at the toe of the slope. Ditches are normally lined with energy-absorbing material, e.g., loose gravel or sand. Where structures cannot be adequately protected, temporary relocation of the personnel and facilities inside the structures may be needed. Where loose boulders or boulder fields are exposed during construction, and/or exist above the proposed final excavation, remedial measures may be required to stabilise them (Grigg and Wong, 1987). This may involve removal of boulders or *in situ* stabilisation. High-level catch fences may be appropriate to provide protection (Chan et al., 1986).

Opportunities to divert traffic may be limited. Nevertheless, having traffic adjacent to roadside excavations should be regarded as an exceptional situation, and will require particular safety precautions, including stringent safety measures such as preventive and protective measures, and supervision by professionally qualified personnel. Rock slopes are often designed to minimise excavation volumes, resulting in steep slopes and a minimum of excavation lifts. This approach may not be appropriate for roadside excavations, where safety is an important consideration. The slope design should also take into account temporary works, short-term and long-term stability, maintenance and aesthetics. The impact of the excavation on the broader area should also be assessed, including the effect on surface run-off and sub-surface hydrogeology.

Rock excavation should commence at the top of a cut slope, after access has been gained by pioneering works, and potential hazards outside and above the working area have been assessed and stabilised as necessary. This requires the mechanism of potential failure to be properly understood, and the adoption of appropriate temporary mitigation measures. For high slopes, temporary benches are often necessary. The excavated material is typically hauled, where possible, to the end of the benches, and dropped down chutes to a levelled stockpiling area, prior to transportation and removal.

The excavation sequence for sections of slope above a highway will require particular attention to ensure that rock is contained on working berms, and does not fall onto the live carriageways. For example, a specified procedure might be adopted for the formation, and subsequent demolition of a rock containment berm on the working slope (Figures 8.35 and 8.36). One hazard associated with such protective berms is that the reinforcement effectively stitches the

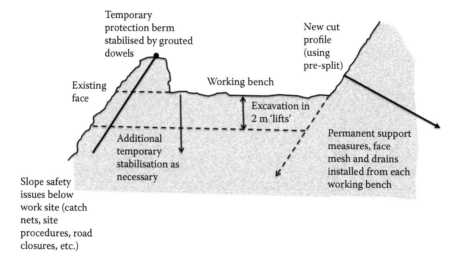

Figure 8.35 Excavating a bench using a temporary, reinforced bund to prevent spillage of debris (and protect personnel). Side view. (Concept by Dr. John Sharpe of Geo-Design, drawing modified from Halcrow China Ltd. 2011. *Study on Methods and Supervision of Rock Breaking Operations and Provision of Temporary Protective Barriers and Associated Measures.* GEO Report No. 260, Geotechnical Engineering Office, Hong Kong Government, 250pp.)

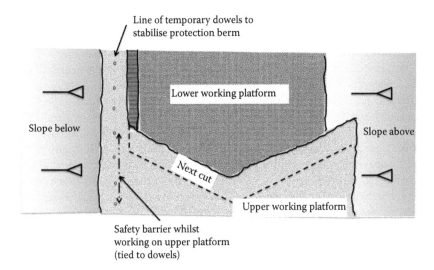

Line of temporary dowels to
stabilise protection berm

Lower working platform

Slope below

Slope above

Next cut

Upper working platform

Safety barrier whilst
working on upper platform
(tied to dowels)

Figure 8.36 Excavating a bench using a temporary, reinforced bund to prevent spillage of debris (and protect personnel). Plan view. (Concept by Dr. John Sharpe, drawing modified from Halcrow China Ltd. 2011. *Study on Methods and Supervision of Rock Breaking Operations and Provision of Temporary Protective Barriers and Associated Measures.* GEO Report No. 260, Geotechnical Engineering Office, Hong Kong Government, 250pp.)

rock masses together, so that any potential rock failure may be larger than would otherwise be the case.

Where appropriate, buffer zones should be provided between toe catch structures and any facility at risk below. Traffic constraints (e.g., partial road closure and work at night to minimise traffic disruption) and workers' safety in erecting the temporary barriers/fences should also be considered in the siting, erection and dismantling of temporary barriers.

8.3.2 Contractual and supervision considerations

All forms of hazard, whether excavation-induced or natural, that are associated with the formation of new rock slopes and/or the upgrading of existing slopes, should be identified, so that necessary precautions can be taken to ensure the safety of both the public and site workers. Not all of these measures can be fully specified in advance for inclusion in the contract documents. Therefore, the contract should not be restrictive on any measures related to safety issues, unless it is clear that the imposed restrictions are achievable. In addition, it is important that the conditions encountered during construction should be closely monitored, so that any changes from the design assumptions can be taken into account during the execution of the works. This is essentially a risk-register approach (Brown, 1999; Hencher, 2012a). Since rapid, unpredicted failures cannot be discounted during blasting, traffic stoppages and personnel clearances should normally be carried out within the danger zone, as specified by the blast designers, and approved by the appropriate authorities.

Designers should carry out site safety inspections and joint safety inspections with contractors. Any unsafe situations or working methods should be rectified during the works. If an immediate danger is identified, the contractor may be instructed to suspend relevant portions of the works until safety measures deemed necessary have been introduced. It is, however, also noted that the issue of an instruction would not normally relieve the contractor of his responsibilities under the contract.

8.3.3 Methods for breakage and removal of rocks

One of the key questions when planning the works is whether rippers, scrapers and grabs can be used to excavate without first breaking the rock using blasting or other splitting techniques. This of course will affect the equipment to be deployed, programming and cost. Excavatability depends mainly on the intact strength of the rock and the degree of open or weak natural fracturing (Franklin et al., 1971) but MacGregor et al. (1994) show that performance can be rather unpredictable. Pettifer and Fookes (1994) revised the chart of Franklin et al. based on numerous case studies and this is used as the background for Figure 8.37. It can be seen according to this graph, that blasting is generally required for strong massive rocks. Superimposed in Figure 8.37 are zones with intact strength characteristic of different weathering grades in granite. In practice it is generally found that Grade IV rock (breakable by hand) can also be ripped and scraped without blasting, even where there are few joints, so the underlying chart is rather pessimistic for such materials. An example is shown in Figure 8.38 where the Grade IV rock contains pervasive microfractures and would be cut relatively easily using mechanical excavators. However, in other uniform rocks of similar strength such as weak mudstone, it may well be found that blasting would be required before the rock could be excavated.

Three principal methods are commonly used for rock-breaking: mechanical, chemical and blasting; these apply to slope-formation, just as they do to general sites. The rock-excavation method should be selected to take account of local geological and geotechnical conditions. Issues will include environmental restrictions, safety, efficiency, speed and cost (see Table 8.2).

Table 8.2 Comparison of rock breaking methods for rock site formation and slopes

Method	Characteristics
Drill and blast	• fast and cheap
	• strict controls and specialist, authorised personnel requirements
	• dangers from flyrock, noise, dust and vibration
	• acceptable control over end result using careful design and presplitting
Mechanical breakers	• generally less hazardous than blasting
	• slow and noisy for slope and site formation
	• ineffective in massive rock, lacking joints
	• requires free face
	• excavation equipment needs to be matched to specific ground condition
Chemical	• using expandable grouts
	• no noise, flyrock or vibration
	• can result in unpredictable breakage
	• impossible to work overhead
	• expensive
	• slow, programming difficult
Additional methods	
Cone fracturing	• non-explosive; using high pressure gas release in drilled holes
	• effective for boulder splitting and minor excavations
	• easy to use and environmentally friendly with low noise, vibration or dust
	• good fragmentation control
	• slow and expensive relative to drill and blast
Diamond saw	• used in dimension stone quarries
	• suitable for architectural finish
	• slow and expensive

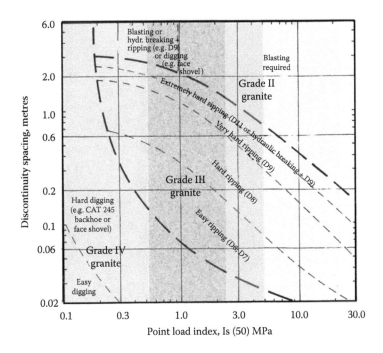

Figure 8.37 Excavatability chart based on Pettifer and Fookes (1994), superimposed with granite weathering grades.

Mechanical breaking (splitters, hydraulic breakers, etc.) is commonly used in urban areas, but these methods may be unacceptable, due to the typically slow excavation rates and noise. Chemical methods, such as expansive grouts (Figure 8.39), are being used more frequently, due to their safety and noise advantages over other methods, albeit at much higher cost. Also, the noise from close drilling might be unacceptable in urban environments. The main drawbacks, when using chemical expansive grouts are uncontrollable seepage and over-break, but these may be overcome to a degree by using expandable cartridges to contain the grout. Chemical grout can be used to form and open up primary fractures prior to mechanical excavation.

In general, lower-energy methods of rock-breaking are generally associated with fewer environmental problems and lower risk, but it needs to be recognised that dislodgment of rock can occur during any rock-slope works, as demonstrated by a fatal rockfall incident where chemical methods were being used for excavation in 1995. More details are given in Halcrow China Ltd. (2011).

Blasting is usually the most effective means of breaking down rock to blocks of less than 0.6 m in diameter, so that excavators can remove them. In blasting operations, the most significant problems are fly-rock, noise, ground vibrations and excessive gas pressures that can dislodge adjacent rock unintentionally.

Blasting works by generating stress waves and gases, which propagate away from the charge into the rock mass. Kutter and Fairhurst (1971) concluded that stress waves initiated cracks, while gas pressures extended them.

In blasting, the principle is to impart a sudden shock to the rock mass as quickly as possible, so that the mass displaces slightly, but not excessively. In this way, the rock is fragmented without over-break or fly-rock.

Figure 8.38 Parallel micro-fracturing in Grade IV granite, in upper part of cutting. Chung Hom Kok, Hong Kong.

There are three basic methods of blasting, namely, bulk blasting, pre-splitting and smooth-wall blasting, and the various techniques are explained by Wyllie and Mah (2004). The pre-split technique is generally accepted as being the most effective for forming stable civil engineering slopes (see Figure 8.40). The principle is to fire the rearmost line of holes first, a few milliseconds before the other holes. This first line forms a continuous fracture (which will be the eventual final slope face), confined by a large over-burden. The first formed fracture protects the rock mass from the stresses, vibrations and gases, as the rest of the holes are fired. It is, however, a misconception to assume that the introduction of this technique is the only step required to control blast damage. Controlling blast damage starts by assessing the state of existing faces. Where poor blasting has already caused fracturing of the rock mass, and loosening and dilation of existing discontinuities behind the line of the final face, it may be too late to remedy the situation by pre-splitting. Where the final rock face is to be formed by blasting, use of controlled blasting techniques should be considered, which employs closely spaced drill holes, and low-velocity explosives. After blasting, inspection and stability assessment of the blasted face and the adjacent slopes, should be carried out by a geotechnical engineer.

Figure 8.39 Chemical grouts used to form slopes in granite. Castle Peak Road, Hong Kong.

Figure 8.40 Slope formed well by pre-split blasting, Hong Kong.

8.3.4 Fly-rock hazards

Blasting in urban environments is inherently risky (Figure 8.41). Rock that flies through the air following a blast is called 'fly-rock'. Fly-rock most frequently arises from the top of benches, but also from front and side faces (Konya and Walter, 1990). Good blast design is the primary protection against flyrock; 83% of 154 incidents investigated in the United

Figure 8.41 Charging the holes. Hazards from flyrock and vibration can be severe in an urban environment.

Kingdom between 1981 and 1988, could have been prevented by better design (HSE, 1989). In some instances, there may be some fundamental geological difficulties. At a granite quarry near Harare, Zimbabwe examined by the author, the operators were having great difficulty in preventing excessive fly-rock, and this seemed to be related to another problem – that the blasted rock fragmentation was very tabular, apparently related to high locked-in *in situ* stresses.

From reported incidents in the 1980s in the United Kingdom and in Hong Kong, significant fly-rock can travel far beyond 200 m, with one instance of up to 800 m. Similar data is reported from Sweden by Lundborg et al. (1975). The maximum 'flying distance' was about 540 m for a 200 mm diameter block. For fragments of 75 mm to 100 mm size (about 2.5 kg), the maximum range was 410–470 m.

Techniques used to minimise fly-rock hazard and risk, include containment of the rock by heavy blasting mats. If the excavation strategy allows, leaving previously blasted rock in place can reduce the incidence of fly-rock. There are extensive requirements (HSC, 1988) for surveying of faces and blast-holes in quarries in the United Kingdom to reduce fly-rock, and this is generally done using ground-based laser, as well as down-hole surveys, to ensure adequate over-burden (Figure 8.42).

Nets are used to provide some protection, as shown in Figure 8.43, but King and Chan (1991) reported that some mesh cages could be pierced by fly-rock. Jeffcock (1995) reports that some nets he tested were of insufficient strength to restrain fly-rock ejected at velocities of up to 900 m/s. Smaller blocks of 0.5 kg were restrained, even when travelling at 70 m/s.

8.4 QUARRYING

8.4.1 Introduction

Quarrying differs from civil engineering in that the working slopes are temporary, and the purpose of blasting is to try to break the rock, not only to allow haulage, but also to break it down to the required end-product aggregate range, and to minimise the need for crushing

Figure 8.42 Laser scanning of slope in working quarry. Dryrigg, Quarry near Horton in Ribblesdale, West Yorkshire, Britain.

Figure 8.43 Blast nets designed to catch flyrock during blasting. Hong Kong.

Figure 8.44 Rock successfully broken by blasting to a manageable size, where an excavator can handle them readily. Anderson Road Quarry, Hong Kong.

(see Figure 8.44). Safety is important, particularly during blasting and excavation works, but the stability of the temporary slopes is a secondary issue.

A computer programme has been developed to aid in the estimation of block-size distribution in a rock mass before and after blasting (Wang and Latham, 1992). In addition, Gama (1995) presented an approach correlating explosive released energy with block-size reduction caused by blasting.

8.5 OPEN-PIT SLOPES

As in quarries, open-pit mine slopes are temporary – until the ore is extracted and site restored (if that is the case). In practice, many sites are essentially abandoned, and gradually collapse. The main difference between open pits and quarries, and most civil slopes, is the large scale, and the determination to cut slopes as steeply as possible to reduce the amount of over-burden (waste rock) that needs to be extracted. There is a fine balance between safety and economy. Failure of individual faces and benches is generally not significant, provided the main haul roads remain available. Because of the temporary nature of the operations, there will be reluctance by management to use support measures other than to protect key routes and structures.

The large scale of open-pit mines means that slope assessment methods, such as wedge-discontinuity analysis, are likely to have limited use, as individual mechanical discontinuities rarely persist for more than a few tens of metres. Analysis tends to be done using large-scale numerical models, such as UDEC which can feature large numbers of discontinuities with different orientations and persistence, expressed statistically. Predictions can be made regarding deformations, and these checked against field measurements made, using inclinometers, total stations and other surveying techniques (Read and Stacey, 2009).

Chapter 9

Underground excavations

Tunnelling is a form of civil engineering construction, carried out in an uncertain and often hostile environment, and relying on the application of special knowledge and resources.

CIRIA (1978)

If it is to your advantage, make a forward move; if not, stay where you are.

Sunzi
The Art of War

9.1 INTRODUCTION

Underground excavation disturbs the pre-existing stress field in terms of the overall rock load and water pressures, if below the water table, as discussed in Chapter 2. Equations and charts for predicting stress changes and concentrations are given in Hoek and Brown (1980) and Brady and Brown (2004), and can be calculated using many commercial software programmes: Phase 2, Examine 3D (Rocscience), FLAC and UDEC and their 3D versions (Itasca) and Plaxis (Plaxis bv). Stresses are concentrated where excavations intersect or where the geometry in the excavation shape is irregular.

Because of the 'sink' produced by the excavation, the pre-existing stress field redistributes to 'flow' around the hole. The rock moves towards the hole created by the excavation until the surrounding rock mass locks up and stabilises naturally, is supported, or collapses.

9.2 DIFFERENCE BETWEEN TUNNELS AND CAVERNS

Tunnels are typically long, often very long, relative to their diameter and this is fundamentally important to the way that the ground can be investigated and the design can be conducted. Almost inevitably, there will be a need to produce a generalised design or sets of designs for tunnels that provide flexibility to cope with all anticipatable conditions. There is a need for preparedness and a strategy for dealing with the worst possible conditions while optimising advance and avoiding wasteful support measures when conditions are good.

By comparison, caverns are larger-diameter structures, in a fixed position relative to the rest of the infrastructure (as in large underground railway stations, crossover structures or hydroelectric powerhouses). Depending on the location and depth, caverns might be investigated in more detail than tunnels and a more realistic design developed prior to construction on the basis of site-specific ground models, testing and numerical modelling. Tunnel design and construction tends to be based on observational techniques and precedent, whereas caverns involve a deterministic design although there will always be an observational element to check that the geology is as anticipated and that deformations during construction are as predicted in the design stage.

9.3 STABILITY CATEGORIES FOR UNDERGROUND EXCAVATIONS

Lunardi (2000) suggests that the design and construction of all tunnels and other underground structures can be split into three categories on the basis of deformation behaviour:

Category A: Stable
Category B: Deforming
Category C: Severe instability

9.3.1 Category A: Stable

The stress in the ground is not sufficiently high to overcome the intrinsic strength of the rock mass. Deformation can be regarded as essentially elastic, occurs rapidly on excavation and is small (of the order of cm). When the rock mass contains rough discontinuities, which is a common situation, then, as the rock deforms towards the excavation, it dilates leading to increased normal stress on rock discontinuities and therefore, increased mass strength. The face and the tunnel periphery are essentially stable; instability only occurs due to the failure of isolated blocks because of localised unfavourable geological structure. Stability is unaffected by water. Stabilisation measures are only necessary to prevent deterioration, to maintain the profile of the excavation in the long-term and as a safety precaution. Lining might still be installed for reasons other than stability, depending on the nature of the project and architectural requirements.

9.3.2 Category B: Deforming

As the tunnel is advanced, the face and tunnel walls deform substantially, as illustrated in Figure 9.1. An 'arch effect' is not formed immediately around the tunnel periphery but at some distance away from the excavation over time. The rock is disturbed and deforms towards the excavation, approximately one tunnel diameter in advance of the face. Stability of the face itself is essential to ensure the overall stability of the excavation.

At 'normal' advance rates, a tunnel is stable in the short term. Rock may break away from the face and around the cavity, but there is sufficient time to install radial support after the face is advanced to stabilise the situation. In some cases, it may be necessary to pre-confine the face, using canopy spiling drilled around the perimeter of the tunnel face or otherwise strengthen the rock in advance of the face. The presence of water can adversely affect the stability, and drainage measures might be required.

9.3.3 Category C: Severe instability

The state of stress is considerably greater than the rock mass strength, even in the face. Natural arching is not possible in such weak rocks relative to the given stress conditions. There may be no time to install support measures before failure occurs. There is a need to improve and support the ground in front of the tunnel face before advancing.

Most rock tunnels fall into either Category A or B, for much of their lengths. Category C applies in highly faulted or weathered zones where the rock mass is especially weak, where the rock mass is prone to swelling or where stresses are particularly high relative to the rock mass strength. The occurrence of Category C conditions in an otherwise benign tunnelling environment often totally outweighs all other aspects of tunnel construction and may result

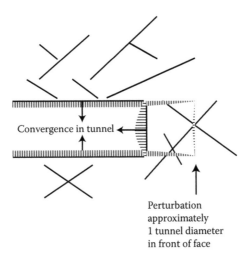

Convergence in tunnel

Perturbation
approximately
1 tunnel diameter
in front of face

Figure 9.1 Deformation at the face and in the tunnel walls. Support is generally constructed after some relaxation of the rock mass has occurred, partly because this is inevitable and also because it reduces the amount of active support that needs to be applied.

in large delays, huge additional costs including remedial works and claims and even abandonment of projects. It is therefore extremely important that, as far as possible, such adverse conditions are anticipated and investigated and that contingencies are put into place to deal with them, if and when encountered.

9.3.4 Other issues

In addition to Lunardi's generalised categories related to broad tunnel behaviour, individual blocks and wedges may become unstable and fail soon after the excavation or in the longer term, because of removal of support, especially in the case of drill-and-blast tunnels where blast-induced stresses and gases have disturbed the rock. It is often a dangerous situation during mucking out and prior to getting temporary support in place.

It is also important to consider other 'environmental' effects that can strongly affect the tunnelling progress; the most important of these is water. Even in Categories A- and B-type rock masses, inflow of groundwater can be a very severe problem and can be difficult to predict and deal with (see Chapter 4). Other environmental hazards include gases, contamination and pollution and active faulting. Anthropogenic hazards for tunnels may include wells, unrecorded boreholes, tunnels and mine workings and, at shallow depths, unexploded bombs and other artefacts.

9.3.5 Overstressing

For shallow excavations, stability in relatively good rock is generally controlled by pre-existing geological variability and anisotropy. The major problems are slab failure from the roof and wedge failure from the tunnel walls. At greater depths, high *in situ* stresses and stress concentrations induced by excavation can lead to severe spalling and even explosive rock bursts; this may be a particular hazard for deep mining and in deep excavations associated with nuclear waste investigations (e.g., Martin and Christiansson, 2009).

Spalling due to overstressing can also occur at shallower depths in excavations in weak rocks; the potential for slabbing failure depends on the ratio of *in situ* stress and intact rock strength. For example, during construction of access tunnels for the Channel Tunnel between England and France, collapses occurred at the UK end due to spalling of what was apparently weak but intact rock on the first exposure. Incipient discontinuities including bedding, that had not been visible during excavation, opened up over days and weeks. The slab failures that ensued forced steel ribs off their supporting foot blocks and necessitated changes in the support measures adopted (Varley, 1996).

9.4 INVESTIGATION

For tunnels and underground excavations, generally, the degree of success hinges on making realistic predictions of the geology, including stress conditions and hydrogeological factors, and then considering the best way to excavate the rock, the requirements for support and the interactions with the proposed finished works. There will always be residual unknowns but if the broad geological conditions are reasonably well anticipated and contingencies are made for any uncertainties, then the project should be constructible within the budget. Sir Harold Harding put it succinctly:

> *Those involved in the design and construction of tunnels should always be prepared to be surprised but not astonished*

(referenced by Brown,1999)

As stated earlier, tunnelling is often fairly routine for good or even relatively poor-quality rocks, but it is the occurrence of zones of Lunardi's Category C, where major collapses can occur, that often outweigh all other conditions. Other considerations that may cause major problems and delays include: abrasive conditions, areas with major and continuing water inflow and sections where the rock is massive and extremely strong and, therefore difficult to break. Site investigation has to be focussed on identifying and characterising such zones.

One process to address the potential hazards is by using the three 'verbal equations' of Knill (1976) to set out the stages of ground investigation and for geotechnical problem analysis, as illustrated by Hencher (1994, 2012a). Each potentially adverse factor, first geological, then environmental and finally relating to the project itself, is considered in turn and its nature and consequence is investigated, as far as possible. Measures and strategies are identified so that each hazard can be mitigated, the risk reduced and the project optimised. This assessment and management procedure is addressed in more detail later in this chapter with reference to risk registers.

9.4.1 Cost of investigation

It is difficult to generalise how much expenditure is justifiable for ground investigation of tunnels and caverns. It really depends on what is already understood from desk and reconnaissance surveys and practical constraints, such as depth of the tunnel and access for deploying a variety of ground investigation techniques, perceived variability of geology and potential consequences. It also depends on the way that tunnels are to be excavated (drill and blast generally being more adaptable to changing conditions), and how robust the design will be. In many cases, forward probing during tunnelling is adopted during construction where poor ground is anticipated.

For shallow structures, quite close spacing of boreholes might be appropriate. For example, for the Toronto Subway Expansion of 25 km of tunnels and 19 stations, the maximum spacing between boreholes, by the final phase of investigation was 50 m for station structures and 75 m for tunnels (Westland et al., 1998). The authors discuss justifiable costs more generally and suggest that beyond a borehole to tunnel length ratio of about 1.5, increased investigation provides little benefit, if any (mainly in terms of reduced claims and other additional costs). Nagel (1992) presents data from a number of projects and concludes that generally, a cost of ground investigation of up to 20% of the project 'contingencies' (engineer's estimate of potential additional costs to the project) is generally justified. In the case of the Dinorwic Diversion Tunnel, discussed later, site investigation costs were almost 2% of the project cost despite little expenditure on boreholes and none on testing. For three of the five tunnels reviewed by West (1983), the cost of site investigation was <0.35% of the construction cost. West identifies facets that could be improved and these are mainly in the design of boreholes (more use of inclined holes especially), sampling, testing and in the use of test results.

9.4.2 Investigation for tunnels: General

Getting the geological model right is very difficult for long tunnels, especially at great depths. Geological mapping and geomorphological interpretation from air photos, satellite images or just from topographic maps can help identify the main geological units, relationships and features, such as major faults. Such interpretation is the standard practice for mapping geologists. Preliminary, ground models are needed at the planning stage for identifying where to locate components of a scheme, such as powerhouses and best routes for tunnels (avoiding major faults as far as possible, in particular). However, despite great geological skills and utilising the 'total geological history' approach advocated by Fookes et al. (2000), extrapolating to the depth of the tunnel is unlikely to give an accurate picture. Even if the overall distribution of lithological units is predicted well, important factors such as openness of discontinuities and nature of any faults can be very difficult to predict accurately. For example, for the Dinorwic Diversion Tunnel in North Wales, West (1983) made a review of geological predictions, based almost solely on surface mapping and found that the overall prediction of rock types was about 64% correct as compared with encountered conditions. Faults identified at the ground surface and anticipated to perhaps cause difficulties were found to be closed and tight at the tunnel level and caused no problems. It is not stated by West whether the less-than-perfect predictions had any particularly adverse effect on the cost or other consequences, but presumably not.

Even where boreholes are put down, for many tunnel projects in hilly areas, they are very widely spaced and sample very small pieces of core, so that the potential for errors in the geotechnical models is high.

9.4.3 Example of geological predictions for a long tunnel

As an example of restrictions on ground investigation, for the Kishanganga hydroelectric power project in Kashmir, India, because of the mountainous terrain and other access restrictions, only three boreholes were put down as part of the investigation for the 24-km headrace tunnel (HRT) linking the dam site to the powerhouse.

The boreholes were located along the downstream portion of the tunnel where weaker rock was anticipated (based on the geological mapping at the ground surface). The tunnel depths at the borehole locations were between 450 and 1050 m but the boreholes were limited to 250 m. Therefore, these did not provide direct geological information at the tunnel level but only general confirmation of the nature of the geological units. Among these three boreholes, only one borehole was actually on the tunnel alignment; the other two were some way off because of access restrictions.

One of the boreholes, where the tunnel depth was 1050 m, was actually located approximately 2 km off the alignment and was used to sample and characterise the rock unit predicted to be at the tunnel level. That borehole was also used to target and characterise specific fault zones and an igneous contact. As the engineering geologist for the project (Mike Palmer of CH2MHILL) says '*Since you only sample a 76 mm diameter section of the tunnel with a borehole, it seems of little consequence whether that 76 mm is exactly on the tunnel route, which is 23 km long, as long as you apply your geological knowledge*'.

The boreholes were largely undertaken to provide information on the rock mass variability, to allow sampling for strength testing of un-weathered material and to undertake *in situ* stress tests; all to assist the numerical modelling work being done for the TBM tunnel. Televiewer logging was used to help understand the geological structure and predict fracture conditions. The rock mass was weathered to depths of about 100 m, but this was not really of interest as it was unlikely to be representative of the majority of tunnel conditions; so, the drilling contractor was instructed to drill an open hole (no recovery) to a 100-m depth.

By comparison several deep holes were undertaken for the powerhouse and for the tailrace tunnel (and other tunnels into the powerhouse). The boreholes for the tailrace tunnel were mainly to determine the bedrock level to maximise the length of construction within andesite rock rather than through the overlying colluvium that was anticipated to be more difficult.

The final geological model largely comprised two main rock units – metasiltstone and metasandstone with a few postulated faults (Figure 9.2). The section was prepared mainly from field mapping of exposure geology. It was originally thought that it was likely that the section of the tunnel under the highest cover might be in granodiorite as encountered on the road leading to the dam site but, on reappraisal, it was considered more probable that the central section would remain in meta-sediments. Some consideration was made regarding tunnel alignment to avoid running parallel to a possible faulted contact between the two units.

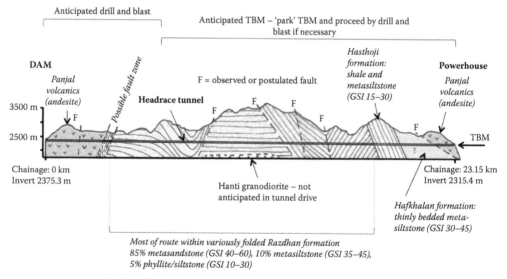

Figure 9.2 Cross section along the line of HRT for Kishanganga. TBM drive from the right. From the TBM end, the andesite/meta-sediment contact was anticipated to be a fault zone comprising black graphitic phyllite with much quartz veining.

Both ends of the tunnel were anticipated to be in andesite. The tunnel was to be constructed in part by TBM (14.5 km), and the rest was to be constructed by drill and blast at a depth of about 1.5 km. There was contractual provision in both the drill and blast and TBM sections of the tunnel to probe ahead and prove geology and investigate hydrogeological conditions, but this was done infrequently. In other tunnels, especially where conditions are worsening, such probing ahead becomes essential, as discussed later.

For the Kishanganga HRT, the main hazards that were anticipated were: poor ground conditions in fault zones with potential collapse, squeezing conditions due to high *in situ* stresses and sudden inflows of groundwater. Clearly, the ground model for the project did not allow foreseeing of hazards' locations with precision; therefore, the TBM was designed robustly to cope with all adverse conditions that might be encountered.

In the event the tunnel was completed 60 days earlier than scheduled, including a record-tunnelling rate (for TBM using segmental liners in the Himalayas) of 816.2 m in November 2012, the success of the tunnelling was largely attributable to close collaboration between the contractor (SELI) and engineer (Halcrow, now CH2M); firstly to recognise the potential hazards, followed by the development of an innovative risk management approach with new modelling techniques and special measures to manage the risks effectively. The special measures were required to prevent compromising the TBM during construction and/or performance of the segmental lining for the completed works.

The special measures adopted were: modifications to the TBM including increased torque on the cutter, increased thrust, clearance of shields, provision for continuous overcutting, forward probing and additional holes for pre-injection grouting.

The purpose of the risk management approach was to provide guidance to the site personnel on the severity of squeezing anticipated along the length of the TBM tunnel, together with monitoring requirements; criteria to be followed during construction, special measures to be implemented in order to allow the TBM to be used to construct as much of the planned length of the tunnel as possible, and with minimum uncertainty and risk to the project.

The contractor for the Kishanganga project tunnelling (SELI) explained the key choices for the TBM in Tunnel Talk (20 May 2014). A double-shield-type machine with high power (~90 kN) was used with a retractable telescopic shield, which enabled the crew to switch from double shield to single-shield operation in those areas where the geology was unstable, to enable excavation to progress in safer conditions. The TBM design also included drill hole portals in the shield that allowed continued TBM excavation while performing probe drilling and ground improvement cycles, extra cutterhead motors, auxiliary thrust rams cylinders and a high conicity of the shield to allow continuous over-excavation by 100 mm to help the machine manoeuvre in squeezing ground. During construction, the crew had to contend with a combination of collapsing ground, cutterhead blockage, the shield getting jammed and therefore, had to hand-mine bypass tunnels over the top of the TBM on three occasions – once in a shear zone where the overburden was 650 m.

Having been prepared for dealing with the unexpected, the contractor had the double shield fitted with top hatches near the rear of the machine to allow the well-experienced crew to hand dig towards the cutterhead. Other measures to mitigate geological challenges included ground consolidation with self-drilling bolts and foam and resin grout injections to fill cavities.

On successful completion of the HRT, it was judged that the geological mapping had been effective in adequately predicting the geological conditions. Two major fault zones were anticipated in the TBM and drill-and-blast tunnels. The chainage, where the fault occurred in the drill-and-blast tunnel was approximately 500 m in error but otherwise, the geology was essentially as predicted. The geological mapping for the TBM tunnelling also identified that the ground conditions would improve over the later stages and, although the preliminary pre-bid information suggested a very weak rock mass (based on test samples collected from the surface that were probably unrepresentative), geological mapping and the boreholes provided information that suggested that the rock mass would in general be better and, so it proved to be.

9.4.4 Directional drilling

Because of the difficulties of drilling to great depths or under water and the general problem of interpolating information between observation points, investigation holes are sometimes drilled to follow the route of the tunnel, particularly to target zones of suspected poor-quality rock. For the 28.3-km-long MetroWest Water Supply Tunnel in Massachusetts, USA, a directional drilling system was used for two sections of the route (Kaplin et al., 1998). Drilling was carried out from the ground surface, bending to follow the tunnel route at a depth of about 63 m, using a flexible core barrel to yield core encased in a clear tube. A programmable survey tool was used to guide the drill string to follow the desired route.

Along one section, the borehole was through granite but also encountered 18.3 m of soft clay gouge. The high total recovery (>95%) allowed the construction risks to be properly evaluated and allowed a TBM to be considered for the work instead of drill and blast, at considerable cost savings. Another borehole was drilled 580 m through a suspected fault zone; in the event, the hole only encountered relatively thin fractured zones of 0.3–2.4 m separated by 2 to 8 m zones of intact rock, which was better than originally anticipated. The borehole was also used to grout the fault zone prior to tunnelling using a TBM and probably contributed to a significant reduction in groundwater ingress and the need for ground support.

The technique has been increasingly used in connection with tunnels in Hong Kong, and indeed, directional drilling is now fairly commonly used, in its own right, to install infrastructure, such as cables and pipes.

One of the early attempts to use directional drilling in Hong Kong was for tunnel F of the sub-sea SSDS project, discussed in more detail later. With 887 m of tunnelling still to complete, the contractor knew that he had to contend with the Tolo Channel Fault, which is a major feature of crustal scale. An attempt was made, very late in the day, to investigate the fault using a borehole drilled backwards along the tunnel line from its target-finishing point on Stonecutter's Island. It was found that the borehole could only advance to 731 m despite three attempts to deviate it. This provided a warning of the difficulties that the tunnellers might expect. In the end, it took 4 months to tunnel through the fault zone and required 750,000 kg of grouting.

9.4.5 Pilot tunnels

Pilot tunnels are commonly used prior to the main tunnelling activities, generally being subsequently used as service tunnels, and as part of the emergency contingency for evacuation or for drainage. Pilot tunnels are of smaller diameter than the main tunnels and therefore, they are likely to be relatively stable. Pilot tunnels allow a better appreciation of the geological conditions along the route than can be gained by other methods.

For the Channel Tunnel, the 'Marine Service Tunnel' acted as a pilot tunnel to establish ground conditions, and was also used to carry out lateral probe drilling and to grout the rock in areas of high permeability to improve conditions for the construction of the larger-diameter-running tunnels (Pollard and Crawley, 1996). For some projects, pilot tunnels are driven first and then used, essentially as top headings to excavate the full circumference of the working tunnel.

9.4.6 Geophysics

As discussed in Chapter 5, various geophysical methods are used to investigate rock masses and to help define properties, such as a dynamic modulus, at the large geotechnical unit scale. For tunnelling, methods that are commonly used include seismic and microgravity.

Seismic reflection and refraction surveys using 'boomer' sources are fairly commonly employed for river and sea crossings. These can reveal the overall structure, depth to bedrock, strata of different velocities and therefore, qualities and importantly any sharp depressions in bedrock that might represent ancient erosion features. Gravity and magnetic surveys can also provide extremely useful data for identifying the presence of major geological structures, as illustrated in Chapter 5.

9.4.7 Investigations for sub-sea tunnels

Inevitably, borehole investigations in marine areas are likely to be of limited extent because of the inherent difficulties in maintaining the location, down time in poor weather and general cost. Often, pilot tunnels are used in such situations.

9.4.7.1 Channel tunnel

Despite the general statement in the previous paragraph, exceptionally large numbers of boreholes were put down for the Channel Tunnel project linking England and France. Harris (1996) describes the site investigation as one of the most thorough ever undertaken. In more than five ground investigation campaigns, 68 boreholes were put down on land (at 1987 comparative costs, £20 thousand each) and 116 marine boreholes (1987 comparative costs, £500 thousand each). However, many of the early boreholes towards the UK end followed a line further north than the route eventually chosen for the tunnel. The boreholes were integrated with, and used to interpret, over 4000-line kilometres of geophysical survey. To establish hydrogeological conditions, there were a considerable number of packer tests and pumping tests performed, though many of these were dismissed as unreliable. On the basis of the ground investigation and interpretations, five Robbins TBMs were selected to excavate most of the tunnels. Because of high-anticipated water inflows towards France, the TBMs were designed to operate in earth pressure balance mode (EPB), that is, with a pressurised chamber to support the face and restrict water inflow where necessary. For the UK end, there were fewer identified faults; so, less water inflow was expected, and open-face machines were adopted. In the event, at the French end, the need to use the EPB machines in closed mode occurred sooner than anticipated. With lower water inflows expected, TBMs designed for the UK end were double-shield machines designed for excavating and supporting broken rock rather than providing support to the face and restricting water inflow. The setup was for essentially 'dry' tunnels with discrete sections (5%) anticipated to be in faulted and broken ground. In fact, the UK TBMs ran into adverse, wet and blocky ground early on, with heavy inflows over a 3.2-km section, which necessitated the TBMs being stopped while grouting was carried out in front of the tunnels.

Despite the extent of the investigation work, in hindsight, Harris (1996) still judged that the ground investigations were inadequate, in the sense that

- There was a substantial underestimation of the length of the tunnel subject to overbreak or identification of the areas subject to overbreak.
- Areas with high water inflow were not adequately identified.
- Variation in structural zones was not identified along the route, and
- The GI did not allow the effect of minor folding to be identified.

Despite early problems with the TBMs, once away from the tunnel section with high water inflow, the Robbins machines advanced quickly with the best month of 1,719 m.

The difficulties arising from poor identification of lengths of difficult ground through ground investigation were compounded by the fact that Rock Mass Classifications (RMR and Q) proved largely inapplicable and gave an inaccurate prediction of the way the ground would behave and were particularly insensitive to predicting overbreak in the Chalk Marl (Harris, 1996). This was perhaps because the classifications were being applied outside their empirical basis where the core parameter, RQD, is not appropriate for Chalk as discussed in Chapter 5.

9.4.7.2 The SSDS tunnels in Hong Kong (later renamed HATS stage 1)

Six tunnels were to be constructed in depth under Hong Kong Harbour in the 1990s, the first such sub-sea tunnels in the Territory (Buckingham, 2003). To establish likely ground conditions, 150 widely-spaced and mostly vertical boreholes were put down. Few of these were in the marine area. In Hong Kong, on land, experience was that usually, where water was encountered in tunnels, conditions dried up quickly (McFeat-Smith et al., 1998). Of course, when tunnelling under standing water and especially the sea, there is potential for unrestricted recharge, which is a major difference.

During the GI, weathered rock was encountered deeper than expected in the boreholes. A decision was made therefore to push the tunnel line deeper, but without any additional boreholes, on the optimistic assumption by the designers that the ground would improve with depth, rock discontinuities would be tight and there would be little water ingress. The contracts for the tunnels were let with all ground risks placed on the contractors.

The contractors, who clearly shared the optimism of the designers, opted for open-face, unshielded-rock TBMs with almost no facility for advance grouting. This proved to be a serious mistake and the contracts came to a halt after excessive inflow of water in tunnels F and C that were only 13.4% and 3.5% complete, respectively, at the time the contract was terminated (Audit Commission HK, 2004). It proved impossible to stem the water flow to meet contractual requirements by post-face grouting and the TBMs were not designed for such wet conditions. The contractor ceased work and the contracts were forfeited with the following arbitrations being won by the owner (all ground risks having been accepted by the contractor).

The contracts were re-let to new contractors without additional ground investigation and largely using the same, refurbished existing unshielded TBMs. The contractual arrangements are unknown but certainly, extensive time was spent carrying out ground treatment in advance of the TBMs in a way that had not been allowed for in the original, forfeited contracts. The most difficult tunnel was tunnel F (see Section 9.4.4). Excavation of the remaining 3098 m still to be completed after the original contract termination was supposed to be finished in 1 year but took nearly 3 years; the advance rate was only 2.9 m per day on average (GEO, 2007). Tunnelling through a single-fault zone of 15-m width that had not been characterised by the ground investigation took 8 months. For the completion of tunnels F and G, the contract sum was $HK356.8 m and additional payments were made, mostly because of the 'difficult ground conditions' in tunnel F, of $HK248.4 m.

On completion, it was concluded that the actual ground conditions were found to be 'much worse than those indicated by the site investigation' and 'major fault zones were more extensive than expected' (Audit Commission HK, 2004). 'In hindsight', the owner (Drainage Service Department) 'considered that the original site investigations were not sufficient to reveal the actual ground conditions'.

9.4.8 Geotechnical baselines and risk registers

9.4.8.1 Geotechnical baselines for tunnels

Reporting of the results from the site investigation, interpretative reporting and setting out Reference Ground Conditions have been addressed in Chapter 5. The concept of contractual Reference Ground Conditions however, takes on particular importance for tunnelling projects where ground conditions are likely to be only broadly predictable along the route.

Aim: Provide a means to judge whether ground conditions are different to those anticipated

The baselines should set out clearly a statement of the best interpretation of likely ground conditions to be encountered during tunnel construction

- Avoid 'soft' baselines with a broad range of material properties or characteristics

- Use unambiguous terminology and descriptions to avoid dispute

- Ensure baseline statements are consistent with payment provision statements and quantities

- Avoid unrealistically conservative or adverse baselines (i.e., 'risk shedding')

- Ensure clarity as the basis of the baselines – risk sharing

- Explain any deviations of baselines from the available database (why interpretations have been made)

Enhance effectiveness of baselines by:

1. Taking care to anticipate conditions realistically using good engineering geological analysis of all available data

The baselines are to be used as a reference if conditions differ

2. Set out likelihoods of occurrence

3. Write clear and concise statements to describe the anticipated conditions

Figure 9.3 Reference ground conditions/baselines.

Reference Ground Conditions or 'Baselines' should be used to set out designers' best estimate of the ground to be encountered, against which some kind of re-measurement can be made, where the contractor considers conditions differing from the Baselines and these have adversely affected his ability to carry out the works and have also led to additional costs. Some guidance on the scope and nature of Baseline documents is given in Figure 9.3.

The British Tunnelling Society (BTS, 2003) defines Geotechnical Baseline Conditions and/or Ground Reference Conditions as follows:

> *Definitive statements about the nature, form, composition and structure of the ground (both artificial and natural) and groundwater together with geotechnical properties of the ground which serve as a basis for construction Contract tendering purposes*

and for the subsequent application of the contract with respect to the conditions actually encountered during Tunnel Works. The Geotechnical Baseline Conditions and/or Ground Reference Conditions represent a contractual definition of 'what is assumed will be encountered'. However, the provision of such Conditions in the Contract is not a warranty that the Conditions will be encountered.

Where the ground is expected to vary, within defined but spatially uncertain limits, Reference Conditions may be expressed in relation to specific bands or zones of characteristics of the ground, or possibly the needs for specific degrees of support, thus providing a basis for estimation and payment.

There is, however, a great risk when defining Reference Ground Conditions for contractual purposes with reference to particular rock mass classifications as shorthand. For example, if one looks at the various components of Q (Chapter 5), it can be seen that there will always be considerable judgement and some kind of averaging, especially when being applied to lengths of tunnel of tens of metres. The description 'poor ground', in a general way, might be readily applied over such lengths (say where there is excessive water inflow or the rock is particularly prone to wedge failure) but the components of Q (such as the number of rock sets, discontinuity quality or RQD) leading to that assignment would not generally stand up to rigorous examination if it came to a legal dispute.

9.4.8.2 Risk registers

For tunnels, a relatively recent management tool is the 'risk register', as used successfully for the Channel Tunnel Rail Link (Brown, 1999).

The British Tunnelling Society, working with the Association of British Insurers prepared the Code of Practice for Risk Management of Tunnel Works in the United Kingdom (BTS, 2003), largely because of unacceptable performance in many underground works. There was a threat that insurers were simply not going to underwrite such projects, if practice did not improve. The Code, whilst targeted at the United Kingdom, has application internationally.

The BTS (2003) defines the Risk Register as follows:

A formalised record of risks identified from the Risk Assessment process including full descriptive details of mitigation and control measures, risk owners and with appropriate cross-references. The Risk Register is the primary means of recording and monitoring the Risk Management process.

The Code of Practice dealt broadly with hazards defining them just *'as an event that has the potential to impact on matters relating to a project'*. For our purposes, we will only consider geotechnical hazards. Hencher (2012a) sets out in some detail, parts of a risk analysis for a long, mountain-loop, rock tunnel in Korea. The tunnel was to be constructed through about 20 km of sedimentary rocks, including, possibly, karstic limestone, coal measures (with mine workings) and mudrocks that had 'squeezed' in previous tunnelling projects. The report ran to many pages individually dealing with the hazards, risks, likelihood, potential mitigation measures (rated according to likely success) and residual risks for excavations either by drill and blast or by TBM. The proposed mitigation measures included investigation ahead of tunnel construction and improving the ground, as well as having detailed strategies in place (and equipment available) to deal with flooding events or instability.

A risk register should be an essential management tool for a project where each risk is assigned to an individual to seek ways to mitigate the potential problems. During construction, regular meetings should be held to check on the progress and to establish the level of residual risks (Brown, 1999).

9.4.9 Investigation for caverns

Investigations for caverns are generally less constrained than for long tunnels, as the construction location is identified early on in the overall design of a project. Caverns are therefore normally designed on some deterministic basis rather than empirically using some prescriptive 'rules'. Because of large spans, caverns are normally sited in good-quality rocks, wherever possible (Sauer et al., 2013). The most important point is to avoid major faults as far as possible, and this can be usually predicted reasonably well from the study of air photographs, satellite images and walk-over surveys by experienced geologists. This is obviously more difficult in urban areas or in areas where the ground has otherwise been markedly altered by man's activities. Seismic refraction surveys or cross-hole seismic tomography may be useful. If the caverns are deep underground, then the number of boreholes that can be put down may be just as restricted by difficulty and cost as for deep tunnels. For example, for the Kishanganga power cavern, only one borehole was put down to the power station depth prior to excavation, and a second borehole was put down for the surge shaft. When there are major uncertainties, then the initial stages of construction, through trial adits and pilot tunnels can serve as part of the confirming investigation. Examples of how severe-encountered conditions are dealt with, are discussed later, when dealing with construction issues.

9.5 DESIGN

9.5.1 Introduction

The design of underground structures requires ground modelling, characterisation, selection of parameters and analysis, so that the design can be optimised in terms of geometry, method of excavation and engineering support measures. Observation and measurement are important during construction to validate ground models and to check that the rock mass is behaving as anticipated. This is the fundamental nature of 'the Observational Method' and/or 'New Austrian Tunnelling Method' (ICE, 1996). In some cases, the original design might be found too conservative so that less support is necessary. Alternatively, and more seriously, ground conditions are encountered that require more support than anticipated or special measures to be adopted. Pells (2011) makes the important comment from review of case studies of major tunnel failures that more than 85% were the result of unexpected or poorly interpreted geology and hydrogeological conditions; the same is, in fact, generally true for many civil-engineering projects (e.g., Burland, 2007).

9.5.2 Design of tunnels

9.5.2.1 Options for tunnelling

For hard rock tunnelling, there are three main options for excavating and the choice of the method is linked to the design, particularly, the amount and type of support required and how this is to be linked to observation during construction:

1. Drill and blast
2. Roadheader
3. Tunnel-boring machine

The advantages and disadvantages are listed in Table 9.1; these factors will assist in judging which method to employ.

Table 9.1 Main methods of tunnelling

Method	Details	Advantages	Disadvantages
Drill and blast	Mechanical excavation using blasting of hard rock combined with mechanical excavators	Rapid deployment Especially useful for short lengths of tunnel but is successfully used for long tunnels Can observe face and tunnel walls easily during excavation (not cramped) Drill holes provide some warning of conditions If necessary can stop to do geological probe drilling Relatively easy to improve ground and apply support in advance of the tunnel Adaptable. Appropriate for the 'Observational Method' Support measures can be selected to meet ground conditions in a way not readily feasible for TBMs	Relatively slow (typically up to 10 m a day with very good equipment and methods) Potentially dangerous period between blasting, mucking and installation of support Vibration and other health and safety restrictions
Roadheader	Use with or without shield Usually mounted on crawler track Similar machines also used in mining	Many of the advantages of drill-blast but with less safety and vibration issues Good for excavating weak and fractured rocks Can be used within a shield to provide support and safety Can have the capacity to excavate and clear muck from the tunnel simultaneously with a linked conveyor system Can cut a range of cross sections, change the diameter and direction quickly and move towards and away from the face under its own power	Not generally applicable to strong and massive rocks Subject to wear in abrasive rock the same way as TBMs Relatively slow advance compared to TBM

(Continued)

Table 9.1 (Continued) Main methods of tunnelling

Method	Details	Advantages	Disadvantages
TBM	Open face or shielded with disc cutters for most rocks (see Figure 9.4) For tunnels where support of face and control of water ingress is required, earth pressure balance mode machine is more appropriate	Rapid advance – can be >100 m a day; cutting and removal of muck carried out simultaneously (but can be delayed in case of unexpectedly poor conditions for which TBM is not designed to cope) TBM body provides support to rock initially Generally clean cut with little overbreak Support can be installed immediately behind the cutting unit either by using a segmental liner installed from the TBM train or by shotcrete/bolting	Expensive initial costs Delay in commencement because of the need to design and manufacture the machine Can get stuck in swelling rock Category C rock (as in very weak ground faults) can lead to collapse and burial of machine Difficult to probe ahead or grout effectively unless the machine is specifically designed to allow this In mixed and weathered rock, it may be difficult to maintain alignment In weak ground, there may be problem with grippers providing thrust Most TBMs are circular; so, wrong shape for many tunnels – that means over-excavation, especially in invert for road tunnels In abrasive ground, slurry causes severe wear to cutters and ancillary equipment Cost and time to replace cutters and picks Inflexible for change in support during construction

Figure 9.4 Open-faced rock TBM with disc cutters.

Usually, more than one method might be appropriate for the same project – it may be a matter of cost, preference or previous experience. For example, for many years in South Korea, there has been a preference for using drill and blast in rock tunnelling despite the rock conditions often being evidently suitable for TBM excavation. This largely stems from poor experience (for whatever reasons) on a number of TBM contracts, which leads to reluctance to use that method. Conversely, in Singapore, TBMs are used very extensively (ironically sometimes with Korean contractors) even though severe weathering and core stones of stronger rock often lead to problems with collapses and settlement in many tunnel (Shirlaw et al., 2000). Elsewhere there may be slow progress because of massive strong rock being encountered even in the same tunnel.

9.5.2.2 Importance of portals

Portals, where the tunnel emerges at the ground surface, are extremely important and can be highly risky areas both during construction (Figures 9.5 and 9.6) and when the works are complete (Figures 9.7 and 9.8). If blocked by a landslide, for example, people may be trapped. If an incident occurs during construction, it may cause major delays to the works. There is a need to consider the stability of the tunnel itself as well as any other hazards, such as rock fall from the surrounding hillside.

Decisions on the level of support required vary from engineer to engineer (Hoek et al., 1995). Judgments are made based on the perceived level of risk, that is, hazard level and potential consequence. This is illustrated in Figures 9.7 and 9.8. For the A55 trunk road through North Wales, it was clearly decided to adopt a very low tolerance level to rock block failure in the portal design. Conversely, in the Italian example in Figure 9.8, it was decided to adopt a minimal support strategy for this one of many similar tunnels through the mountainous terrain adjacent to Lake Maggiore.

Figure 9.5 Temporary construction access to Young Dong mountain-loop railway tunnel, South Korea. Comprehensive slope support measures were adopted that include soil nails to improve slope stability above the portal, anchored beams and thick shotcrete.

Figure 9.6 Formwork for constructing a reinforced concrete canopy prior to commencing a drill-and-blast tunnel (Chongqing, China). Slope protection above the canopy comprises steel mesh with shotcrete.

9.5.2.3 Water inflows

Water inflow is a major hazard for tunnelling works, especially, beneath the sea or other standing water. Methods to predict inflows and deal with water are addressed in Chapter 4.

9.5.2.4 Support based on RMCs

RMCs, such as RMR, Q and RMi, introduced in Chapter 5 have specific application to tunnel construction where they allow some informed judgement over the amount of support

Figure 9.7 Cautious design with numerous instrumented anchors, bolts and dowels used to support rock at the tunnel portal (A55 North Wales, UK).

Figure 9.8 Minimal support of netting and localised shotcrete for a short tunnel, Lake Maggiore, Italy.

required. They are linked to case histories and experience and often work reasonably well within that context in similar ground. Examples are often found however where they do not work too well, generally on a site-specific basis. Pells (2011) compares predictive support requirements based on the Q system in the Sydney region of Australia with that actually employed. In several cases more support was actually used than predicted as necessary from the Q charts but elsewhere, failures occurred. As noted earlier, for the Channel Tunnel construction, the use of both RMR and Q led to the conclusion that the ground would behave better than it actually did.

RMC approaches have been often criticised as being over-simplistic or incapable of representing the rock mass properly and this is of course true to a point. When dealing with real ground conditions with variable geology, how can one reduce that to a single number? Muir Wood in his Harding Lecture (2004) points out, with respect to the concept of the application of RMCs to 'stand-up time' that a single RMC rating might be derived for very different rock masses – some with inherent unstable mechanisms, and others without, – that will behave differently during construction and require quite different support measures. Nevertheless, the value of RMCs is in their empirical databases that allow precedent practice to be applied quickly and with some degree of confidence as discussed below.

9.5.2.5 Use of classification systems for 'precedent design'

Despite criticisms, the use of one or more RMCs – preferably at the same time – will help decision making and it is usual current practice for tunnel designers to attempt to predict the quality of rock along tunnel routes linked to classifications. Of course, the amount of investigation data that will be available for any tunnel (even when using directional drilling along the tunnel route), preclude any realistic prediction of quality. Generally, broad judgements are made. For example, when a tunnel is anticipated to run close to a major fault, or near the Earth's surface, in terrain where weathering is a common issue, it will be assumed that the rock will be of relatively poor quality, even if there is little direct evidence. Conversely, if driving through the centre of an igneous pluton, it might be assumed that the rock quality will be excellent. The judged categories might be quantitatively supported from knowledge of seismic velocities – low velocities being associated with poorer ground quality.

Among the commonly used RMCs, RMR is relatively easy to use, albeit heavily biased towards joint frequency with inherent problems regarding discontinuity definition, as discussed in Chapter 5. It is set out in Figure 9.9 below, together with instructions on adjustments appropriate for tunnelling and guidance on support measures to be installed in Figures 9.10 and 9.11, respectively.

The Q system and RMi system of Palmström (2009) can be used in a similar manner to RMR to estimate rock support requirements using charts based on precedent.

9.5.2.6 Support in squeezing ground

Hoek (2000) concentrates on the construction of tunnels in especially poor ground, citing cases where particular difficulties have been encountered, including the complete destruction of tunnel-boring machines. Some of his recommendations are summarised in Figure 9.12. In very poor rock, multiple headings will often be used, and the face will be supported, as discussed earlier, with reference to Category C rock of Lunardi (2000). It may also be necessary to reinforce outside the tunnel perimeter using spiles or columns of jet grout before advancing. In really poor ground, pre-grouting, drainage and even freezing might be required to allow the tunnel to advance.

For the Yacambu water supply tunnel in the Venezuelan Andes, originally TBMs were used, but one was crushed and the other was withdrawn (Wallis, 1984). The tunnels were repaired to some extent and then, excavation progressed using drill and blast and a roadheader. Initially, the procedure adopted was to apply shotcrete before mucking out. Then, fully grouted 3-m-long rock bolts were installed at 1-m centres followed by another 20-cm shotcrete. Despite these efforts, within 48 h, the invert blew by 50–60 cm and plates and nuts were forced off the rock bolts. The project engineer commented, 'This rock is plastic.

A: *Classification Parameters*								
1	Strength of intact rock (MPa)	Point load PLI_{50}	>10	10–4	4–2	2–1	\multicolumn{2}{l}{PLI_{50} Not applicable}
		UCS	>250	250–100	100–50	50–25	25–5	5–1	<1
	Rating		**15**	**12**	**7**	**4**	**2**	**1**	**0**
2	RQD		90–100%	75–90%	50–75%	25–50%	<25%		
	Rating		**20**	**17**	**8**	**3**			
3	Spacing of discontinuities		>2 m	0.6–2 m	0.2–0.6 m	60–20 mm	<60 mm		
	Rating		**20**	**15**	**10**	**8**	**5**		
4	Condition of discontinuities		Very rough, impersistent, no aperture, unweathered.	Sl. rough, aperture <1 mm, sl. weathered	Sl. rough, aperture <1 mm, highly weathered	Slicken-sided or gouge <5 mm or aperture 1–5 mm, continuous	Soft gouge >5 mm thick, separation > 5 mm, continuous		
	Rating		**30**	**25**	**20**	**10**	**0**		
5	Groundwater	Inflow per 10 m tunnel length	None	<10 litres/min	10–25 litres/min	25–125 litres/min	>125 litres/min		
		Joint water pressure/ major principal stress, σ_1	0	0–0.1	0.1–0.2	0.2–0.5	>0.5		
		General conditions	Completely dry	Damp	Wet	Dripping	Flowing		
	Rating		**15**	**10**	**7**	**4**	**0**		

Base RMR = $\sum$ Ratings 1–5 (Range 100 to 8) [Actually Bienawski (2011) has clarified that RMR can extend down to zero – the lower bound 3 and 5 values above are apparently averages]

Figure 9.9 RMR system for classifying rock masses. (After Bieniawski, Z.T. 1989. *Engineering Rock Mass Classifications*, New York: Wiley, 251pp.)

It never stabilises'. The tunnelling technique was modified to allow the rock to decompress slowly. '*What it requires is a tolerant, flexible intermediate support and a very heavy duty final support'*. Decompression continued, through the invert, for 3–4 months and then the rock was graded flat; a closing-steel invert was then fitted to each arch and embedded in concrete as a solution to the problems.

9.5.2.7 Support measures and internal liners including pressure tunnels

The shotcrete and bolt systems recommended for most tunnels, and incorporated in the RMC charts, balance the excess rock load. For many tunnels, this system of support will be relied upon for the lifetime of the structure with no additional liner installed. For some projects, however, an inner concrete liner with an outer, waterproof-lining sheet fixed to the primary shotcrete liner, is subsequently constructed. This may be for safety reasons, aesthetics or to improve flow in water-carrying tunnels. It is often assumed for the design that the rockbolts and shotcrete will deteriorate in the long term so that a reinforced concrete, cast *in situ* liner, will be designed to carry the entire rock load as well as any other forces, such as from water pressure.

B: *Rating adjustment for discontinuity orientations*

Guidelines for assessment of adversity of discontinuities							
	Strike perpendicular to tunnel axis				Strike parallel to tunnel axis		Dip almost horizontal (0–20), irrespective of strike
	Drive with dip		Drive against dip				
Dip°	45–90	20–45	45–90	20–45	45–90	20–45	
	Very favourable	Favourable	Fair	Unfavourable	Very unfavourable	Fair	Fair

Rating adjustments for RMR						
Strike and dip orientations		Very favourable	Favourable	Fair	Unfavourable	Very unfavourable
Rating	Tunnels and mines	0	−2	−5	−10	−12

C: *Rock mass classes*

Class	I	II	III	IV	V
Description	Very good	Good	Fair	Poor	Very poor
Rating	100–81	80–61	60–41	40–21	<20

D: *Interpretation of rock classes for underground openings*

Class	I	II	III	IV	V
Description	Very good	Good	Fair	Poor	Very poor
Average stand-up time	20 years for 15 m span	1 year for 10 m span	1 week for 5 m span	10 h for 2.5 m span	30 min for 1 span

Figure 9.10 Guidelines for judging adversity of discontinuities in tunnel construction and adjustments to be made to basic RMR. (After Bieniawski, Z.T. 1989. *Engineering Rock Mass Classifications*, New York: Wiley, 251pp.)

When the internal liner is formed (generally cast against the primary liner, with or without a separating waterproofing membrane), initially, the external loading on that liner from the rock mass and from groundwater will be zero. Fairly quickly, following construction of the liner, water pressures that have been disturbed by construction will build up to their original preconstruction levels in the case of fully waterproofed tunnels. The excavation might be designed to be self-draining, so that water pressure will not develop on the liner; but this is only appropriate where drawdown of the water table is not an issue. It is usually an issue in urban areas where dewatering can cause settlement of structures nearby and even at great distances (km) away from the line of the tunnel. It is also an issue where water is extracted from boreholes for irrigation or water supply. Because of the potential for environmental impacts and damage as well as health and safety reasons during construction, maximum acceptable inflows are often specified for tunnelling. The allowable limits are expressed as litres per minute over an arbitrary length of a tunnel. The limits might be specified for pre-bores drilled in advance of the tunnel face as well as for inflow to the tunnel as a whole. Often, controlling water inflow requires grouting the rock mass in front of the advancing tunnel face (Chapter 4). In critical situations, it may be necessary to resort to other measures such as ground freezing or constructing physical impermeable barriers.

In hydroelectric power schemes, lengths of tunnel (especially penstocks) will be periodically subjected to very high differential pressures from the tunnel, outwards into the ground, and from groundwater, into the tunnel when empty, and in this case, steel liners are often used that can resist the cyclic compressive and tensile stresses.

Rock Mass Class	Excavation		Rock bolts (20 mm diameter, fully grouted)	Shotcrete	Steel sets
I – Very Good Rock	Full face 3 m advance		Generally no support required except spot bolting for adverse blocks		
II – Good Rock	Full face 1 to 1.5 m advance. Complete support 20 m from face		Local bolts in crown 3 m long, spaced 2.5 m with occasional wire mesh	50 mm in crown locally	Not required
III – Fair Rock	Top heading and bench 1.5 to 3 m advance in top heading Commence support after each blast Complete support 10 m from face	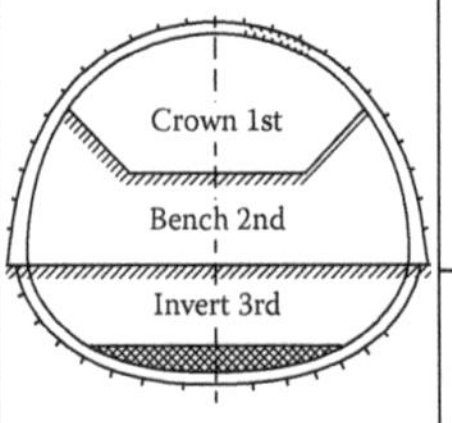	Systematic 4m long bolts, spaced at 1.5 to 2 m in crown and walls with wire mesh in crown	50–100 mm in crown and 30 mm in sidewalls	Not required
IV – Poor Rock	Top heading and bench 1 to 1.5 m advance in top heading Install support concurrently with excavation, 10 m from face		Systematic bolts 4–5 m long, spaced 1–1.5 m in crown and sidewalls with wire mesh	100–150 mm in crown and 30 mm in sidewalls	Light to medium ribs spaced 1.5 m where required
V – Very Poor Rock	Multiple drifts 0.5 to 1.5 m advance in top heading Install support concurrently with excavation. Shotcrete asap	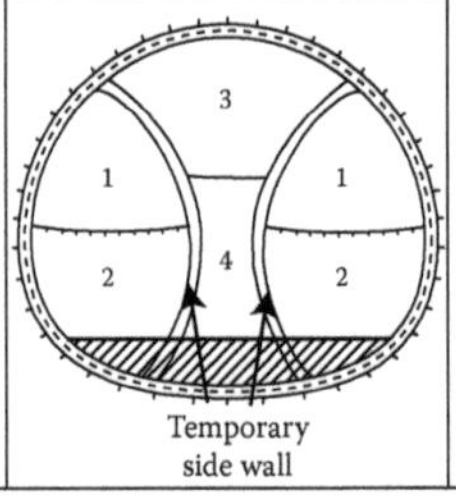	Systematic bolts 5–6 m long, spaced 1–1.5 m in crown and sidewalls with wire mesh. Bolts in invert.	150–200 mm in crown, 150 mm in sidewalls and 50 mm in face	Medium to heavy ribs spaced 0.75 m with steel lagging and possibly forepoling. Close invert to prevent closure.

Figure 9.11 Guidelines for excavation and support of 10-m-span rock tunnel on the basis of RMR assessment. (After Bieniawski, Z.T. 1989. *Engineering Rock Mass Classifications*, New York: Wiley, 251pp.)

9.5.2.8 Tunnels designed for TBM excavation

Choice of a machine, method of excavation and anticipation and planning for how different conditions can be dealt with, is key. Barla and Perlizza (2000) provide a good summary of the options and advantages of different types of TBM available, although this industry is in constant development with new solutions to the challenges in tunnelling. Open-face unshielded TBMs are used where good-quality rock is anticipated throughout the excavation and support is erected as for drill-and-blast tunnels. Shielded machines have pre-cast concrete segment liners that are erected immediately behind the cutter section of the TBM

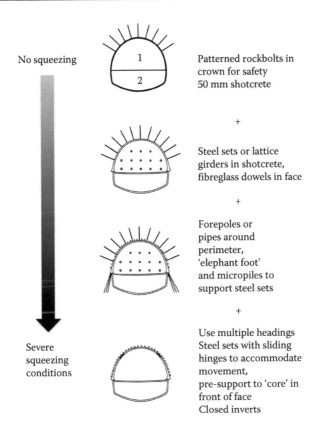

No squeezing

1
2

Patterned rockbolts in
crown for safety
50 mm shotcrete

+

Steel sets or lattice
girders in shotcrete,
fibreglass dowels in face

+

Forepoles or
pipes around
perimeter,
'elephant foot'
and micropiles to
support steel sets

+

Severe
squeezing
conditions

Use multiple headings
Steel sets with sliding
hinges to accommodate
movement,
pre-support to 'core' in
front of face
Closed inverts

Figure 9.12 Methods adopted for tunnel support in squeezing conditions. (Modified from Hoek, E. 2000. Big tunnels in bad rock. 2000. Terzaghi Lecture. *ASCE Journal of Geotechnical and Geoenvironmental Engineering*, **127**, 726–740.)

train (see Figure 9.13). Single-shield TBMs advance by pushing against the concrete segments. Double-shield TBMs have two modes. When the conditions are good, they can grip against the tunnel walls to advance in the same way as for unshielded TBMs. In poorer-quality rock, the thrust is shifted to thrust cylinders that push off against the tunnel segments behind the machine.

Advance rates are predicted mainly based on intact strength or 'strain energy' of the rock and degree of fracturing and the power and torque of the proposed machine, but it is a complex matter. The commonly used relationships are those proposed by the Colorado School of Mines and by the University of Trondheim (1994).

Figure 9.14 shows a hypothetical advance rate curve in which the advance rate should increase markedly, the weaker the rock encountered, both in terms of intact strength and degree of fracturing. In fact, TBMs tend to work best in 'average rock'. TBM advance rates are affected by orientation of prominent discontinuities. Thuro and Plinniger (2003) report how best progress in schistose-foliated rock was achieved where the foliation was at right angles to the tunnel advance. This rather surprising result is explained because the disc cutters act in compression, forming new tensile fractures in line with advance but the rotation of the cutter head takes advantage of the low shear strength on the foliation resulting in an effective breakage mechanism. Where foliation is in line with the tunnel, the advance rate is much lower.

Figure 9.13 Large-diameter launch chamber with pre-cast concrete segments stacked and awaiting transport to the tunnel face for installation behind the TBM. Kishanganga project, India.

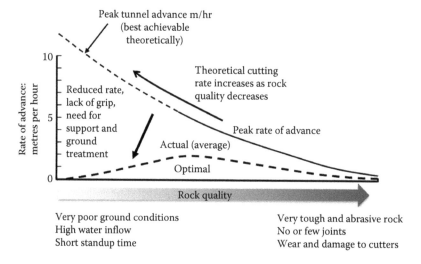

Figure 9.14 Schematic comparison between theoretical advance rate (cuttability of rock) and actual. (Modified from Barton, N.R. 2000. *TBM Tunnelling in Jointed and Faulted Rock*, Rotterdam: Balkema, 173pp.)

If the rock is extremely strong and without fractures, the TBM may find it very difficult to advance (Figure 9.15). Conversely, if the ground conditions deteriorate, there will be delays in supporting the rock and perhaps more caution on behalf of the operators. In really poor ground, the TBM may be stopped, buried and pushed back by inflows or even destroyed.

RMC Q (Q_{TBM}) can be used for predicting advance rates (Barton, 2000, 2002, 2003, 2005). Bieniawski (2007) makes similar claims for a rock mass excavatability (RME) index mainly based on RMR. It is suggested that a 3.5-m TBM tunnel might be excavated in excellent rock, with full site efficiency at between 120 and 60 m a day, on average. These rates then need to be adjusted downwards for a range of factors including learning curve, quality of manpower and other restrictions. The correlations presented for two projects in Ethiopia over the period of monitoring and for other projects in Spain are impressive. However, for one of the Ethiopian projects, there were serious problems and delays due to collapses, as

 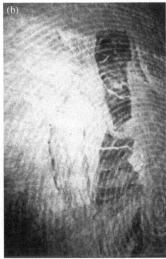

Figure 9.15 (a) 10-m diameter full-face-unshielded Robbins TBM used for Manapouri HRT, New Zealand, and (b) grooves from disc cutters in very strong, massive rock at Manapouri.

discussed later, both before the period of monitoring RME and afterwards. There is no doubt that these events would have seriously impacted the advance rates compared to those achieved for relatively good and more 'forgiving' sections of the tunnel.

9.5.2.9 Tunnelling in weathered rock

In weaker ground, such as mixed weathered rocks, the face may need to be supported by compressed air, earth balance or slurry pumped into a chamber at the front of the machine (Maidl et al., 2014). The use of compressed air involves strict health and safety requirements, and the sudden loss of pressure is always a risk especially where the tunnel runs into some old borehole or well unexpectedly. For EPB and slurry machines, there is always a delicate balance between maintaining sufficient pressure to hold up the face while preventing blow-outs, especially for shallow urban tunnels. There are also inherent difficulties in tunnelling through heterogeneous mixes of hard rock and soil that might even flow and many major failures are reported in such conditions. Furthermore, it may be difficult to steer the TBM on the desired path because it will be deviated away from stronger strata. In fact, it is often preferable to either tunnel well above or below rock head rather than design the tunnel to pass through very variable rock although the location is often dictated by architectural or other constraints. Details and examples of some of the difficulties encountered in tunnelling through weathered rock are given by Shirlaw et al. (2000).

9.5.3 Design of caverns

Caverns are large underground excavations and have many uses including large power-houses and transformer galleries for hydroelectricity projects (HEPs), large underground stations and crossovers, for storage and for sports facilities, to name a few.

For large power and transformer caverns, the key issues are the locations of these caverns to each other and to the ground surface. With respect to ground conditions, other particularly important aspects are the initial field stresses and the orientation of discontinuity systems. Generally, caverns are sited in blocks of good-quality rock wherever possible, without major faults and with the main dimensions oblique to major discontinuity set strikes.

9.5.3.1 Cavern shape

In all underground openings, the intention is that the surrounding rock mass should transfer the *in situ* stress around the perimeter of an excavation without inducing too much deformation. The optimum shape of a cavern will be the one that minimises the moments in the concrete lining. The common shape adopted is an arched roof with vertical sidewalls. In some situations, especially with horizontally bedded, massive strata, the roof shape adopted might be trapezoidal or even flat roofed (Pells, 2007), but this can be very risky where there are adversely oriented joints. Peck et al. (2013) analyse various situations and show that the least displacement and potential for instability is for a roof shaped as a tight arch with a small radius of curvature. The displacement and potential for collapse increases as the roof is flattened. At greater depths, sidewalls may be slightly curved to reduce tensile stress concentration and support requirements (Hoek and Moy, 1993). Generally, the roof and sidewalls are designed with pattern bolting, essentially at right angles to the perimeter. The higher the sidewalls, the longer the anchorages employed.

9.5.3.2 Case example: Preliminary design for large-span underground station

The ground investigation for a proposed station at Taegu in South Korea, illustrated in Figure 9.16, included geophysical characterisation as well as about 20 boreholes spaced on an average at about 50 m. Most boreholes extended to the station depth at 50 m and proved strong to very strong mudstones with RQD typically >50% and often approaching 100%, other than in a faulted zone of up to 100-m width where the RQD was much lower at excavation level. For preliminary design purposes, the engineers assumed the rock to be of good quality and essentially isotropic with parameters derived using the Hoek–Brown

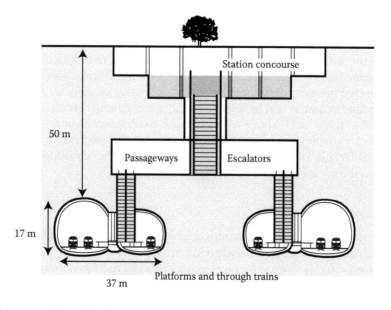

Figure 9.16 Conceptual design for Taegu, high-speed train, underground station.

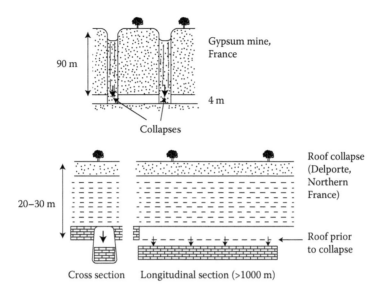

Figure 9.17 Mechanisms for roof failure in rock with vertical orthogonal joints. (After Maury, V. 1993. *Comprehensive Rock Engineering*, Vol. 4, Pergamon Press, 369–412.)

criterion (Chapter 6). The overall span of each station excavation was to be 32.6 m and the height was 16.9 m.

However, on reviewing the ground model, it was identified that many of the boreholes had near-vertical joints, orthogonal to the bedding that was often incipient but sometimes open. The frequency of occurrence of joints in boreholes indicated that the mass was pervasively jointed in this way. The vertical joints were often infilled with calcite and were evidently tensile fractures. This model was confirmed through discussions with local structural geologists as well as from mapping in quarries and road cuttings nearby the site. From observation in boreholes and field exposure, many of the vertical joints were likely to be persistent and 'weak', following the classification given in Chapter 5 (retaining some tensile strength such that the joint might be opened up by disturbance during drilling and handling). It was also noted that the joints were persistent for many metres – that is, the scale of the project. Field mapping showed that the rocks were also intruded with occasional dykes and that fault surfaces could be planar and polished. On the basis of this review, it was identified that further analysis was necessary to allow design against specific failure mechanisms that were not covered in the Hoek–Brown model approach assuming homogeneity and isotropic rock.

The designers were aware that failure on vertical joint systems could be quite extensive even for relatively narrow openings, as described by Maury (1993) and illustrated in Figure 9.17, and that this would require specific design measures to counteract.

Various failure mechanisms could be envisaged, made worse by the complexity of the series of tunnels for cross passageways, lift shafts and escalators, let alone the major platform excavations with spans exceeding 32 m and lengths approaching 1 km, as illustrated in Figure 9.18.

9.5.3.3 Rock load

There are several methods that might be used to estimate rock loads for the design of support systems, including final liners. Many of these are reviewed by Choquet and Hadjigeorgiou (1993) and Hoek et al. (1995) and all were used for the project illustrated in Figures 9.16 and 9.18, to assess the scale of the potential problems.

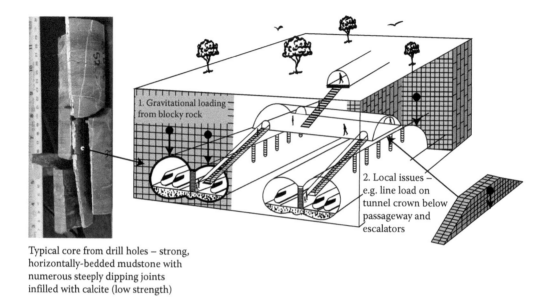

Typical core from drill holes – strong,
horizontally-bedded mudstone with
numerous steeply dipping joints
infilled with calcite (low strength)

Figure 9.18 Potential failure mechanisms for the underground station (preliminary concept).

One of the earliest methods for estimating rock load is that of Terzaghi (1946) and this is still useful for prediction of rock load on liners based on a broad appreciation of rock mass qualities. One of the major attractions of this method is that the terminology is essentially recognisable as geological as is the GSI of Marinos and Hoek (2000).

Terzaghi's conceptual model, with a loosened zone of width B_1 and height H_p above the tunnel is illustrated in Figure 9.19. For rock with a fairly random fracture network, much of the weight of the rock above the tunnel opening will be transferred by the rock adjacent to the tunnel through arching.

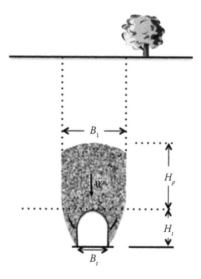

Figure 9.19 Terzaghi's concept of rock load on a tunnel (for steel arches).

Table 9.2 Terzaghi's prediction for rock load (on steel arches) (simplified)

Rock condition	Rock load (multiply by unit weight of rock)	Remarks
1. Hard and intact (no joints)	0	Light lining required only if spalling occurs.
2. Massive or stratified, moderately jointed	0 to 0.5 B_t	Light support, mainly for protection against spalling and localised wedge and block failure. Little or no side pressure.
3. Moderately blocky and seamy	0.25 B to 0.35 $(B_t + H_t)$	
4. Very blocky and seamy (smaller blocks and looser)	0.35 to 1.0 $(B_t + H_t)$	
5. Completely fragmented, brecciated or even sand size	1.10 $(B_t + H_t)$	Considerable side pressure. Softening by water may require either continuous support for feet of ribs or circular ribs.
6. Squeezing rock at moderate depth	1.1 to 2.1 $(B_t + H_t)$	Heavy side pressure, invert struts required. Circular ribs recommended.
7. Squeezing rock at great depth	2.1 to 4.5 $(B_t + H_t)$	
8. Swelling rock	Up to 76 m irrespective of tunnel dimensions	Circular ribs required; in extreme cases, use yielding support.

This rock load depends on the width of the tunnel, B_t and its height H_t. It is assumed that the tunnel is below the water table; if not, Terzaghi advises that the rock load can be reduced by 50% for rock classes 4–6 in Table 9.2.

Using Terzaghi's table as a scoping calculation for the Taegu station example, with $B_t = 32.6$ m and $H_t = 16.9$ m, and where the rock might be considered moderately blocky and seamy, with RQD typically 75%–80%:

The vertical load then $= 0.25\,B_t$ to $0.35(B_t + H_t) \times$ unit weight of rock

$$= 8 \text{ to } 17\,\text{m} \times 27\ \text{kN/m}^3$$

$$= 216 \text{ to } 468\ \text{kPa (kN per square metre of tunnel crown)}$$

Little or no side pressure would be expected.

Again, as a scoping calculation, this compares to a worst-case scenario where the rock load might extend to the ground surface, as in Maury's case examples in Figure 9.17. In this case, the vertical load on a liner might be up to $50 \times 27\ \text{kN/m}^3$, that is, well over 1 MPa.

The U.S. Corps of Engineers (1980) provide a method of estimating loads based on observations in a range of tunnels and openings of widths 4.5 to 30 metre span, heights 4 to 60 metre and at depths shallower than 150 m. Therefore, this method is relevant to the example under discussion here. Using this approach, the average support pressure is estimated as:

Above the spring line (where the roof arch meets the walls of the tunnel):

(Vertical rock load) $= 0.2 \times$ span $\times$ unit weight of rock

$$= 6.52 \times 27\ \text{kN/m}^3$$

$$= 176\ \text{kPa}$$

Below the spring line:

(Horizontal rock load) = 0.1 × height × unit weight of rock

$$= 46 \text{ kPa}$$

Rock load can also be estimated using RMCs. Using the Rock Structure Rating (RSR) approach of Wickham et al. (1972), then the predicted vertical load is $W_r = 284$ kPa which is close to the lower bound prediction from Terzaghi's table, which is not surprising as Choquet and Hadjigeorgiou (1993) point out that the RSR method draws heavily on Terzaghi's work.

Barton et al. (1974) provide guidance on predicting rock load based on Q (Chapter 5). For this case, assuming the number of joint sets >2, then the support pressure in the tunnel roof (crown) is calculated as

$$P_{roof} = (0.2 \times Q^{-1/3})/J_r$$

where J_r is a joint roughness number.

The estimated Q value for the best-quality rock in the example under discussion was calculated as 33 and, taking $J_r = 3$ (also for a good-quality rock mass),

$$P_{roof} = (0.2 \times 33^{-1/3})/3 = 21 \text{ kPa}$$

which is very low compared to other methods and considerations.

For the relatively good-quality rock anticipated, the maximum span (unsupported) is estimated from

$$2 \text{ ESR } Q^{0.4}$$

where ESR ranges from 3 to 5 for a temporary mine opening to 0.8 for public facilities. For $Q = 33$, the maximum span would be about 8 m (times ESR); hence, for the multiple-heading sequence envisaged for this structure, the predicted span ties in quite well with requirements (temporary excavations) if the Q approach was to be considered appropriate for this geological situation.

Grimstad and Barton (1993) gave an alternative equation for estimating rock load:

$$P_{roof} = 2(J_n^{1/2})Q^{-1/3}/3 J_r$$

Taking $J_n = 4$ (two joint sets), $Q = 33$ and $J_r = 3$

$$P_{roof} = 140 \text{ kPa}$$

If the bedding was allowed for as a third set, then $J_n = 9$ and if J_r was taken as smooth planar ($J_r = 1$), keeping $Q = 33$, $P_{roof} = 624$ kPa; the range 140 to 624 kPa is not far off the range predicted using Terzaghi's approach.

9.5.3.4 Conclusions regarding Taegu calculation of rock load

This section demonstrates that rock load might be predicted in a number of ways, based on precedent and broad assumptions. The predicted range is large but even the highest-predicted rock load would be much less than that imposed by a large vertical block failure mechanism of the sort illustrated by Maury (1993). Similarly, it would be unwise to assume that the calculated unsupported span of 8 m was reliable, given the evident potential for discrete block failure. In this case, there is no point in taking this any further analytically, and certainly not by numerical modelling. The correct procedure is to carry out further investigation, using non-vertical boreholes and trial adits to establish the extent, persistence and strength of the sets of vertical joints. The alternative is to design all the structures for very persistent vertical discontinuities and to ensure safety during excavation by using inclined rock anchorages to pin the rock blocks together.

9.5.4 Numerical modelling

Numerical modelling is commonly used in the design of underground excavations, particularly for large-span caverns. As is always emphasised, the results will only ever be as good as the input parameters and validity of the geological model. In particular, the variables of intact rock strength, discontinuity network geometry and *in situ* stresses will strongly affect outcomes. Given the normal state of ignorance of these parameters, it rarely warrants detailed 3D modelling. It is usually more productive and instructive to carry out series of 2D models with sensitivity analysis, varying the parameters according to the predictions of the ground model, however refined that is.

For many projects nowadays, rock loads are predicted using numerical modelling, although such modelling can be complex especially where underground openings are asymmetric and where the geometry changes along the line of the excavation. Furthermore, it may lead to overconfidence where rock mass and pre-existing stress conditions are largely unknown. This is especially true where there is a preferential adverse geological structure for which the use of the standard isotropic assumed rock mass strengths and stiffness parameters are inappropriate. Suitable software packages include FLAC, Phase2 and Plaxis. An example using UDEC is shown in Figure 9.20. Within the numerical models, a degree of relaxation of the rock will need to be assumed prior to the installation of the primary support (shotcrete and rock bolts). The degree of relaxation depends on the nature of the rock mass but in the modelling, it will be, largely, a matter of judgement. The next stage of modelling will be to construct the inner liner up against the face (possibly with low friction interface to account

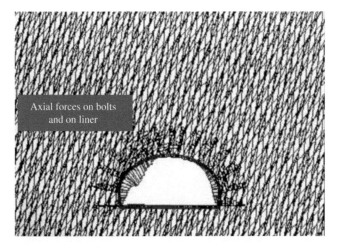

Figure 9.20 UDEC model of underground excavation with rock bolts and concrete liner. (After Swannell, N.G. and Hencher, S.R. 1999. Cavern design using modern software. *Proceedings of the 10th Australian Tunnelling Conference*, Melbourne, March.)

for a waterproofing liner). The final stage of modelling needs to account for the loss of the primary support (either fully or partially – if partial loss of support can be justified logically after considering the likely deterioration processes). The load will then be transferred onto the inner lining and the liner needs to be designed for this load. Clearly, the predictions will be partly subjective and very much dependent on assumed parameters, which in turn depend on assumed rock mass conditions, often with very little actual knowledge of the rock away from the excavation boundaries – certainly before excavation is actually carried out.

Muir Wood (2004) urges caution in the use of numerical models as follows:

'Numerical modelling of an advanced nature is an invaluable tool for research. For practical application it must be kept simple. Tunnel failures have resulted from excessive reliance on numerical models manipulated unwittingly to indicate spurious safety – an expensive form of uncertainty'.

9.6 CONSTRUCTION

9.6.1 Construction of tunnels by drill and blast or roadheader

As illustrated earlier with respect to the use of RMR to anticipate support requirements for tunnels (Figure 9.11), the choice of the construction sequence is largely a matter of the ground conditions anticipated. If the rock is of good quality with long anticipated stand-up time, then the tunnel may be advanced full face, maybe, by drilling and charging up to 5-m-long blast holes, depending on the diameter, complexity, skill of operators and availability of excellent equipment. When the rock is weaker and there is potential instability, then the tunnel or cavern may be excavated in multiple excavations ('drifts'), generally with the crown excavated first.

An advantage of tunnelling by drill and blast/roadheader is that rock conditions can be mapped continuously, monitoring can be carried out and decisions can be made on-site, as to the support necessary, which is fundamental to the observation method, discussed in the next section. That said, there will be a tendency to continue with the same support system unless conditions change drastically, partly because the contractor will have certain equipment available and established methodologies, and hence will be reluctant to alter these for relatively minor changes in rock quality.

9.6.2 The observational method

The observational method is a formalised approach, whereby measurements are made during construction to compare them to anticipated conditions and performance. Peck (1969) in his Rankine lecture, sets out the methodology primarily for soil, but the same principles largely apply to rock conditions albeit that rock commonly fails rather quickly and in a brittle manner; this somewhat limits the opportunity for observation, analysis and judgement over what support measures to introduce, based on measurements.

According to Peck, the application embodies the following 'ingredients' (modified here with relevance to rock tunnels):

1. Knowledge of the broad engineering geological conditions along the route, including geotechnical properties, but not necessarily in detail.
2. Assessment of the most probable conditions to be encountered along the route and the most unfavourable deviation from the conditions (nowadays as assessed by a risk register).
3. Design based on probable behaviour for the most likely conditions to be encountered along the tunnel route.

4. Identification of measurements (quantities) and observations to be made during construction and prediction of their likely values based on a working hypothesis (preliminary design).
5. Calculation of values of the same quantities under the most unfavourable conditions. Comparison of actual measurements during construction with those in points 4 and 5 above will serve to inform the construction team of the seriousness of the situation (provided the original design models were reasonable).
6. Established course of action for every foreseeable condition.
7. Measurement of quantities during construction and evaluation of actual conditions.
8. Modification of the design to suit the actual conditions.

Muir Wood (2004) reviews the methodology and identifies that it has many applications to tunnel construction, including assessing stability and judging the need for support, comparing actual water ingress with that predicted (and the need to take action) and control of settlement. He emphasises, as did Peck, that inherent to its adoption is a need for preparation of counter-actions for the worst-possible conditions, where initial assumptions are shown to be inadequate. The intention to use the Observation Method should be set out in the contractual arrangements before construction. Alternatively, where the method is introduced later to mitigate sudden surprises during construction, Muir Wood emphasises that the basic rules still need to be observed.

Powderham and Nicholson (1996) note that with respect to the design stage, as an adjunct to using the Observational Method, that there needs to be clarity regarding a number of issues including

1. The basis of the design.
2. What are the criteria to establish the most probable conditions?
3. What FoS has been adopted?
4. What is the acceptable level of risk?

These need to be understood by those making decisions during construction, interpreting tunnel behaviour and instrument data, so that they know what actions to take. It was a delayed response to the instrumented data (partly because of poorly defined roles and management) that was one of the main factors highlighted as important to the collapse of tunnels for the Heathrow Express in London (Muir Wood, 2000; Hencher, 2012a).

9.6.3 Mapping

Mapping of geological conditions throughout tunnelling is vitally important. This is done to anticipate poor ground that might affect the excavation (e.g. if about to drill and blast into a zone of poor-quality rock that might collapse and chimney upwards) or support requirements. When approaching fault zones or hydrothermal alteration, there are often strong clues, such as increased fracturing, discolouration, presence of minerals, such as epidote, calcite and talc and veining that can be recognised and interpreted by an experienced geologist. Quite often, poor ground appears first in the corner of an excavated face and provides a warning to be cautious in advancing the tunnel. If poor ground is anticipated, then there may well be a need for forward probing in advance of the face. Examination of the rate of drilling and arisings from the drilling process will aid the interpretation. The same is true when drilling blast holes.

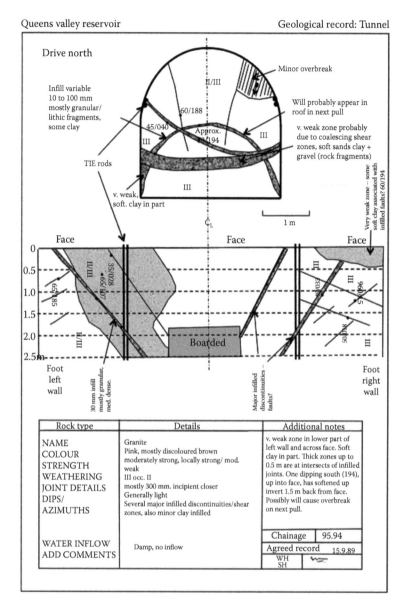

Figure 9.21 Mapping the tunnel face and newly excavated walls and roof following a 2.5-m 'pull'. (After Hencher, S.R. 2012a. *Practical Engineering Geology. Applied Geotechnics*, Vol. 4, Spon Press, 450pp.)

Mapping is usually done using a proforma onto which ground conditions can be sketched (Figure 9.21). Such records should be agreed between parties as a true record, to facilitate any later discussion on claims.

9.6.4 Monitoring

Monitoring of displacements is usually done in tunnels to compare with predicted values. Sometimes, levels are set for action under the headings: Alert, Action and Alarm.

The typical instruments include multipoint borehole extensometers, strain gauges, tilt metres and precise settlement points. Load cells may be installed within liners to monitor rock loads. In urban situations, it will also be important to monitor settlement caused by tunnelling and probably piezometric levels.

9.6.5 Investigating in front of the tunnel during construction

When there is a potential for poor ground ahead, either estimated from the pre-existing ground model or from observations (geological and measurement) during construction, then special measures may need to be taken. Sometimes, these measures are set out in the contractual documents, especially where there is an environmental or other perceived risk to health and safety. An important trigger might be that water inflows are greater than some value allowed for the project even though, sometimes, designers set values at a level where they are really unachievable (Pells, 2004). The commonly used methods include

- Probe drilling ahead of, and to the side of, the tunnel, using blow-out preventers in case of intersecting water under high pressure. Drilling is usually by percussive methods and requires close geological supervision. In some cases, core drilling may be required.
- Tunnel seismic profiling (TSP) from the tunnel face and/or sidewalls. This is carried out using vibration sources that could be small blasts or, for a TBM tunnel, the vibrations from the excavation itself. This method allows some prediction of fault zones (low-velocity horizons) and caves and other openings in karstic terrain.

9.6.6 Installation of support

In drill and blast tunnels installation of rock support should usually be undertaken as soon as possible, as illustrated in Figure 9.22. In many instances, shotcrete can be installed over the top of the muck pile, before it is removed. Remote methods should be used, if possible, to avoid the need to work under unsupported roofs.

Rapidly curing shotcrete can bind the rock blocks together to provide some degree of security as other support measures are installed. Steel ribs or lattice girders are erected and bolted together and blocked off the surrounding rock with hardwood blocks and wedges. Care must

Figure 9.22 Shotcreting the tunnel face above a partially excavated muck pile to try to stabilise the situation.

be taken to ensure that lattice girders are fully encased in shotcrete and that steel ribs are also adequately embedded. Steel lagging can also be used between steel ribs, especially where shotcrete application is difficult due to wet conditions or where the ground has unravelled. Where there is overbreak, it might be possible to partially fill the hole with rock blocks and then grout between the lagging plates and the rock; otherwise, voids should be concreted.

The tunnel invert can be closed with steel ribs infilled with shotcrete and/or concrete, to improve support and to prevent the invert heaving, provided this does not clash with the final design requirements. As noted earlier, in a poor ground, it may be necessary to allow the ground to relax for some time before re-profiling and applying support. Obviously, this requires a balanced approach and engineering judgement hopefully linked to numerical analysis and feedback from instruments (the Observation Approach). A concreted floor surface is good for construction traffic, improves safety (and reduces the risk of overtopping wellington boots) and can incorporate drainage measures.

The use of steel fibre-reinforced shotcrete is preferred for speed and safety, because, then, there is no need to install a welded wire mesh below an unsupported roof. Shotcrete can be applied as dry mix or wet mix. The wet-mix process provides better control on the mix proportions (with better potential strength outcomes) and reduces the hazard of dust in the tunnel. To avoid the problem of wet-mix shotcrete 'going off', it can be stored close to the tunnel face by using hydration control additives. Further, additives can be added later to 'reawaken' the shotcrete, which can then be used in the normal manner. This allows the shotcrete to be readily available at the face, which is particularly important if a rapid response is required to poor ground conditions or the travel time from the shotcrete-batching plant is great. The use of microsilica in modern shotcrete mixes aids 'stickiness' and cohesion, as well as providing other benefits.

In fault zones, or heavily sheared or fractured rock, a friction-anchored bolt may be more appropriate than the fully grouted rock bolts used in other sections of the tunnel. Friction-anchored bolts include the 'Swellex' bolt. This type of bolt has the advantage of very quick installation time and provision of immediate support along the full-bolt length. This type of bolt is more expensive than the grouted rebar type of rock bolt, but shows significant benefits in terms of speed and safety, and is often appropriate.

In areas of poor ground and where there are concerns over larger-scale mechanisms of failure or squeezing ground, then more substantial support is used, such as lattice girders (incorporated into a shotcrete liner) and steel arches (Figure 9.23).

Generally, the need for such measures is judged during the works by experienced tunnel engineers rather than through a process of investigation and numerical analysis, although more measured approaches based on displacements are possible, as discussed in Section 9.6.2.

9.6.7 Support in advance of the tunnel

As discussed at the beginning of this chapter, Lunardi (2000) emphasises the importance of the stability of the face to the overall tunnel stability (not just circumferential deformation). For Category C ground (as in severely faulted or weathered material), when it is not possible to form a natural or partially supported arch, then the ground will need to be improved in advance of tunnelling.

9.6.7.1 Reinforcing spiles

Reinforcing spiles can be drilled above the tunnel to form an umbrella prior to advancing the tunnel. The spiles may be reinforcing bars or larger-diameter steel tubes. This technique can be combined with grouting and/or other measures. Some local enlargement of the

Figure 9.23 Steel ribs and shotcreting to support a zone where tunnel collapse occurred when blasting into weak, hydrothermally altered granite at a depth of about 200 m. Black Hills Tunnel, Hong Kong.

tunnel may be necessary to allow installation of the spiles. Large-diameter spiles are usually installed using a drilling jumbo equipped to install large-diameter (100 mm) casing tubes. Proprietary systems are available. Large-diameter pipe spiles are typically installed at 500-mm centres, possibly in two rows in severe conditions. The pipes are usually perforated to allow grout injection.

9.6.7.2 Other methods

Other methods that might be employed in poor ground include jet grouting to form structural elements and ground freezing.

Another technique is to install a pattern of horizontal glass fibre dowels in holes drilled ahead of the face. These dowels reinforce the face, so that it can be excavated as a vertical full face. The dowels are cut easily as the face advances.

In some instances, when the problems are severe and localised, then it may be necessary to drive additional tunnels and adits around the problem area, and perhaps use these to install reinforcement or to drain excessive water. This is a particularly important technique commonly used where TBMs get stuck.

9.6.8 TBM excavation

For unshielded TBMs, where there is no immediate lining of pre-cast concrete segments, it is possible to log the geological conditions in detail and to apply only the support that is necessary, given the quality of rock. Part of the tailrace tunnel at Manapouri, excavated by an unshielded TBM is illustrated in Figure 9.24. Clearly, the rock shown was of an excellent quality but, still, there was a minimum requirement of pattern bolting. In many areas, this was perhaps of questionable necessity, but it was installed for reasons of ease of construction as well as to meet requirements regarding health and safety by the governing authorities (Charlie Watts, personal communication).

For the Manapouri tunnel, correlation between defect spacing and advance rate was strong and fitted with the Norwegian NTH methodology (University of Trondheim, 1994)

Figure 9.24 Tailrace tunnel Manapouri scheme, New Zealand.

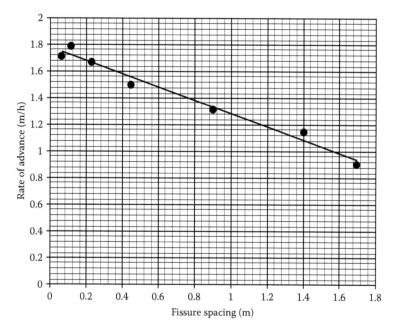

Figure 9.25 Relationship between advance rate and fissure spacing for Manapouri tailrace tunnel. (Data from Macfarlane, D.F., Watts, C.R. and Nielsen, B. 2008. Field application of NTH fracture classification at the second Manapouri tailrace tunnel, New Zealand. *Proceedings of the North American Tunnelling, 2008,* 236–242.)

(Figure 9.25). Where fracturing was closer, progress was dictated by the speed of support installation.

TBMs while offering many advantages in terms of speed, especially in relatively uniform ground, can be inadaptable when poor conditions are encountered, unexpectedly (Hoek, 2000; Barla and Pelizza, 2000; Carter, 2011). As discussed in Section 9.5.2.8, for one of the two tunnels in Ethiopia (Gibe II headrace), when things were progressing well, it was possible to develop a good relationship between advance rate and rock mass quality factors (Bieniawski, 2007). However, the overall excavation achievements were overshadowed by a major collapse in October 2006 with inflows of hot mud at 40-bar pressure from a fault zone. One TBM was pushed backwards and seven segment rings were crushed (Wallis, 2010). There was a 2-year delay for that section of the tunnel. Following the completion of the tunnel, during the opening ceremony, an estimated 8000 m³ of rock fell into the tunnel. According to the contractor, the 'faulty' material that collapsed was not detected during excavation. This illustrates one of the difficulties with TBM construction; where lining segments are erected immediately behind the cutting train of a shielded TBM, it is very difficult to log geological conditions so that major zones that might affect the long-term integrity of the tunnel might not be observed or recorded.

Similarly, it is often a difficult and disruptive exercise to drill ahead of a shielded TBM, especially where progress is acceptable, however much an engineering geologist might wish to do so; instead, the geologist is limited to examine the rock excavated as chippings and to try to interpret this in terms of the likely *in situ* ground conditions.

9.7 CAVERN CONSTRUCTION

For tall caverns as used in HEP powerhouses, generally, excavation begins with a top heading and the installation of rock support in the roof arch followed by a concrete liner.

Generally, if the ground conditions are quite good, without major zones of fault gouge or weathered or hydrothermally altered rock, then suitable solutions can be found to deal with local instability. For example, in the Kishanganga powerhouse excavation, a steeply dipping fault zone extended essentially parallel to one sidewall and an unexpected set of joints dipping into the excavation were found on the other side wall. These problems were simply dealt with by installing additional rock bolts, re-oriented to deal with the potential wedge failures (Figure 9.26a and b).

When conditions are more severe or the rock mass is generally weaker, then, more drastic remedies will need to be adopted. For the Mingtan cavern in Taiwan, the presence of infilled faults and steep bedding necessitated a series of cable anchors to be installed from

Figure 9.26 (a) Rock bolting to the sidewall of power cavern where fractured ground (a fault) runs almost vertically. (b) Rock bolting to cope with specific failure mechanisms of steeply dipping joints in the sidewall, Kishanganga powerhouse, Kashmir, India.

galleries to support the roof of the cavern before excavation (Hoek and Moy, 1993). For the Pergau dam project, where *in situ* stress levels were found to be lower than anticipated (see Chapter 3 and Murray and Gray, 1997), the site was anticipated to be in reasonably good granite with weathering limited to about 80 m, below ground surface. However, two major discontinuities were encountered during the excavation of tunnels at depth. Two particular major 'fissures' were associated with hydrothermal alteration and were largely infilled with completely altered granite (sand) and calcite. Elsewhere, the infill included epidote and zeolites, typical of hydrothermal alteration in granite. Locally, there were voids up to 800 mm. Despite grouting and drainage, it was then determined that there was potential for a wedge failure of 22,900 tonnes into the power cavern (at a depth of about 350 m). The wedge could not be supported by standard design detail of dowels, mesh and shotcrete. This necessitated the construction of a purpose-built gallery to 82 cable anchors, each of 225-tonne capacity, incrementally as the cavern was excavated (Varley et al., 1997).

There are no hard-and-fast-rules for how construction is to be achieved – it depends on the geometry, the strength of the rock mass, orientation of the fracture network as well as water and *in situ* stress conditions. It also depends on other factors, such as the construction sequence and access. This is well illustrated by the Channel Tunnel Project.

For the crossover cavern towards the UK side of the Channel Tunnel, the cavern was constructed in weak chalk marl using roadheader machines in multiple headings, as illustrated in Figure 9.27. Tape and optical measurement of deformation allowed the efficiency of rock dowels and load-bearing behaviour of the shotcrete shell to be determined in line with NATM principles (John and Allen, 1996).

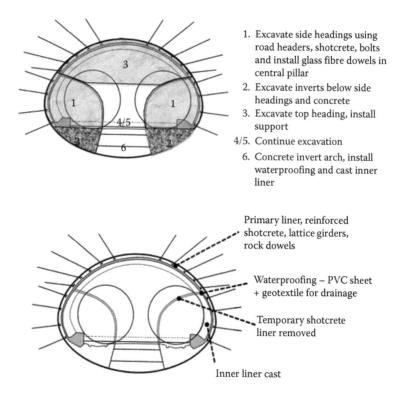

1. Excavate side headings using road headers, shotcrete, bolts and install glass fibre dowels in central pillar
2. Excavate inverts below side headings and concrete
3. Excavate top heading, install support
4/5. Continue excavation
6. Concrete invert arch, install waterproofing and cast inner liner

Primary liner, reinforced shotcrete, lattice girders, rock dowels

Waterproofing – PVC sheet + geotextile for drainage

Temporary shotcrete liner removed

Inner liner cast

Figure 9.27 Construction sequence and the final support design for English crossover for the Channel Tunnel. (Modified from John, M. and Allen, R. 1996. *Engineering Geology of the Channel Tunnel*, London: Thomas Telford, 310–336.)

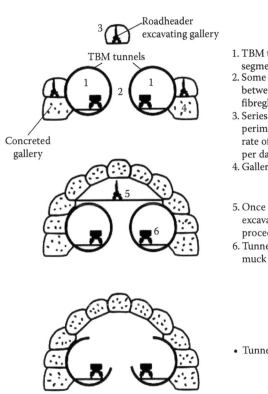

1. TBM tunnels completed with segmental liners
2. Some reinforcement of pillar between tunnels using mostly fibreglass dowels
3. Series of galleries excavated around perimeter using road headers with rate of advance between 5 and 12 m per day
4. Galleries filled with concrete

5. Once outer galleries concreted, excavate top heading and then proceed downwards
6. Tunnels used in part to take out muck

• Tunnel segments dismantled

• Invert excavation and concrete

Figure 9.28 Construction sequence for the French crossover cavern for the Channel Tunnel. (Modified from Leblais, Y. and Leblond, L. 1996. *Engineering Geology of the Channel Tunnel*, London: Thomas Telford, 349–360.)

At the French end, a different approach was adopted. In this case, the two TBM tunnels had already been constructed; so, the crossover cavern needed to be constructed around these existing tunnels. The excavation procedure adopted was as illustrated in Figure 9.28. Firstly, an arch was formed around the periphery of the final cavern structure using a series of individual galleries excavated using roadheader machines. Each gallery was then fully concreted. Once the arch was formed, the cavern chamber was excavated, existing tunnel liners were removed and the invert arch closed (Leblais and Leblond, 1996).

Appendix: Conversion factors (to two decimal places)

To convert from	To	Multiply by
Length		
Inches	Millimetres	25.4
Feet	Metres	0.30
Metres	Feet	3.28
Kilometres	Miles	0.62
Miles	Kilometres	1.61
Area		
Square metres	Square feet	10.76
Square feet	Square metres	9.29×10^{-2}
Volume		
Cubic metres	Cubic feet	35.31
Cubic feet	Cubic metres	0.02
Litres	Cubic metres	0.001 m^3
Cubic metres	Litres	1000
Gallons (UK)	Litres	4.55
Litres	Gallons (UK)	0.22
Cubic metres	Gallons (US)	264.17
Cubic metres	Gallons (UK)	219.97
Force		
Pounds	Kilograms	0.45
	Tonnes (metric)	4.53×10^{-4}
	Newtons	4.44
Tonnes (metric)	Kilograms	1000
	Pounds	2204.62
	Kilonewtons	9.80
Kilonewtons	Pounds	224.81
	Tonnes (metric)	0.10
	Kilograms	101.97
Stress or Pressure		
Pounds/square foot	Kilonewtons/square metre	0.04
Pounds/square inch	Kilonewtons/square metre	6.89
Kilograms/square centimetre	Pounds/square inch	14.22

(Continued)

To convert from	To	Multiply by
	Kilonewtons/square metre	98.06
Tonnes (metric)/square metre	Kilograms/square centimetre	0.10
	Kilonewtons/square metre	9.80
Kilonewtons/square metre	Pounds/square foot	20.88
	Kilograms/square centimetre	0.01
	Tonnes (metric)/square metre	0.10
	1 bar = 100 kPa = 0.1 MPa	
Unit weight		
Tonnes (metric)/cubic metre	Grams/cubic centimetre	1.00
	Megagrams/cubic metre	1.00
	Pounds/cubic foot	62.42
Pounds/cubic foot	Tonnes (metric)/cubic metre	0.01
	Kilonewtons/cubic metre	1.57×10^{-1}
Kilonewtons/cubic metre	Tonnes (metric)/cubic metre	0.10
	Pounds/cubic foot	6.36

References

AASHTO. 2007. *LRFD Bridge Design Specification*, 4th edition. Washington, DC: American Association of State Highway and Transportation Officials.

Aksoy, E., Unceöz, M., and Koçyigit, A. 2007. Lake Hazar Basin: A negative flower structure on the East Anatolian Fault System (EAFS), SE Turkey. *Turkish Journal of Earth Sciences*, 16, 319–338.

Alejano, L.R., Bedi, A., Bond, A., Ferrero, A.M., Harrison, J.P., Lamas, L., Migliaza, M.R. et al. 2013. Rock engineering design and the evolution of Eurocode 7. In *Rock Mechanics for Resources, Energy and Environment*, Vigo, Spain (eds., M. Kwasniewski and D. Lydzba), Boca Raton, FL: Taylor & Francis, 777–782.

Al-Harthi, A.A. and Hencher, S.R. 1993. Modelling stability of the Pen-Y-Clip tunnel, UK. *Safety and Environmental Issues in Rock Engineering, Eurock '93*, Lisbon: Balkema, 451–459.

Ameen, M.S. 1995. Fractography: Fracture topography as a tool in fracture mechanics and stress analysis. An introduction. In *Fractography: Fracture Topography as a Tool in Fracture Mechanics and Stress Analysis* (ed., M.S. Ameen), London: Geological Society, Special Publication No. 92, 1–10.

American Society for Testing and Materials. 2008. *Standard Test Method for Performing Laboratory Direct Shear Tests of Rock Specimens Under Constant Normal Force*. West Conshohocken: ASTM International.

American Society for Testing and Materials. 2013. ASTM D5313/D5313M-12 *Standard Test Method for Evaluation of Durability of Rock for Erosion Control under Wetting and Drying Conditions*. West Conshohocken: ASTM International.

Amontons, G. 1699. De la résistance causée dans les machines. *Memoires de l'Academie Royale*, A, 257–282.

Anderson, E.M. 1951. *The Dynamics of Faulting*. Edinburgh: Oliver and Boyd, 206pp.

Anon. 1970. The logging of rock cores for engineering purposes. *Quarterly Journal of Engineering Geology*, 3, 1–14.

Anon. 1972. The preparation of maps and plans in terms of engineering geology. *Quarterly Journal of Engineering Geology*, 5, 293–382.

Anon. 1995. The description and classification of weathered rocks for engineering purposes. Geological Society Engineering Group Working Party Report. *Quarterly Journal of Engineering Geology*, 28, 207–242.

Anon. 1997. *Tropical Residual Soils*. Geological Society Engineering Group Working Party Revised Report, 184pp.

Archambault, G., Verreault, N., Riss, J., and Gentier, S. 1999. Revisiting Fecker–Rengers–Barton's methods to characterize joint shear dilatancy. In *Rock Mechanics for Industry* (eds., Amadei, Kranz, Scott, and Smeallie), Rotterdam: Balkema, 423–430.

Archard, J.F. 1958. Elastic deformation and the laws of friction. *Proceedings of the Royal Society, Section A 243*, 190–205.

Archard, J.F. 1974. Surface topography and tribology. *Tribology*, 7, 213–220.

Atkinson, B.K. 1987. Introduction to fracture mechanics and its geophysical applications. *Fracture Mechanics of Rock*, London: Academic Press, pp. 1–26.

Atkinson, J.H. 1990. Discussion. In *Field Testing in Engineering Geology* (eds., F.G. Bell, M.G. Culshaw, J.C. Cripps, and J.R. Coffey), London: Geological Society, Engineering Geology Special Publication No. 6, 235–235.

Atkinson, J.H. 2002. What is the matter with geotechnical engineering? *Proceedings of the Institution of Civil Engineers, Geotechnical Engineering*, 155, 155–158.

Attewell, P.B. 1993. The role of engineering geology in the design of surface and underground structures. *Comprehensive Rock Engineering*, Vol. 1, *Fundamentals*. Oxford: Pergamon Press, (Volume Editor, Brown, E.T.), 111–154.

Audit Commission. H.K. 2004. Chapter 3 Harbour Area Treatment Scheme Stage I, pp. 52–79. http://www.legco.gov.hk/yr03-04/english/pac/reports/42/ch3.pdf

Australian Standards. 1993. *Geotechnical Site Investigations*. AS-1726.

Azzoni, A., La Barbera, G., and Zaninetti, A. 1995. Analysis and prediction of rockfalls using a mathematical mode. *International Journal of Rock Mechanics and Mining Science & Geomechanical Abstracts*, 32(7), 709–724.

Azzoni, A., Rossi, P.P., Drigo, E., Giani, G.P., and Zaninetti, A. 1992. *In situ* observation of rockfall analysis parameters. *Proceedings of the 6th International Symposium on Landslides*, Christchurch, 307–314.

Baecher, G.B. and Christian, J.T. 2003. *Reliability and Statistics in Geotechnical Engineering*. The University of Michigan: John Wiley & Sons Ltd., 605pp.

Bandis, S.C. 1990. Scale effects in the strength and deformability of rocks and rock joints. In *Scale Effects in Rock Masses* (ed., Pinto da Cunha), Rotterdam: Balkema, 59–76.

Bandis, S.C., Lumsden, A.C., and Barton, N.R. 1981. Experimental studies of scale effects on the shear behaviour of rock joints. *International Journal of Rock Mechanics and Mining Sciences and Geomechanics Abstracts*, 18, 1–21.

Barla, G. and Pelizza, S. 2000. TBM tunnelling in difficult conditions. *Proceedings of GeoEng 2000*, Melbourne, Australia, 329–354.

Barton, N. 2002. Some new Q-value correlations to assist in site characterization and tunnel design. *International Journal of Rock Mechanics and Mining Science*, 39, 185–216.

Barton, N.R. 1982. Shear strength investigations for surface mining. In *Stability in Surface Mining, Proceedings 3rd International Conference* (ed., C.O. Brawner), Vancouver, 171–196.

Barton, N.R. 2000. *TBM Tunnelling in Jointed and Faulted Rock*. Rotterdam: Balkema, 173pp.

Barton, N.R. 2003. TBM or drill-and-blast. *Tunnels & Tunnelling International*, June, 20–22.

Barton, N.R. 2005. Comments on "A critique of QTBM." *Tunnels & Tunnelling International*, July, 16–19.

Barton, N.R. and Bandis, S.C. 1990. Review of predictive capabilities of JRC-JCS model in engineering practice. In *Rock Joints, Proceedings International Symposium on Rock Joints*, Loen, Norway (eds., N. Barton and O. Stephansson), Rotterdam: Balkema, 603–610.

Barton, N.R. and Choubey, V. 1977. The shear strength of rock joints in theory and practice. *Rock Mechanics*, 12(1), 1–54.

Barton, N.R., Lien, R., and Lunde, J. 1974. Engineering classification of rock masses for the design of tunnel support. *Rock Mechanics and Rock Engineering*, 6, 189–236.

Beale, G. and Read, J. 2014. *Guidelines for Evaluating Water in Pit Slope Stability*. Boca Raton, FL: CRC Press, Taylor & Francis, 680pp.

Bear, J. 1972. *Dynamics of Fluids in Porous Media*. American Elsevier.

Beer, A.J., Stead, D., and Coggan, J.S. 2002. Estimation of the joint roughness coefficient (JRC) by visual comparison. *Rock Mechanics & Rock Engineering*, 35, 65–74.

Beitnes, A. 2005. Lessons to be learned from Romeriksporten. *Tunnels & Tunnelling International*, June, 36–38.

Bergbauer, S. and Martel, S.J. 1999. Formation of joints in cooling plutons. *Journal of Structural Geology*, 21, 821–835.

Bieniawski, Z.T. 1976. Rock mass classification in rock engineering. In *Exploration for Rock Engineering, Proceedings of Symposium*, Cape Town (ed., Z.T. Bieniawski), Vol. 1, 97–106.

Bieniawski, Z.T. 1978. Determining rock mass deformability—Experience from case histories. *International Journal of Rock Mechanics and Mining Sciences Geomechanical Abstracts*, 15, 237–247.

Bieniawski, Z.T. 1989. *Engineering Rock Mass Classifications*. New York: Wiley, 251pp.

Bieniawski, Z.T. 2007. Predicting TBM excavatability—Part II. *Tunnels & Tunnelling International*, December, 15–18.

Bieniawski, Z.T. 2011. Misconceptions in the applications of rock mass classifications and their corrections. *ADIF Seminar on Advanced Geotechnical Characterization for Tunnel Design*, Madrid, Spain, June 29, 2011, 32pp.

Bigger, J.C. 1995. Rock fall mitigation program at Nantahala Dam using wire rope nets in conjunction with other techniques. *46th Highway Geology Symposium*, May 14–18, Charleston, West Virginia.

Black, J.H. 2010. The practical reasons why slug tests (including falling and rising head tests) often yield the wrong value of hydraulic conductivity. *Quarterly Journal of Engineering Geology*, 43, 345–358.

Black, J.H., Barker, J.A., and Woodman, N.D. 2007. *An investigation of 'sparse channel networks'*. Report No. R-07-35. Svensk Kärnbränslehantering AB (SKB), 123pp.

Blythe, F.G.H. 1967. *A Geology for Engineers*, 5th edition, London: Arnold, 351pp.

Bowden, F.P. and Tabor, D. 1956. *The Friction and Lubrication of Solids. Methuen's Monographs on Physical Subjects*. London: Methuen & Co. Ltd.

Bowles, J.E. 1996. *Foundation Analysis and Design*, 5th edition. New York: McGraw-Hill, 1175pp.

Brady, B.H.G. and Brown, E.T. 2004. *Rock Mechanics for Underground Mining*, 3rd edition. Amsterdam: Springer.

Brand, E.W. 1981. Some thoughts on rain-induced slope failure. *Proceedings 10th International Symposium on Soil Mechanics and Foundation Engineering*, Stockholm, Vol. 3, 373–376.

British Geological Survey. 1997. *Geology of the West Cumbria District*. Memoir for 1:50 000 Geological Sheets 28 Whitehaven, 37 Gosforth and 47 Bootle (England and Wales), 138pp.

British Standards Institution. 1986. BS 8004:1986 *Code of Practice for Foundations*. 186pp.

British Standards Institution. 1989. BS 8081:1989 *Code of Practice for Ground Anchorages*.

British Standards Institution. 1999. BS 5930:1999 *Code of Practice for Site Investigation*. 206pp.

British Standards Institution. 2002. EN ISO 14688-1. *Geotechnical Investigation and Testing—Identification and Classification of Soil—Part 1: Identification and Description*.

British Standards Institution. 2003. EN ISO 14689-1. *Geotechnical Investigation and Testing—Identification and Classification of Rock—Part 1: Identification and Description*.

British Tunnelling Society. 2003. *The Joint Code of Practice for Risk Management of Tunnel Works in the UK*, 18pp.

Brown, E.T. (ed.) 1981. Suggested methods for determining shear strength. In *Rock Characterisation Testing and Monitoring*. Oxford: Pergamon Press.

Brown, E.T. and Hoek, E.T. 1978. Trends in relationships between measured rock *in-situ* stresses and depth. *International Journal of Rock Mechanics and Mining Sciences*, 15, 211–215.

Brown, R.H.A. 1999. The management of risk in the design and construction of tunnels. *Korean Geotechnical Society, Tunnel Committee Seminar, 21st Seminar*, Seoul, September 21, 22pp.

Buckingham, R.J. 2003. *Problems Associated with Water Ingress into Hard Rock Tunnels*. Unpublished MSc dissertation, the University of Hong Kong.

Budkewitsch, P. and Robin, P.-Y. 1994. Modelling the evolution of columnar joints. *Journal of Volcanology and Geothermal Research*, 59, 219–239.

Burland, J.B. 2007. Terzaghi: Back to the future. *Bulletin Engineering Geology and Environment*, 66, 29–33.

Burland, J.B. 2008. Discussion to Schofield, 2006. Interlocking, and peak and design strengths. *Géotechnique*, 58, 527.

Burland, J.B. and Burbidge, M.C. 1985. Settlement of foundations on sand and gravel. *Proceedings of the Institution of Civil Engineers*, 78, 1325–1381.

Byerlee, J.D. 1978. Friction of rocks. *Pure Applied Geophysics*, 116, 615–626.

California Division of Mines and Geology. 1981. Recommended guidelines for preparing engineering geologic reports. *Engineering Geologist*, GSA, 16, No. 2, May 3–5. (Reference from lecture notes of Professor Fred Kulhawy.)

CALTRANS. 2010. *Soil and Rock Logging, Classification, and Presentation Manual*, State of California: Department of Transportation, 81pp.

Carter, T.G., 2011. Himalayan ground conditions challenge innovation for successful TBM tunnelling. Invited paper in *Proc. Hydrovision India 2011*, SESSION 5c: (Risk Management in Tunnelling), 20pp.

Carter, T.G., de Graaf, P., Booth, P., Barrett, S., and Pine, R. 2002. Integration of detailed field investigations and innovative design key factors to the successful widening of the Tuen Mun Highway. *Proceedings of the 22nd Annual Seminar*, Geotechnical Division of the Hong Kong Institution of Engineers, May 8, 2002, 187–201.

Chan, Y.C., Chan, C.F., and Au, S.W.C. 1986. Design of a boulder fence in Hong Kong. *Proceedings of the International Conference on Rock Engineering and Excavation in an Urban Environment*, Hong Kong, 87–96.

Chau, K.T., Wong, R.H.C., and Lee, C.F. 1998. Rockfall problems in Hong Kong and some new experimental results for coefficients of restitution. *International Journal of Rock Mechanics and Mining Sciences & Geomechanical Abstracts*, 35(4–5), 662–663.

Cho, S.-M., Ahn, S.-S., Lee, Y.-K., and Oh, S.-T. 2009a. Incheon Bridge: Facts and the state of the art. *Proceedings of the International Commemorative Symposium for the Incheon Bridge*, Inchon, South Korea, 18–28.

Cho, S.-M., Kim, J.-H., Lee, H.-G., Kim, Z.-C., Shin, S.-H., Song, M.-J., Kim, D.-J., and Park, Y.-H. 2009. Pile load tests on the Incheon Bridge Project. *Proceedings of the International Commemorative Symposium for the Incheon Bridge*, Inchon, South Korea, 488–499.

Choquet, P. and Hadjigeorgiou, J. 1993. The design of support for underground excavations. *Comprehensive Rock Engineering* (chief editor, J. Hudson), Vol. 4, 313–348.

Clayton, C.R.I. 1995. *The Standard Penetration Test (SPT): Methods and Use*. CIRIA Report 143, 144pp.

Clayton, C.R.I., Simons, N.E., and Mathews, M.C. 1995. *Site Investigation. A Handbook for Engineers*, 2nd edition. Granada, 424pp.

Coggan, J.S., Stead, D., Howe, J.H., and Faulks, C.I. 2013. Mineralogical controls on the engineering behavior of hydrothermally altered granites under uniaxial compression. *Engineering Geology*, 16, 89–102.

Combault, J., Morand, P., and Pecker, A. 2000. Structural response of the Rion Antirion Bridge. *Proceedings of the 12th World Conference on Earthquake Engineering*, Auckland, Australia.

Cook, K.L., Turowski, J.M., and Hovius, N. 2014. River gorge eradication by downstream sweep erosion. *Nature Geoscience*, 7, 682–686. doi:10.1038/ngeo2224.

Cooper, A.H. and Waltham, A.C. 1999. Subsidence caused by gypsum dissolution at Ripon, North Yorkshire. *Quarterly Journal of Engineering Geology*, 32, 305–310.

Culshaw, M.G. and Waltham, A.C. 1987. Natural and artificial cavities as ground engineering hazards. *Quarterly Journal of Engineering Geology and Hydrogeology*, 20, 139–150.

Davis, G.H. and Reynolds, S.J. 1996. *Structural Geology of Rocks and Regions*, 2nd edition. John Wiley & Sons, 776pp.

Davy, P., Bour, O., De Dreuzy, J.-R., and Darcel, C. Flow in multiscale fractal fracture networks. *Fractal Analysis for Natural Hazards*, Geological Society, London, Special Publications, 261, 31–35.

Dearman, W.R. 1986. Discussion on fracture state. In *S I Practice: Assessing BS5930, Geological Society, Engineering Geology Special Publication No. 2* (ed., A.B. Hawkins), 69–70.

Dearman, W.R. and Fookes, P.G. 1974. Engineering geological mapping for civil engineering practice in the United Kingdom. *Quarterly Journal of Engineering Geology*, London: Geological Society, 7, 223–256.

Deere, D.U. 1968. Geological considerations. In *Rock Mechanics in Engineering Practice* (eds., K.G. Stagg and O.C. Zienkiewicz), Chapter 1, New York: Wiley, 1–20.

Deere, D.U. and Deere, D.W. 1989. *Rock Quality Designation (RQD) after Twenty Years*. Contract Report GL-89-1, US Army Corps of Engineers, 67pp. plus Appendix.

DeGraff, J.M. and Aydin, A. 1987. Surface morphology of columnar joints and its significance to mechanics and direction of joint growth. *Geological Society of America Bulletin*, 99, 605–617.

Dering, C.J. 2003. Ground conditions risk: A conspiracy theory. *Proceedings 14th SE Asian Geotechnical Conference on Geotechnical Engineering*, Vol. 3, 161–164.

Dershowitz, W. and LaPointe, P. 1994. Discrete fracture approaches for oil and gas applications. In *Proceedings 1st North American Rock Mechanics Symposium, Rock Mechanics* (eds., Nelson and Laubach), Rotterdam: Balkema, 19–30.

Dershowitz, W., Lee, G., Geier, J., Foxford, T., LaPointe, P., and Thomas, A. 1996. *FracMan Interactive Discrete Feature Data Analysis, Geometric Modeling, and Exploration Simulation, User Documentation, Version 2.5*. Redmond, Washington: Golder Associates.

Dixon, N. and Spriggs, M. 2007. Quantification of slope displacement rates using acoustic emission monitoring. *Canadian Geotechnical Journal*, 44(8), 966–976.

Domenico, P.A. and Schwartz, F.W. 1990. *Physical and Chemical Hydrogeology*, New York: Wiley.

Douglas, P.M. and Voight, B. 1969. Anisotropy of granites: A reflection of microscopic fabric. *Géotechnique*, 19, 376–398.

Dowrick, D.J. 2009. *Earthquake Resistant Design and Risk Reduction*, 2nd edition, United Kingdom: Wiley, 548pp.

Dunnicliff, J. 1993. *Geotechnical Instrumentation for Monitoring Field Performance*. New York: Wiley, 577pp.

Dyke, C.G. and Dobereiner, L. 1991. Evaluating the strength and deformability of sandstones. *Quarterly Journal Engineering Geology*, 24, 123–134.

Eberhardt, E. 2012. The Hoek–Brown failure criterion. *Rock Mechanics & Rock Engineering*, 45, 981–988.

Ebuk, E.J. 1991. *The Influence of Fabric on the Shear Strength Characteristics of Weathered Granites*. Unpublished PhD thesis, the University of Leeds, 486pp.

Ebuk, E.J., Hencher, S.R., and Lumsden, A.C. 1993. The influence of structure in the shearing mechanisms of weakly bonded soils derived from granites. In *The Engineering Geology of Weak Rock* (eds., Cripps et al.), Rotterdam: Balkema, 207–215.

Edmonds, C.N. 2008. Karst and mining geohazards with particular reference to the Chalk outcrop, England. *Quarterly Journal of Engineering Geology and Hydrogeology*, 41, 261–278.

Engelder, J.T. and Scholtz, C.H. 1976. The role of asperity indentation and ploughing in rock friction—II. *International Journal of Rock Mechanics and Mining Sciences & Geomechanics Abstracts*, 13, 155–163.

Engelder, T. 1985. Loading paths to joint propagation during a tectonic cycle: An example from the Appalachian Plateau, USA. *Journal of Structural Geology*, 7, 459–476.

Engelder, T. 1999. Transitional-tensile fracture propagation: A status report. *Journal of Structural Geology*, 21, 1049–1055.

Engelder, T., Lash, G.G., and Uzcátegui, R.S. 2009. Joint sets that enhance production from Middle and Upper Devonian gas shales of the Appalachian Basin. *AAPG Bulletin*, 93, 857–889.

ERM-Hong Kong Ltd. 1998. *Quantitative Risk Assessment of Boulder Fall Hazards in Hong Kong: Phase 2 (GEO Report No. 80)*. Report prepared for the Geotechnical Engineering Office, Hong Kong, 61pp.

Farmer, I.W. 1983. *Engineering Behaviour of Rocks*, 2nd edition, Chapman & Hall, 208pp.

Farrell, N.J.C., Healy, D., and Taylor, C.W. 2014. Anisotropy of permeability in faulted porous sandstone. *Journal of Structural Geology*, 63, 50–67.

Fecker, E. and Rengers, N. 1971. Measurement of large scale roughness of rock planes by means of profilograph and geological compass. *Proceedings Symposium on Rock Fracture*, Nancy, France, Paper 1–18.

Fekete, S., Diederichs, M., and Lato, M. 2010. Geotechnical and operational applications for 3-dimensional laser scanning in drill and blast tunnels. *Tunnelling and Underground Space Technology*, 25, 614–628.

Fell, R., Ho, K.K.S., Lacasse, S., and Leroi, E. 2005b. A framework for landslide risk assessment and management. In *Landslide Risk Management* (eds., Hung, Fell, Couture and Eberhardt), New York: Balkema, 3–25.

Fell, R., MacGregor, P., Stapledon, D., Bell, G. 2005a. *Geotechnical Engineering of Dams*. London: Taylor & Francis.

Fleischmann, K.H. 1990. Rift and grain in two granites of the White Mountain magma series. *Journal Geophysical Research*, 95(B13), 21463–21474.

Fogg, G.E., Noyes, C.D., and Carle, S.F. 1998. Geologically based model of heterogeneous hydraulic conductivity in an alluvial setting. *Hydrogeology Journal*, 6(1), 131–143.

Fookes, P.G., Baynes, F.J., and Hutchinson, J.M. 2000. Total geological history: A model approach to the anticipation, observation and understanding of ground conditions. *GeoEng 2000*, Melbourne, **1**, 370–460.

Fookes, P.G. and Sweeney, M. 1976. Stabilisation and control of local rockfalls and degrading rock slopes. *Quarterly Journal of Engineering Geology*, **9**, 37–55.

Franklin, J.A., Broch, E., and Walton, G. 1971. Logging the mechanical character of rock. *Transactions of the Institution of Mining and Metallurgy*, 80A, 1–9.

Fry, N. 1984. *The Field Description of Metamorphic Rocks*. Milton Keynes, UK: The Open University Press, 110pp.

Fujii, Y., Takemura, T., Takahashi, M., and Lin, W. 2007. Surface features of uniaxial tensile fractures and their relation to rock anisotropy in Inada granite. *International Journal of Rock Mechanics and Mining Science*, **44**, 98–107.

Fumagalli, E. 1968. Model simulation of rock mechanics problems. In *Rock Mechanics in Engineering Practice* (eds., K.G. Stagg and O.C. Zienkiewicz), New York: John Wiley & Sons, 358–384.

Fyfe, J.A., Shaw, R., Campbell, S.D.G., Lai, K.W., and Kirk, P.A. 2000. *The Quaternary Geology of Hong Kong*. Geotechnical Engineering Office, Civil Engineering Department, the Government of the Hong Kong Special Administrative Region, 209p plus maps.

Gama, C.D. 1995. A model for rock mass fragmentation by blasting. *Proceedings of the Eighth International Congress on Rock Mechanics*, Tokyo, Vol. 1, pp. 73–76.

Gannon, J.A., Masterton, G.G.T., Wallace, W.A., and Muir Wood, D. 1999. *Piled Foundations in Weak Rocks*. CIRIA Report 181, 140pp.

Gawthorpe, R.L. and Leeder, M.R. 2000. Tectono-sedimentary evolution of active extensional basins. *Basin Research*, **12**, 195–218.

Geotechnical Control Office. 1984a. *Geotechnical Manual for Slopes*, 2nd edition. Hong Kong: Geotechnical Control Office, 295pp.

Geotechnical Control Office. 1984b. *CHASE: Cutslopes in Hongkong Assessment of Stability by Empiricism*. Unpublished Report, Geotechnical Control Office, Hong Kong, 5 Vols.

Geotechnical Control Office. 1988. *Guide to Rock and Soil Descriptions* (Geoguide 3). Geotechnical Control Office, Hong Kong, 189pp.

Geotechnical Engineering Office. 2000. *Highway Slope Manual*. Geotechnical Engineering Office, Hong Kong, 114pp.

Geotechnical Engineering Office. 2006. *Foundation Design and Construction*. GEO Publication 1/2006, 376pp.

Geotechnical Engineering Office. 2007. *Engineering Geological Practice in Hong Kong*. GEO Publication 1/2007, 278pp.

Geotechnical Engineering Office 2009. *Prescriptive Measures for Man-made Slopes and Retaining Walls*. GEO Publication No. 1/2009, Geotechnical Engineering Office, Hong Kong, 76pp.

Geotechnical Engineering Office. 2011. *Technical Guidelines on Landscape Treatment for Slopes*. GEO Publication No. 1/2011, Geotechnical Engineering Office, Hong Kong, 217pp.

Gnirk, P. 1993. *Overview Volume II. Natural Barriers*. OECD/NEA International Stripa Project 1980–1992, SKB, Stockholm, Sweden, 327pp.

Goehring, L., and Morris, S.W. 2008. Scaling of joints in basalt. *Journal of Geophysical Research*, **113**, 10203, 18.

Golder Associates (HK) Ltd. 1998. Tuen Mun Road Widening at Tai Lam Section. Design and Construction Consultancy. Design Report: Temporary and Permanent Rockfall Fences. Agreement No. CE 18/97.

Goodman, R., Moye, D., Schalkwyk, A., and Javendel, I. 1965. Groundwater inflow during tunnel driving. *Bulletin Association Engineering Geology*, **2**, 39–56.

Goodman, R.E. 1995. Block theory and its application. *Géotechnique*, **45**(3), 383–423.

Goodman, R.E. and Shi, G. 1985. *Block Theory and Its Application to Rock Engineering*. New Jersey: Prentice-Hall Inc.

Gordon, J.E. 1978. *Structures or Why Things Don't Fall Down*. London: Penguin Books, 395pp.

Grasselli, G. and Egger, P. 2003. Constitutive law for the shear strength of rock joints based on three-dimensional surface parameters. *International Journal of Rock Mechanics & Mining Sciences*, **40**, 25–40.

Griffith, A.A. 1921. The phenomena of rupture and flow in solids. *Philosophical Transactions of the Royal Society*, **A221**, 163–198.

Grigg, P.V. and Wong, K.M. 1987. Stabilisation of boulders at a hill slope site in Hong Kong. *Quarterly Journal of Engineering Geology*, **20**, 5–14.

Grimstad, E. and Barton, N. 1993. Updating of the Q-system for NMT. *Proceedings of the International Symposium on Sprayed Concrete. Modern Use of Wet Mix Sprayed Concrete for Underground Support*, Fagernes. Norwegian Concrete Association, Oslo.

Gupta, S., Collier, J.S., Palmer-Felgate, A. et al. 2007. Catastrophic flooding origin of shelf valley systems in the English Channel. *Nature*, **448**, 342–345.

Hack, R. 1998. *Slope Stability Probability Classification, SSPC*, 2nd edition. Emschede, Netherlands: International Institute for Aerospace Survey and Earth Sciences (ITC), Publication No. 43, 258pp.

Hack, R. and Huisman, M. 2002. Estimating the intact rock strength of a rock mass by simple means. In *Engineering Geology for Developing Countries—Proceedings of 9th Congress of the International Association for Engineering Geology and the Environment*, Durban, South Africa, September 16–20 (eds., J.L. van Rooy and C.A. Jermy), 1971–1977.

Halcrow China Ltd. 2002. *Investigation of Some Selected Landslides in 2000* (Vol. 2). GEO Report No. 130, Geotechnical Engineering Office, Hong Kong Government, 173pp.

Halcrow China Ltd. 2003. *Interim Report on Detailed Hydrogeological Study of the Hillside near Yee King Road, Tai Hang*. Landslide Study Report LSR 4/2003: 123pp.

Halcrow China Ltd. 2011. *Study on Methods and Supervision of Rock Breaking Operations and Provision of Temporary Protective Barriers and Associated Measures*. GEO Report No. 260, Geotechnical Engineering Office, Hong Kong Government, 250pp.

Hancock, P.L. 1985. Brittle microtectonics: Principles and practice. *Journal of Structural Geology*, 7, 437–457.

Hardee, H.C. 1980. Solidification in Kilauea Iki lava lake. *Journal of Volcanological and Geothermal Research*, 7, 211–223.

Harris, C.S. 1996. Tunnelling difficulties: Working towards an explanation. In *Engineering Geology of the Channel Tunnel* (eds., C.S. Harris, M.B. Hart, P.M. Varley, and C.D. Warren). London: Thomas Telford, 381–397.

Hawkins, A.B. and McConnell, B.J. 1991. Sandstones as geomaterials. *Quarterly Journal Engineering Geology*, **24**, 135–142.

Hazzard, J.F. and Damjanac, B. 2013. Further investigations of microseismicity in bonded particle models. *Proceedings of the 3rd International FLAC/DEM Symposium*, Hangzhou, China, October 22–24. Minneapolis, MN: Itasca Consulting Group.

Heathcote, J.A., Jones, M.A., and Herbert, A.W. 1996. Modelling groundwater flow in the Sellafield area. *Quarterly Journal of Engineering Geology*, **29**, S59–S81.

Heidbach, O., Tingay, M., Barth, A., Reinecker, J., Kurfeß, D., and Müller, B. 2010. Global crustal stress pattern based on the World Stress Map database release 2008, *Tectonophysics*, **462**, 1–4.

Hencher, S.R. 1976. A simple sliding apparatus for the measurement of rock friction. *Géotechnique*, 26(4), 641–644.

Hencher, S.R. 1977. *The Effect of Vibration on the Friction between Planar Rock Surfaces*. Unpublished PhD thesis, Imperial College Science and Technology, London University.

Hencher, S.R. 1981. *Report on Slope Failure at Yip Kan Street (11SW-D/C86) Aberdeen on 12th July 1981*. Geotechnical Control Office Report No. GCO 16/81, Geotechnical Control Office, Hong Kong, 26pp. plus 3 appendices (unpublished).

Hencher, S.R. 1983. *Landslide Studies 1982 Case Study No. 4 South Bay Close*. Special Project Report No. SPR 5/83, Geotechnical Control Office, Hong Kong, 38pp. (Unpublished).

Hencher, S.R. 1984. *Three Direct Shear Tests on Volcanic Rock Joints*. Geotechnical Control Office Information Note, IN 5/84, Hong Kong Government, 33pp. (unpublished).

Hencher, S.R. 1985. Limitations of stereographic projections for rock slope stability analysis. *Hong Kong Engineer*, **13**(7), 37–41.

Hencher, S.R. 1987. The implications of joints and structures for slope stability. In *Slope Stability—Geotechnical Engineering and Geomorphology* (eds., M.G. Anderson and K.S. Richards), Chichester, UK: John Wiley & Sons Ltd, 145–186.

Hencher, S.R. 1994. Recognising the significance of complex geological conditions. Tunnelling in difficult conditions. In *Proceedings of Short Course, Associazione Geotecnica, Italiana, MIR, '94* (ed., G. Barla), Turin: Polytecnico di Torino, pp. 1–1 to 1–8m.

Hencher, S.R. 1996a. Fracture flow modelling: Proof of evidence. In *Radioactive Waste Disposal at Sellafield, UK: Site Selection, Geological and Engineering Problems* (eds., R.S. Haszeldine and D.K. Smythe), Department of Geology and Applied Geology, University of Glasgow, 349–358.

Hencher, S.R. 1996b. Fracture flow modelling: Supplementary proof of evidence. In *Radioactive Waste Disposal at Sellafield, UK: Site Selection, Geological and Engineering Problems* (eds., R.S. Haszeldine and D.K. Smythe), Department of Geology and Applied Geology University of Glasgow, 359–370. http://www.foe.co.uk/archive/nirex/

Hencher, S.R. 2008. The "new" British and European standard guidance on rock description. A critique by Steve Hencher. *Ground Engineering*, **41**(7), 17–21.

Hencher, S.R. 2009. Technical trends for long span bridge foundations. Invited Paper. *Proceedings of International Commemorative Symposium for the Incheon Bridge*, ROTREX 2009, Inchon, South Korea, 456–465.

Hencher, S.R. 2010. Preferential flow paths through soil and rock and their association with landslides. *Hydrological Processes*, **24**, 1610–1630.

Hencher, S.R. 2012a. *Practical Engineering Geology*. Applied Geotechnics, Vol. 4, London and New York: Spon Press, 450pp.

Hencher, S.R. 2012b. Discussion of comparison of different techniques of tilt testing and basic friction angle variability assessment by Alejano, Gonzalez and Muralha. *Rock Mechanics and Rock Engineering*, **45**, 1137–1139.

Hencher, S.R. 2014. Characterising discontinuities in naturally fractured outcrop analogues and rock core: The need to consider fracture development over geological time. *Advances in the Study of Fractured Reservoirs, Geological Society of London Special Publication*, **374**, 113–123.

Hencher, S.R. and Acar, I.A. 1995. The Erzincan Earthquake, Friday, 13 March 1992. *Quarterly Journal of Engineering Geology*, **28**, 313–316.

Hencher, S.R. and Knipe, R.J. 2007. Development of rock joints with time and consequences for engineering. *Proceedings of the 11th Congress of the International Society for Rock Mechanics*, **1**, 223–226.

Hencher, S.R., Lee, S.G., Carter, T.G., and Richards, L.R. 2011. Sheeting joints—Characterisation, shear strength and engineering. *Rock Mechanics and Rock Engineering*, **44**, 1–22.

Hencher, S.R., Liao, Q.H., and Monighan, B. 1996. Modelling slope behaviour for open pits. *Transactions of the Institution of Mining & Metallurgy*, **105**, A37–A47.

Hencher, S.R., Massey, J.B., and Brand, E.W. 1985. Application of back analysis to some Hong Kong landslides. *Proceedings of the 4th International Symposium on Landslides*, Toronto, Vol. 1, 631–638.

Hencher, S.R. and McNicholl, D.P. 1985. Engineering in weathered rock. *Quarterly Journal of Engineering Geology*, **28**, 253–266.

Hencher, S.R. and Richards, L.R. 1982. The basic frictional resistance of sheeting joints in Hong Kong Granite. *Hong Kong Engineer*, **11**(2), 21–25.

Hencher, S.R. and Richards, L.R. 1989. Laboratory direct shear testing of rock discontinuities. *Ground Engineering*, **22**(2), 24–31.

Hencher, S.R. and Richards, L.R. 2015. Assessing the shear strength of rock discontinuities at laboratory and field scales. *Rock Mechanics & Rock Engineering*, **48**, 883–905.

Hencher, S.R., Sun, H.W., and Ho, K.K.S. 2008. The investigation of underground streams in a weathered granite terrain in Hong Kong. In *Geotechnical and Geophysical Site Characterization* (eds., Huang and Mayne), London: Taylor & Francis, 601–607.

Hencher, S.R., Toy, J.P., and Lumsden, A.C. 1993. Scale dependent shear strength of rock joints. In *Scale Effects in Rock Masses 93* (ed., Pinto da Cunha), Rotterdam: Balkema, 233–240.

Hencher, S.R., Tyson, J.T., and Hutchinson, P. 2005. Investigating substandard piles in Hong Kong. *Proceedings 3rd International Conference on Forensic Engineering*, London: Institution of Civil Engineers, 107–118.

Hill, S.J. and Wallace, M.I. 2001. Mass modulus of rock for use in the design of deep foundations. In *Proceedings 14th Southeast Asian Geotechnical Conference*, (eds., K.K.S. Ho and K.S. Li), Lisse: Swets & Zeitlinger, Vol. 1, 333–338.

Hobbs, N.B. 1974. General report and state-of-the-art-review. In *Proceedings of a Conference on Settlement of Structures*, Cambridge: Pentech Press, 579–609.

Hoek, E. 1968. Brittle failure of rock. In *Rock Mechanics in Engineering Practice* (eds., Stagg and Zienkiewicz), London: Wiley & Sons, 99–124.

Hoek, E. 1999. Putting numbers to geology—An engineer's viewpoint. The 2nd Glossop Lecture. *Quarterly Journal of Engineering Geology*, **32**, 1–20.

Hoek, E. 2000. Big tunnels in bad rock. 2000 Terzhagi Lecture. *ASCE Journal of Geotechnical and Geoenvironmental Engineering*, **127**, 726–740.

Hoek, E. 2014a. The development of rock engineering. Hoek's Corner. https://www.rocscience.com/

Hoek, E. 2014b. History of rock engineering. Hoek's Corner. https://www.rocscience.com/

Hoek, E. 2014c. In situ and induced stress. Hoek's Corner. https://www.rocscience.com

Hoek, E. 2014d. Rock mass properties. Hoek's Corner. https://www.rocscience.com/

Hoek, E. and Bieniawski, Z.T. Brittle fracture propagation in rock under compression. *International Journal Fracture Mechanics*, **1**(3), 137–155.

Hoek, E. and Bray, J.W. 1974. *Rock Slope Engineering*. Institution of Mining and Metallurgy, 309pp.

Hoek, E. and Bray, J.W. 1981. *Rock Slope Engineering* (revised 3rd edition). Institution of Mining and Metallurgy, 358pp.

Hoek, E. and Brown, E.T. 1980. *Underground Excavations in Rock*. London: Institution of Mining and Metallurgy, 527pp.

Hoek, E. and Brown, E.T. 1988. The Hoek–Brown failure criterion—A 1988 update. *Proceedings of 15th Canadian Rock Mechanics Symposium*, Toronto, 31–38.

Hoek, E. and Brown, E.T. 1997. Practical estimates of rock mass strength. *International Journal of Rock Mechanics & Mining Sciences*, **34**, 1165–1186.

Hoek, E. and Diederichs, M.S. 2006. Empirical estimation of rock mass modulus. *International Journal of Rock Mechanics & Mining Sciences*, **43**, 203–215.

Hoek, E., Kaiser, P.K., and Bawden, W.F. 1995. *Support of Underground Excavations in Rock*. Rotterdam: Balkema, 215pp.

Hoek, E. and Marinos, P. 2000. Predicting tunnel squeezing. *Tunnels and Tunnelling International*, Part 1, **32/11**, 45–51, November 2000; Part 2, **32/12**, 33–36, December 2000.

Hoek, E. and Martin, C.D. 2014. Fracture initiation and propagation in intact rock—A review. *Journal of Rock Mechanics and Geotechnical Engineering*, **6**, 287–300.

Hoek, E. and Moy, D. 1993. Design of large powerhouse caverns in weak rock. *Comprehensive Rock Engineering*, Vol. 5, *Surface and Underground Project Case Histories* (editor in chief, J.A. Hudson; Vol. editor, E. Hoek), Oxford: Pergamon Press, 85–110.

Holmes, A. 1965. *Principles of Physical Geology*, 2nd edition. Nelson & Sons, 1288pp.

Holmøy, K.H. and Nilsen, B. Significance of geological parameters for predicting water inflow in hard rock tunnels. *Rock Mechanics & Rock Engineering*, **47**, 853–868.

Hong Kong Buildings Department. 2004. *Code of Practice for Foundations*. The Government of the Hong Kong Special Administrative Region, 67pp.

Houlsby, A.C. 1976. Routine interpretation of the Lugeon water-test. *Quarterly Journal Engineering Geology*, **9**, 303–313.

HSC. 1988. *Explosives at Quarries, Quarries (Explosives) Regulations—Approved Code of Practice*. Health and Safety Commission, UK, Her Majesty's Stationary Office.

HSE. 1989. *Flyrock Projection from Quarry Blasting*. Health and Safety Executive, Quarries Topic report, 1–9.

Hubbert, M.K. 1951. Mechanical basis for certain familiar geologic structures. *Geological Society of America Bulletin*, **62**, 355–372.

Hubbert, M.K. and Rubey, W.W. 1959. Role of fluid pressure in mechanics of overthrust faulting: 1. Mechanics of fluid-filled porous solids and its application to overthrust faulting. *Geological Society of America Bulletin*, **70**, 115–166.

Hubbert, M.K. and Willis, D.G. 1957. Mechanics of hydraulic fracturing. *Trans. AIME*, **210**, 153–166.

Hudson, J.A. 1989. *Rock Mechanics Principles in Engineering Practice*. CIRIA, London: Butterworths, 72pp.

Hudson, J.A. and Harrison, J.P. 1997. *Engineering Rock Mechanics. An Introduction to the Principles.* Oxford: Pergamon, 444pp.

Hungr, O. and Evans, S.G. 1989. *Engineering Aspects of Rockfall Hazard in Canada.* Geological Survey of Canada, Open File 2061, 102pp.

Hunt, R.E. 2005. *Geotechnical Engineering Investigation Handbook*, 2nd edition. Boca Raton, Florida: Taylor & Francis, 1066p.

Institution of Civil Engineers. 1996. *Sprayed Concrete Linings (NATM) for Tunnels in Soft Ground. ICE Design and Practice Guide*, London: Thomas Telford, 88pp.

Institution of Civil Engineers. 2005. *NEC3 Engineering and Construction Contract.* London: Thomas Telford.

International Society for Rock Mechanics. 1978. Suggested methods for the quantitative description of discontinuities in rock masses. *International Journal of Rock Mechanics and Mining Sciences & Geomechanics Abstracts*, **15**, 319–368.

International Society for Rock Mechanics. 1981. Basic geotechnical description of rock masses. *International Journal of Rock Mechanics Mining Sciences and Geomechanics Abstracts*, **18**, 85–110.

Itasca. 2004. *UDEC Version 4.0 User's Guide*, 2nd edition, Minnesota, USA: Itasca Consulting Group.

Iverson, R.M. 2000. Landslide triggering by rain infiltration. *Water Resources Research*, **36**, 1897–1910.

Jaeger, J.C. and Cook, N.G.W. 1969. *Fundamentals of Rock Mechanics*, 1st edition. London: Methuen & Co. Ltd.

Jaeger, J.C., Cook, N.G.W., and Zimmerman, R.W. 2007. *Fundamentals of Rock Mechanics*, 4th edition. Malden, Massachusetts; Oxford: Blackwell Publishing, 475pp.

Jeffcock, A.E. 1995. Simulation tests to determine the suitability of nets for preventing blast projected flyrock in quarries. *Explosives Engineering*, **1**, 4–10.

John, M. and Allen, R. 1996. NATM on the channel tunnel. In *Engineering Geology of the Channel Tunnel* (eds., C.S. Harris, M.B. Hart, P.M. Varley, and C.D. Warren), Chapter 21. London: Thomas Telford, 310–336.

Jones, J.S. 1999. *Almost Like a Whale.* London: Doubleday, 499pp.

Kaplin, J., Dowdell, R., and Tokle, V. 1998. Directional core drilling for the Metro West Water Supply Tunnel. In *Proceedings North American Tunnelling '98* (ed., Ozdmir), Rotterdam: Balkema, 95–103.

Käsling, H. and Thuro, K. 2010. Determining rock abrasivity in the laboratory. In *Proceedings of the 11th IAEG Congress, Auckland, New Zealand, Geologically Active* (eds., Williams et al.), London: Taylor & Francis Group, 1973–1980.

Keefer, D.K. 1984. Rock avalanches caused by earthquakes: Source characteristics. *Science*, **223**, 1288–1290.

Keen, D.H. 1995. Raised beaches and sea-levels in the English Channel in the Middle and Late Pleistocene: Problems of interpretation and implications for the isolation of the British Isles. *Geological Society, London, Special Publications*, **96**, 63–74.

King, U.U. and Chan, W.P. 1991. Hong Kong surface blasting accidents and mitigation methods employed to ensure site safety. *Hong Kong Engineer, August*, 18–25. Mines Division, Civil Engineering Services Department.

Knapett, J.A. and Craig, R.F. 2012. *Craig's Soil Mechanics.* London and New York: Taylor & Francis, 552pp.

Knill, J.L. 1976. Cow Green revisited. Inaugural Lecture, Imperial College of Science and Technology, University of London.

Knill, J.L. 2002. Core values: The first Hans Cloos Lecture. *Proceedings 9th AEG Congress*, Durban, 1–45.

Kong, W.K. 2011. Water ingress assessment for rock tunnels: A tool for risk planning. *Rock Mechanics & Rock Engineering*, **44**, 755–765.

Konya, C.J. and Walter, E.J. 1990. *Surface Blast Design.* New Jersey: Prentice-Hall.

Kruseman, G.P. and deRitter, N.A. 2000. *Analysis and Evaluation of Pumping Test Data*, 2nd edition. International Institution for Land Reclamation and Improvement, Publication 47, 377pp.

Krynine, D.P. and Judd, W.R. 1957. *Principles of Engineering Geology and Geotechnics*. New York: McGraw-Hill Book Company, 730pp.

Kulander, B.R. and Dean, S.L. 1995. Observations on fractography with laboratory experiments for geologists. In *Fractography: Fracture Topography as a Tool in Fracture Mechanics and Stress Analysis* (ed., M.S. Ameen), London: Geological Society London Special Publication, 92, 59–82.

Kulatilake, P.H.S.W., Um, J., Panda, B.B., and Nghiem, N. 1999. Development of new peak shear strength criterion for anisotropic rock joints. *Journal of Engineering Mechanics*, 125, 1010–1017.

Kutter, H.K. and Fairhurst, C. 1971. On the fracture process in blasting. *International Journal of Rock Mechanics and Mining Sciences*, 8, 181–202.

Kveldsvik, V., Nilsen, B., Einstein, H.H., and Nadim, F. 2008. Alternative approaches for analyses of a 100,000 m³ rock slide based on Barton-Bandis shear strength criterion. *Landslides*, 5, 161–176.

Ladeira, F.L. and Price, N.J. 1981. Relationship between fracture spacing and bed thickness. *Journal of Structural Geology*, 3, 179–183

Lambe, T.W. and Whitman, R.V. 1979. *Soil Mechanics*. New York: John Wiley & Sons, 553pp.

Lanaro, F., Sato, T., and Nakama, S. 2009. Depth variability of compressive strength test results of Toki Granite, from Shobasama and Mizunami construction sites, Japan. *Rock Mechanics and Rock Engineering*, 42, 611–629.

Lancaster-Jones, P.F.F. 1975. The interpretation of the Lugeon test. *Quarterly Journal of Engineering Geology*, 8, 151–154.

Lang, P.A. 1988. Rook 209 excavation response test in the underground research laboratory. *Proceedings of an NEA Workshop on excavation Response in Geological Repositories for Radioactive Waste*, Winnipeg, 295–329.

Leblais, Y. and Leblond, L. 1996. French undersea crossover: Design and construction. In *Engineering Geology of the Channel Tunnel* (eds., C.S. Harris, M.B. Hart, P.M. Varley, and C.D. Warren), London: Thomas Telford, 349–360.

Lee, S.G. and Hencher, S.R. 2013. Assessing the stability of a geologically complex slope where strong dykes locally act as reinforcement. *Rock Mechanics and Rock Engineering*, 46, 1339–1351.

Leeder, M. 1999. *Sedimentology and Sedimentary Basins: From Turbulence to Tectonics*. Oxford: Blackwell Publishing, 592p.

Littlejohn, G.S. and Bruce, D.A. 1975. Rock anchors—state-of-the-art Part 1: Design, *Ground Engineering*, 8(3), 25–32; 8(4), 41–48. Part 2: Construction. *Ground Engineering*, 8(6), 36–45.

Liu, Y.C. and Chen, C.S. 2007. A new approach for application of rock mass classification on rock slope stability assessment. *Engineering Geology*, 89, 129–143.

Loew, S., Barla, G., and Diederichs, M. 2010. Engineering geology of Alpine tunnels: Past, present and future. In *Geologically Active* (ed., Williams et al.), London: Taylor & Francis Group, 201–254.

Londe, P. 1987. The Malpasset dam failure. *Engineering Geology*, 24, 295–329.

Long, J.C.S., Remer, J.S., Wison, C.R., and Witherspoon, P.A. 1982. Porous media equivalents for networks of discontinuous fractures. *Water Resources Research*, 18, 645–658.

Lord, J.A., Clayton, C.R.I., and Mortimore, R.N. 2002. *Engineering in Chalk*. Construction Industry Research and Information Association Report, C574.

Lovitt, M.B. and Collins, A.H. 2012. Improved tunnelling performance through smarter drilling and design. In *Tunnelling in Rock by Drilling and Blasting* (eds., Spathis and Gupta), London: Taylor & Francis, 7–14.

Lu, P. and Latham, J.P. 1999. Developments in the assessment of in-situ block size distributions of rock masses. *Rock Mechanics and Rock Engineering*, 32, 29–49.

Lunardi, P. 2000. The design and construction of tunnels using the approach based on the analysis of controlled deformation in rocks and soils. *Tunnels and Tunnelling International ADECO-RS Approach*, May 2000, 30p.

Lundborg, N., Persson, A., Ladegaard-Pedersen, A., and Holmberg, R. 1975. Keeping the lid on flyrock. *Engineering and Mining Journal*, 176(5), 95–100.

Lyle, P. 2000. The eruption environment of multi-tiered columnar basalt lava flows. *Journal of the Geological Society, London*, 157, 715–722.

Lysmer, J. and Duncan, J.M. 1969. *Stresses and Deflections in Pavements and Foundations*, 4th edition. Berkeley: Department of Civil Engineering, University of California.

MacCulloch, J. 1815. Geological description of Staffa. *The Monthly Magazine*, **39**, 240–243.

Macfarlane, D.F., Watts, C.R., and Nielsen, B. 2008. Field application of NTH fracture classification at the second Manapouri tailrace tunnel, New Zealand. *Proceedings North American Tunnelling*, 2008, 236–242.

MacGregor, F., Fell, R., Mostyn, G.R., Hocking, G., and McNally, G. 1994. The estimation of rock rippability. *Quarterly Journal of Engineering Geology*, **27**, 123–144.

Maidl, B., Herrenknecht, Maidl, U., Wehrmeyer, G., and Sturge, D.S. 2014. *Mechanised Shield Tunnelling*, 2nd edition. Berlin Heidelberg New York: Wiley, 490p.

Maidl, B., Schmid, L., Ritz W., Herrenknecht, M., and Sturge, D.S. 2008. *Hardrock Tunnel Boring Machines*. Berlin: Wiley, 356p.

Mak, N. and Blomfield, D. 1986. Rock trap design for pre-splitting slopes. *Conference on Rock Engineering in an Urban Environment*, Institution of Mining and Metallurgy, Hong Kong, 263–270.

Mandl, G. 2005. *Rock Joints. The Mechanical Genesis*. Oxford: Springer, 221p.

Marinos, P. and Hoek, E. 2000. GSI—A geologically friendly tool for rock mass strength estimation. *Proceedings of GeoEng 2000*, Melbourne, Australia, 1422–1442.

Marinos, V., Marinos, P., and Hoek, E. 2005. The geological strength index: Applications and limitations. *Bulletin Engineering Geology & Environment*, **64**, 55–65.

Martin, C.D., Davison, C.C., and Kozak, E.T. 1990. Characterising normal stiffness and hydraulic conductivity of a major shear zone in granite. In *Proceedings International Symposium on Rock Joints*, Loen (eds., Barton and Stephansson), Rotterdam: Balkema, 549–556.

Martin, D.C. 1994. Quantifying drilling-induced damage in samples of Lac du Bonnet granite. *Proceedings 1st North American Rock Mechanics Symposium*, Austin, Texas, 419–427.

Martin, D.C. and Christiansson, R. 2009. Estimating the potential for spalling around a deep nuclear waste repository in crystalline rock. *International Journal of Rock Mechanics and Mining Sciences*, **46**, 219–228.

Martin, R.P. and Hencher, S.R. 1986. Principles for description and classification of weathered rocks for engineering purposes. *SI Practice: Assessing BS5930, Geological Society, Engineering Geology Special Publication No. 2*, London: Geological Society, 299–308.

Masset, O. and Loew, S. 2010. Hydraulic conductivity distribution in crystalline rocks, derived from inflows to tunnels and galleries in the Central Alps, Switzerland. *Hydrogeology Journal*, **18**, 863–891.

Matsuo, S. 1986. Tackling floods beneath the sea. *Tunnels & Tunnelling*, March, 42–45.

Maurenbrecher, M. 2008. Analysing the analysis of the Malpasset arch dam failure of 1959. *GEOinternational*, July, 58–62.

Maury, V. 1993. An overview of tunnel, underground excavation and boreholes collapse mechanisms. *Comprehensive Rock Engineering*, Vol. 4 (editor in chief, J. Hudson). London: Pergamon Press, 369–412.

McClay, K.R. 1987. *The Mapping of Geological Structures*. Geological Society of London Handbook, Chichester: John Wiley & Sons, 161p.

McClay, K.R. and Coward, M.P. 1981. The Moine thrust zone: An overview. Thrust and nappe tectonics. *Geological Society of London Special Publication*, **9**, 241–260.

McFeat-Smith, I., MacKean, R., and Waldmo, O. 1998. Water inflows in bored rock tunnels in Hong Kong: Prediction, construction issues and control measures. *ICE Conference on Urban Ground Engineering*, Hong Kong.

McNicholl, D.P., Pump, W.L., and Cho, G.W.F. 1985. Groundwater control in large scale slope excavations—five case histories from Hong Kong. In *Groundwater in Engineering Geology* (eds., J.C. Cripps, F.G. Bell, and M.G. Culshaw), London: Geological Society, Engineering Geology Special Publications, Vol. 3, 533–540.

Megaw, T.M. and Bartlett, J.V. 1981. *Tunnels. Planning, Design, Construction*, Vol. 1. London: Ellis Horwood Limited, 284pp.

Mellor, M. and Hawkes, I. 1971. Measurement of tensile strength by diametral compression of discs and annuli. *Engineering Geology*, **5**, 173–225

Miller, K.G., Kominz, M.A., Browning, J.V., Wright, J.D., Mountain, G.S., Katz, M.E., Sugarman, P.J. et al. 2005. The Phanerozoic record of global sea-level change. *Science*, **310**, 1293–1298.

Miller, K.G., Mountain, G.S., Wright, J.D., and Browning J.V. 2011. A 180-million-year record of sea level and ice volume variations from continental margin and deep-sea isotopic records. *Oceanography*, 24(2), 40–53.

Milodowski, A.E., Gillespie, M.R., Naden, J., Fortey, N.J., Shepherd, T.J., Pearce, J.M., and Metcalfe, R. 1998. The petrology and paragenesis of fracture mineralization in the Sellafield area, West Cumbria. *Proceedings of the Yorkshire Geological Society*, 52, 215–241.

Mortimore, R.N. 2012. Making sense of Chalk—A total-rock approach to its engineering geology. 11th Glossop Lecture. *Quarterly Journal of Engineering Geology and Hydrogeology*, 45, 252–334.

Morton, K.L., Bouw, P.C., and Connelly, R.J. 1988. The prediction of minewater inflows. *Journal South African Institution Mining Metallurgy*, 88(July), 219–226.

Moye, D.G. 1955. Engineering Geology for the Snowy Mountains scheme. *Journal of the Institution of Engineers*, Australia, 27, 287–298.

Muir Wood, A. 2000. *Tunnelling: Management by Design*. London: E & FN Spon, 320pp.

Muir Wood, A. 2004. *Ahead of the face*. The 2004 Harding Lecture. London: British Tunnelling Society, 21pp.

Muralha, J., Grasselli, G., Tatone., B., Blumel, M., Chryssanthakis, P., and Jiang, Y.-J. 2013. ISRM suggested method for laboratory determination of the shear strength of rock joints: Revised version. *Rock Mechanics and Rock Engineering*, 47, 291–302.

Murray, A.D. and Gray, A.M. 1997. The Pergau Hydroelectric Project Part 3: Civil engineering design. *Proceedings institution Civil Engineers Water, Maritime & Energy*, 124, 173–188.

Nagel, K.-H. 1992. Limits of the geological predictions constructing the Samanalawera pressure tunnel, Sri Lanka. *Bulletin of the International Association Engineering Geology*, 45, 97–110.

National Highway Institute 1993. *Rockfall Hazard Rating System - Participant's Manual*. NHI Course No. 130220, Publication No. FHWA SA-93-057, National Highway Institute, U.S. Department of Transport, 102 p.

Nelson, R.A. 2001. *Geological Analysis of Naturally Fractured Reservoirs*, 2nd edition, Gulf Professional Publishing, Boston: Elsevier, 332pp.

Nicholson, D.T., Lumsden, A.C., and Hencher, S.R. 2000. Excavation-induced deterioration of rock slopes. *Proceedings of Conference on Landslides in Research, Theory and Practice*. Cardiff: Thomas Telford, Vol. 3, 1105–1110.

Nilsen, B. 2014. Characteristics of water ingress in Norwegian subsea tunnels. *Rock Mechanics & Rock Engineering*, 47, 933–945.

Nirex 1996. Nuclear science and technology testing and modelling of thermal, mechanical and hydro-geological properties of host rocks for deep geological disposal of radioactive waste. *Proceedings of a Workshop, Brussels*, January 12–13, 127–140.

Nirex 2007. *Geosphere Characterisation Report: Status Report*, October 2006, 198pp.

Norbury, D. 2010. *Soil and Rock Description in Engineering Practice*. Boca Raton, Florida: CRC Press, 288pp.

Odling, N.E. 1997. Scaling and connectivity of joint systems in sandstones from western Norway. *Journal of Structural Geology*, 19, 1257–1271.

Ohnishi, Y., Herda, H., and Yoshinaka, R. (1993). Shear strength scale effect and the geometry of single and repeated rock joints. *Proceedings of the 2nd International Workshop on Scale Effects in Rock Masses*, Lisbon, 167–173.

Ollier, C.D. 1975. *Weathering*, 2nd impression. Longman Group Limited, 304p.

Ollier, C.D. 2010. Very deep weathering and related landslides. In *Weathering as a Predisposing Factor to Slope Movements* (eds., D. Calcaterra and M. Parise), London: Geological Society, Engineering Geology Special Publications, 23, 5–14.

Olsson, O. and Gale, J.E. 1995. Site assessment and characterization for high-level nuclear waste disposal: Results from the Stripa Project, Sweden. *Quarterly Journal of Engineering Geology and Hydrogeology*, 28(Suppl. 1), S17–S30.

Palmström, A. 1982. The volumetric joint count—A useful and simple measure of the degree of rock jointing. *Proceedings 4th Congress International Association Engineering Geologists*, Delhi, 221–228.

Palmström, A. 2000. Recent developments in rock support estimates by the RMi. *Journal of Rock Mechanics and Tunnelling Technology*, 6, 1–19.

Palmström, A., 2009. Combining the RMR, Q, and RMi classification systems. *www.rockmass.net*, 25p.

Palmström, A. and Broch, E. 2006. Use and misuse of rock mass classification systems with particular reference to the Q-system. *Tunnelling and Underground Space Technology*, 21, 575–593.

Papaliangas, T.T. (1996). *Shear behaviour of rock discontinuities and soil-rock interfaces*. Unpublished PhD thesis, the University of Leeds.

Papaliangas, T.T., Hencher, S.R., and Lumsden, A.C. 1994. Scale independent shear strength of rock joints. *Proceedings IV CSMR/ Integral Approach to Applied Rock Mechanics*, Santiago, Chile, 123–133.

Patton, F.D. and Deere, D.U. 1970. Significant geological factors in rock slope stability. *Proceedings Symposium on Planning Open Pit Mines*, Johannesburg, A.A. Balkema, 143–151.

Peacock, D.C.P. 2001. The temporal relationship between joints and faults. *Journal of Structural Geology*, 23, 329–341.

Peck, R.B. 1969. Advantages and limitations of the observational method in applied soil mechanics. Ninth Rankine Lecture. *Géotechnique*, 19, 171–187.

Peck, R.B., Hanson, W.E., and Thornburn, T.H. 1974. *Foundation Engineering*, 2nd edition. New York: John Wiley and Sons, 514pp.

Peck, W.A., Sainsbury, D.P., and Lee, M.F. 2013. The importance of geology and roof shape on the stability of shallow caverns. *Australian Geomechanics*, 48(3), September, 1–14.

Pecker, A. 2004. Design and construction of the Rion Antirion Bridge. In *Geotechnical Engineering for Transportation Projects, Geotechnical Special Publication No. 126, ASCE Geo-Institute* (eds., M.K. Yegian and E. Kavazanjian), Reston, VA, 216–240.

Peckover, F.L. and Kerr, J.W.G. 1976. Treatment of rock falls on transportation routes. *Proceedings of the 29th Conference*, Canadian Geotechnical Society, Vancouver, B.C., Canada.

Pells, P.J.N. 2004. Rock mass grouting to reduce inflow. *Tunnels and Tunnelling*, March, 34–37.

Pells, P.J.N. 2007. Limitations of rock mass classification systems. *Tunnels & Tunnelling*, April, 33–37.

Pells, P.J.N. 2008. What happened to the mechanics in rock mechanics and the geology in engineering geology? *Journal of the Southern African Institute of Mining and Metallurgy*, 108, 309–323.

Pells, P.J.N. 2011. Against limit state design in rock. *Tunnels & Tunnelling*, February, 34–38.

Pettifer, G.S. and Fookes, P.G. 1994. A revision of the graphical method for assessing the excavatability of rock. *Quarterly Journal of Engineering Geology*, 27, 145–164.

Phillips, F.C. 1973. *The Use of Stereographic Projection in Structural Geology*, 3rd edition. London: Edward Arnold, 90p.

Pine, R.J., Coggan, J.S., Flynn, Z.N., and Elmo, D. 2006. Development of a new numerical modelling approach for naturally fractured rock masses. *Rock Mechanics and Rock Engineering*, 39(5), 395–419.

Pine, R.J., Owen, D.R.J., Coggan, J.S., and Rance, J.M. 2007. A new discrete fracture modelling approach for rock masses. *Geotechnique*, 57(9), 757–766.

Pine, R.J. and Roberds, W.J. 2005. A risk-based approach for the design of rock slopes subject to multiple failure modes—Illustrated by a case study in Hong Kong. *International Journal of Rock Mechanics & Mining Sciences*, 42, 261–275.

Pirazzoli, P.A. 1996. *Sea-Level Changes. The Last 20000 Years*. Chichester: John Wiley & Sons, 211p.

Piteau, D.R. 1973. Characterising and extrapolating rock joint properties in engineering practice. *Rock Mechanics*, Supplement 2, 2–31.

Pollard, C. and Crawley, J. 1996. UK tunnels: Ground treatment. In *Engineering Geology of the Channel Tunnel* (eds. C.S. Harris, M.B. Hart, P.M. Varley, and C.D. Warren), London: Thomas Telford, 261–269.

Pollard, D.D. and Aydin, A. 1988. Progress in understanding jointing over the past century. *Geological Society America Bulletin*, 100, 1181–1204.

Post, G. and Bonazzi, D. 1987. Latest thinking on the Malpasset accident. *Engineering Geology*, 24, 339–353.

Powderham, A.J. and Nicholson, D.P. 1996. The way forward. In *The Observational Method in Geotechnical Engineering*. (eds., D. Nicholson and A. Powderham), London: Thomas Telford, 195–204.

Power, C.M. 1998. *Mechanics of Modelled Rock Joints under True Stress Conditions Determined by Electrical Measurements of Contact Area*. Unpublished PhD thesis, University of Leeds, 293p.

Power, C.M. and Hencher, S.R. 1996. A new experimental method for the study of real area of contact between joint walls during shear. *Rock Mechanics, Proceedings 2nd North American Rock Mechanics Symposium*, Montreal, 1217–1222.

Price, N.J. 1966. *Fault and Joint Development in Brittle and Semi-brittle Rock*. Oxford; New York: Pergamon Press, 176p.

Price, N.J. and Cosgrove, J. 1990. *Analysis of Geological Structures*. Cambridge: Cambridge University Press.

Priest, S.D. 1993. *Discontinuity Analysis for Rock Engineering*. London; New York: Chapman & Hall, 473p.

Priest, S.D. and Hudson, J.A. 1981. Estimation of discontinuity spacing and trace length using scanline surveys. *International Journal for Rock Mechanics and Mining Science*, 18, 183–197.

Ramsay, J.G. 1967. *Folding and Fracturing of Rocks*. New York; London: McGraw-Hill, 568p.

Ramsay, J.G. 1980. Shear zone geometry: A review. *Journal of Structural Geology*, 2, 83–99.

Rawnsley, K.D. 1990. *The influence of joint origin on engineering properties*. Unpublished PhD thesis, University of Leeds, 388p.

Rawnsley, K.D., Rives, T., Petit, J.-P., Hencher, S.R., and Lumsden, A.C. 1992. Joint development in perturbed stress fields near faults. *Journal of Structural Geology*, 14, 939–951.

Raymer, J.H. 2001. Predicting groundwater inflow into hard-rock tunnels: Estimating the high-end of the permeability distribution. In *Proceedings of Rapid Excavation and Tunnelling Conference*, Littleton, Colorado: SME, 1027–1038.

Rayner, D.H. 1967. *The Stratigraphy of the British Isles*. Cambridge: Cambridge University Press, 453p.

Read, J. and Stacey, P. 2009. *Guidelines for Open Pit Slope Design*. Collingwood, Victoria: CSIRO Publishing, 512p.

Read, S. and Richards, L.R. 2007. Design inputs for stability assessment of dams on New Zealand greywackes. *Dams, Securing Water for Our Future: Proceedings of NZSOLD ANCOLD 2007 Conference*, Queenstown, 21 November 2007, 319–329.

Reeves, A., Ho, K.K.S., and Lo, D.O.K. 1999. Interim risk criteria for landslides and boulder falls from natural terrain. Geotechnical Risk Management. *Proceedings of the 18th Annual Seminar of the Hong Kong Institution of Engineers*, Hong Kong, 129–136.

Reid, H.F. 1910. *The Mechanics of the Earthquake: The California Earthquake of April 18, 1906*. Report of the State Investigation Commission, Vol. 2, Washington, DC: Carnegie Institution of Washington.

Richards, L.R. and Cowland, J.W. 1982. The effect of surface roughness on the field shear strength of sheeting joints in Hong Kong granite. *Hong Kong Engineer*, 10, 39–43.

Richards, L.R. and Cowland, J.W. 1986. Stability evaluation of some urban rock slopes in a transient groundwater regime. *Proceedings Conference on Rock Engineering and Excavation in an Urban Environment*, IMM, Hong Kong, 357–363, (Discussion 501–6).

Ritchie, A.M. 1963. Evaluation of rockfalls and its control. *Highways Research Record*, 17, 14–28.

Rives, T., Rawnsley, K.D., and Petit, J.-P. 1994. Analogue simulation of natural orthogonal joint set formation in brittle varnish. *Journal of Structural Geology*, 16, 419–429.

Rives, T., Razack, M., Petit, J.-P., and Rawnsley, K.D. 1992. Joint spacing: Analogue and numerical simulation. *Journal of Structural Geology*, 14, 925–937.

Roberts, J.C. 1995. Fracture surface markings in Liassic limestone at Lavernock Point, South Wales. In *Fractography: Fracture Topography as a Tool in Fracture Mechanics and Stress Analysis* (ed., M.S. Ameen), London, Geological Society London Special Publication No. 92, 175–186.

Robertshaw, C. and Tam, T.K. 1999. Tai Po to Butterfly Valley treated water transfer. *World Tunnelling*, 12(8), 383–385.

Rouleau, A. and Raven, K.G. 1995. Site specific simulation of groundwater flow and transport using a fracture network model. Proceedings of Conference on Fractured and Jointed Rock Masses, Lake Tahoe, (eds., Meyer, Cook, Goodman and Tsang), June 1992, Rotterdam: Balkema, 567–572.

Ruxton, B.P. and Berry, L. 1957. The weathering of granite and associated erosional features in Hong Kong. *Bulletin of the Geological Society of America*, 68, 1263–1292.

Saeb, S. and Amadei, B. 1992. Modelling rock joints under shear and normal loading. *Int. J. Rock Mech. Min. Sci. & Geomech. Abstr.*, **29**(3), 267–278.

Sauer, E., Marcher, T., and John, M. 2013. Decisive design basis and parameters for power plant caverns. *World Tunnel Congress, Geneva, Underground—The Way to the Future.* (eds., G. Anagnostou and H. Ehrbar), London: Taylor & Francis Group, 1858–1864.

Schofield, A.N. 2006. Interlocking, and peak and design strengths. *Géotechnique*, **56**, 357–358.

Scholz, C.H. 1968. Microfracturing and the inelastic deformation of rock in compression. *Journal of Geophysical Research*, **73**, 1417–1432.

Scholz, C.H. 1990. *The Mechanics of Earthquakes and Faulting.* Cambridge: Cambridge University Press, 439pp.

Seers, T. and Hodgetts, D. 2014. Comparison of digital outcrop and conventional data collection approaches for the characterization of naturally fractured reservoir analogues. *Advances in the Study of Fractured Reservoirs, Geological Society of London Special Publication*, **374**, 51–77.

Seidel, J.P. and Haberfield, C.M. 2002. A theoretical model for rock joints subjected to constant normal stiffness direct shear. *International Journal of Rock Mechanics and Mining Sciences*, **39**, 539–553.

Selby, M.J., 1980. A rock mass strength classification for geomorphic purposes: With tests from Antarctica and New Zealand. *Zeitschrift für Geomorphologie*, **23**, 31–51.

Selby, M.J. 1993. *Hillslope Materials and Processes*, 2nd edition. Oxford: Oxford University Press, 451p.

Serafim, J.L. and Pereira, J.P. 1983. Consideration of the geomechanical classification of Bieniawski. *Proc. Int. Symp. Eng Geol Underground Construction* (Lisbon), **1**(II), 33–44.

Shirlaw, J.N., Hencher S.R., and Zhao, J. 2000. Design and construction issues for excavation and tunnelling in some tropically weathered rocks and soils. *Proceedings of GeoEng 2000*, Melbourne, Australia, Volume 1: Invited Papers, 1286–1329.

Sibson, R.H. 1977. Fault rocks and fault mechanisms. *Journal Geological Society London*, **133**, 191–213.

Siddle, H.J. 1985. Groundwater control by drainage gallery at Aberfan, South Wales. In *Groundwater in Engineering Geology* (eds., J.C. Cripps, F.G. Bell, and M.G. Culshaw), London: Geological Society, Engineering Geology Special Publications, **3**, 533–540.

Siegesmund, S., Weiss, T., and Vollbrecht, A. (eds.) 2002. *Natural Stone, Weathering Phenomena, Conservation Strategies and Case Studies.* London: Geological Society, Special Publications, 205.

Simons, N., Menzies, B., and Mathews, M. 2001. *A Short Course in Soil and Rock Slope Engineering.* London: Thomas Telford, 432p.

Skempton, A.W. 1966. Some observations on tectonic shear zones. *Proceedings 1st Congress International Society Rock Mechanics*, Lisbon, **1**, 329–335.

Slob, S. 2010. *Automated Rock mass characterisation using 3-D terrestrial laser scanner.* Unpublished PhD Thesis, Technical University of Delft, 287p.

Smally, I.I. 1966. Contraction crack networks in a basalt flow. *Geological Magazine*, **103**, 110–113.

Smith, M.R. and Collis, L. (eds.) 2001. *Aggregates.* London: Geological Society, Engineering Geology Special Publications No. 17, 339 p.

Snow, D.T. 1968. Rock fracture spacings, openings, and porosities. *Journal of Soil Mechanics of the Foundations Division of the American Society of Civil Engineers*, **94**, 73–91.

Soper, N.J. and Dunning, F.W. 2005. Structure and sequence of the Ingleton Group, basement to the central Pennines of northern England. *Proceedings of the Yorkshire Geological Society*, **55**, 241–261.

Sowers, G.I. 1988. Foundation problems in residual soils. *Proceedings Int. Conf. Engineering Problems of Regional Soils.* Beijing, pp. 154–171.

Spang, K. and Egger, P. 1990. Action of fully-grouted bolts in jointed rock and factors of influence. *Rock Mechanics and Rock Engineering*, **23**, 201–229.

Spörli, K.B. and Rowland, J.V. 2006. "Column on column" structures as indicators of lava/ice interaction, Ruapehu andesite volcano, New Zealand. *Journal of Volcanology and Geothermal Research*, **157**, 294–310.

Starfield, A.M. and Cundall, P.A. 1988. Towards a methodology for rock mechanics modelling. *International Journal of Rock Mechanics and Mining Sciences & Geomechanics Abstracts*, **25**, 99–106.

Starr, D., Dissanayake, A., Marks, D., Clements, J., and Wijeyakulasuriya, V. 2010. South West transport corridor landslide. *Queensland Roads*, 8, 57–73.

Stauffer, D. 1985. *Introduction to Percolation Theory*. London: Taylor and Francis.

Storms, J.E.A., Hoogendoorn, R.M., Dam, R.A.C., Hoitink, A.J.F., and Kroonenberg, S.B. (2005). Late-Holocene evolution of the Mahakam delta, East Kalimantan, Indonesia. *Seidmentary Geology*, 180, 149–166.

Stow, D.A.V. 2005. *Sedimentary Rocks in the Field. A Colour Guide*. London: Manson Publishing Ltd., 320p.

Strozzi, T., Delaloye, R., Poffet, D., Hansmann, J., and Loew, S. 2011. Surface subsidence and uplift above a headrace tunnel in metamorphic basement rocks of the Swiss Alps as detected by satellite SAR interferometry. *Remote Sensing of Environment*, 8p.

Sturzenegger, M., Stead, D., and Elmo, D. 2011. Terrestial remote sensing-based estimation of mean trace length, trace intensity and mean block size/shape. *Engineering Geology*, 119, 96–111.

Swales, M.J. 1995. *Strength characteristics of potential shear planes in Coal Measures strata of South Wales*. Unpublished PhD thesis, the University of Leeds, 304p, plus appendices.

Swannell, N.G. and Hencher, S.R. 1999. Cavern design using modern software. *Proceedings of 10th Australian Tunnelling Conference*, Melbourne, March.

Tapley, M.J., West, B.W., Yamamoto, S., and Sham, R. 2006. Challenges in construction of Stonecutters Bridge and progress update. *International Conference on Bridge Engineering*, Hong Kong.

Tchalenko, J.S. 1968. The evolution of kink bands and the development of compression textures in sheared clays. *Tectonophysics*, 6, 159–174.

Terzaghi, K. 1925. Principles of soil mechanics. *Engineering News Record*, 95, 19–27.

Terzaghi, K. 1946. Rock defects and loads on tunnel supports. In *Rock Tunnelling with Steel Supports* (eds., R.V. Proctor and T. White), Youngstown, Ohio: Commercial Shearing and Stamping Co., 15–99.

Thuro, K. and Plinninger, R.J. 2003. Hard rock tunnel boring, cutting, drilling and blasting: Rock parameters for excavatability. *ISRM 2003-Technology roadmap for rock mechanics*, South African Institute of Mining and Metallurgy, 1227–1233.

Todd, D.K. 1980. *Groundwater Hydrology*, 2nd edition. New York; Chichester: Wiley, 535p.

Tomlinson, M.J. 2001. *Foundation Design and Construction*, 7th edition. Harlow: Prentice Hall, 569p.

Tottergill, B.W. 2006. *FIDIC Users' Guide: A Practical Guide To The 1999 Red And Yellow Books: Incorporating Changes And Additions To The 2005 MDB Harmonised Edition*. London: Thomas Telford.

Toy, J.P. 1993. *An appraisal of the effects of sample size on the shear strength behaviour of rock discontinuities*. Unpublished MSc dissertation, the University of Leeds.

Tucker, M.E. 1982. *The Field Description of Sedimentary Rocks*.Milton Keynes, UK: The Open University Press, 112p.

Ulusay, R. and Hudson, J.A. (eds.) 2006. *The Complete ISRM Suggested Methods for Rock Characterisation, Testing and Monitoring* 1974–2006, International Society for Rock Mechanics, 628p.

University of Trondheim, 1994. *Hard Rock Tunnel Boring*, Project Report 1–94, The Norwegian Institute of Technology, Norway.

U.S. Corps of Engineers. 1980. *Engineering and Design. Rock Reinforcement*. Engineer Manual No. 1110-2-2907, February, 1980.

Varley, P.M. 1996. The 1974 Channel tunnel project. In *Engineering Geology of the Channel Tunnel* (eds., C.S. Harris, M.B. Hart, P.M. Varley, and C. Warren), Chapter 9, London: Thomas Telford, 118–128.

Varley, P.M., Parkin, R.J.H., and Patel, R. 1997. Stabilisation of a large rock wedge in the underground powerhouse of the Pergau Hydroelectric Project, Malaysia. *Ground Anchorages and Anchored Structures*, London: Thomas Telford, 159pp.

Vervoort, A. and De Wit, K. 1997. Correlation between dredgeability and mechanical properties of rock. *Engineering Geology*, 47, 259–267.

Walker, F. and Mathias, M. 1946. The petrology of two granite-slate contacts at Cape Town, South Africa. *Quarterly Journal of the Geological Society*, 102, 499–521.

Wallis, P. 2010. Collapse of headrace tunnel after grand opening. *TunnelTalk*, Feb., 2.

Wallis, S. 1984. Pinching and punching the mighty Andes at Yacambu. *Tunnels & Tunnelling*, July, 19–23.

Wang, H. and Latham, J.P. 1992. *Design of fragmentation blasting in surface rock excavation*. ISRM Symposium: Eurock '92, Rock Characterization. Edited by J.A. Hudson, Chester UK, 233–238.

Warner, J. 2004. *Practical Handbook of Grouting—Soil, Rock and Structures*. New York: John Wiley & Sons.

Warren, J.E. and Root, P.J. 1963. The behaviour of naturally fractured reservoirs. *Society Petroleum Engineers Journal*, 3, 245–255.

Wentzinger, B., Starr, D., Fidler, S., Nguyen, Q., and Hencher, S.R. 2013. Stability analyses for a large landslide with complex geology and failure mechanism using numerical modelling. *Proceedings International Symposium on Slope Stability in Open Pit Mining and Civil Engineering*, Australia: Brisbane, 733–746.

West, G. 1983. Comparisons between real and predicted geology in tunnels: Examples from recent cases. *Quarterly Journal of Engineering Geology and Hydrogeology*, 16, 113–126.

West, I. 2014. http://www.southampton.ac.uk/~imw/Kimmeridge-Bay.htm

Westland, J., Busbridge, J.R., and Ball, J.G. 1998. Managing subsurface risk for Toronto's Rapid Transit Expansion Program. In *Proceedings North American Tunnelling '98* (ed., Ozdmir), Rotterdam: Balkema, 37–45.

Whitten, D.G.A. and Brooks, J.R.V. 1972. *The Penguin Dictionary of Geology*. Harmondsworth: Penguin Books, 495p plus appendices.

Wickham, G.E., Tiedemann, H.R., and Skinner, E.H. 1972. Support determination based on geologic predictions. In *Proc. 1st North American Rapid Excavation & Tunnelling Conference (RETC), Chicago* 1. (eds., K.S. Lane and L.A. Garfield), New York: American Institute of Mining, Metallurgical and Petroleum Engineers (AIME), 43–64.

Williamson, I.T. and Bell, B.R. 2012. The Staffa Lava Formation: Graben-related volcanism, associated sedimentation and landscape character during the early development of the Palaeogene Mull Lava Field, NW Scotland. *Scottish Journal of Geology*, 48, 1–46.

Wise, D.U. 1964. Microjointing in basement, Middle Rocky Mountains of Montana and Wyoming. *Geological Society of America Bulletin*, 75, 287–306.

Woodworth, J.B. 1896. On the fracture system of joints, with remarks on certain great fractures. *Proceedings of the Boston Society of Natural History*, 27, 163–184.

Wyllie, D.C. 1999. *Foundations on Rock: Engineering Practice*, 2nd edition. London; New York: E & FN Spon.

Wyllie, D.C. and Mah, C.W. 2004. *Rock Slope Engineering*, 4th edition. London; New York: E & FN Spon, 431p.

Wyllie, D.C. and Norrish, N.I. 1996. Rock strength properties and their measurement. In *Landslides Investigation and Mitigation*, Special Report 247 (eds., A.K. Turner and R. Schuster), Washington, D.C.: National Academy Press, 372–390.

Yu, Y.F., Siu, C.K., and Pun, W.K. 2005. *Guidelines on the Use of Prescriptive Measures for Rock Slopes*. GEO Report 161, 34pp.

Zhang, L. and Einstein, H.H. 2010. The planar shape of rock joints. *Rock Mechanics & Rock Engineering*, 43, 55–68.

Zhang, L., Einstein, H.H., and Dershowitz, W.S. 2002. Stereological relationship between trace length and size distribution of elliptical discontinuities. *Geotechnique*, 52(6), 419–433.

Index

Milton Keynes UK
Ingram Content Group UK Ltd.
UKHW052026141024
449569UK00016B/716